ENGINEERING DESIGN GRAPHICS

ENGINEERING DESIGN GRAPHICS

SKETCHING, MODELING, AND VISUALIZATION

Third Edition

James M. Leake

Department of Industrial & Enterprise Systems Engineering
University of Illinois at Urbana-Champaign

Molly Hathaway Goldstein

Department of Industrial & Enterprise Systems Engineering
University of Illinois at Urbana-Champaign

with special contributions by

Jacob L. Borgerson

Wiss, Janney, Elstner Associates, Inc.
Houston, Texas

WILEY

SENIOR VICE PRESIDENT	Smita Bakshi
SENIOR DIRECTOR	Don Fowley
EDITORIAL ASSISTANT	Molly Geisinger
SENIOR MANAGING EDITOR	Judy Howarth
PRODUCTION EDITOR	Vinolia Benedict Fernando
COVER PHOTO CREDIT	© courtesy of James M. Leake and Molly Hathaway Goldstein, created by the following students:

(Pruning Shears) Tommy Chavez, Blake Lesser, Chaitanya Maroju, Nicholas Sandora
(Camp Stove) Jinhi Lee, Victor Wang, Jake Wolff / (Quadcopter) Chengbin (Simon) Zhang
(Face Mask) Josie Suter, David Du, Yalin Li, Harshey Aggarwal
(Business Instructional Facility) Li Ding, Sara Patttison, Qifa Shi, Wenjing Wu
(Campus Instructional Facility) Grant Imhof, Patrick Kundzicz, Johnny Park, Arannya Roy, John Sadek
(ECE Building) Alene Dressler, Yihui Dong, Calvin Ro, Clarence Boyd
(Bike Helmet) Joon Eun Park, Zachary Junk, Alexandra Brown, Riti Shah

This book was set in 10/12pt TimesTenLTStd by Straive™.

Founded in 1807, John Wiley & Sons, Inc. has been a valued source of knowledge and understanding for more than 200 years, helping people around the world meet their needs and fulfill their aspirations. Our company is built on a foundation of principles that include responsibility to the communities we serve and where we live and work. In 2008, we launched a Corporate Citizenship Initiative, a global effort to address the environmental, social, economic, and ethical challenges we face in our business. Among the issues we are addressing are carbon impact, paper specifications and procurement, ethical conduct within our business and among our vendors, and community and charitable support. For more information, please visit our website: www.wiley.com/go/citizenship.

Evaluation copies are provided to qualified academics and professionals for review purposes only, for use in their courses during the next academic year. These copies are licensed and may not be sold or transferred to a third party. Upon completion of the review period, please return the evaluation copy to Wiley. Return instructions and a free of charge return shipping label are available at: www.wiley.com/go/returnlabel. If you have chosen to adopt this textbook for use in your course, please accept this book as your complimentary desk copy. Outside of the United States, please contact your local sales representative.

ISBN: 978-1-119-49043-2 (PBK)
ISBN: 978-1-119-49261-0 (EVALC)

Library of Congress Cataloging-in-Publication Data

Names: Leake, James M., author. | Goldstein, Molly Hathaway, author. | Borgerson, Jacob L., author.
Title: Engineering design graphics : sketching, modeling, and visualization / James Leake, Molly Hathaway Goldstein, Jacob Borgerson.
Description: 3rd edition. | Hoboken : Wiley, [2022] | Includes index.
Identifiers: LCCN 2021054648 (print) | LCCN 2021054649 (ebook) | ISBN 9781119490432 (paperback) | ISBN 9781119492603 (adobe pdf) | ISBN 9781119492382 (epub)
Subjects: LCSH: Engineering design. | Engineering graphics.
Classification: LCC TA174 .L425 2022 (print) | LCC TA174 (ebook) | DDC 620/.0042—dc23/eng/20211123
LC record available at https://lccn.loc.gov/2021054648
LC ebook record available at https://lccn.loc.gov/2021054649

The inside back cover will contain printing identification and country of origin if omitted from this page. In addition, if the ISBN on the back cover differs from the ISBN on this page, the one on the back cover is correct.

SKY10090471_111124

Jim dedicates this work to
those who complete us, to that which we shared,
to the spirit of place, the sacred quest,
to Stephanie Jo.

Molly dedicates this work
to Matt—What a gift to dream in the same direction.

☐ PREFACE

As in the earlier editions of this book, the goal remains to provide a clear, concise treatment of the essential topics included in a modern engineering design graphics course. Projection theory continues to provide the instructional framework, and freehand sketching the means for learning the important graphical concepts at the core of this discipline.

This book, though, is based on teaching two University of Illinois at Urbana-Champaign (UIUC) courses over the past 20 years, a first-year engineering design graphics course and a 400 level CAD technology and design thinking course. Thus, additional goals are to present a cornerstone to capstone treatment of computer-aided design and to provide a solid foundation in engineering design. The cornerstone component includes engineering graphics, freehand sketching, CAD modeling, spatial visualization, and an introduction to design using reverse engineering and product dissection. The capstone phase (2nd, 3rd, 4th year, senior design) includes the different kinds of CAD (parametric vs direct, solid vs NURBS surface, freeform, BIM), additive manufacturing, 3D scanning and reality capture, simulation and generative design, as well as engineering design, human-centered design, and design thinking.

Since the release of the second edition in 2012 (and even before), important CAD trends include the move toward cloud-based CAD, the use of freeform modeling techniques (based on T-splines or subdivision surfaces) to create sculpted, organic shapes, the proliferation of additive manufacturing, the incorporation of scanners and mesh manipulation tools, the growing acceptance of a model-based digital definition of products, the maturation of building information modeling (BIM), and the increasing use of generative design. Within this same time frame, design thinking and human-centered design have gradually gained acceptance in engineering education. These then, are the currents driving this new edition.

The content of the core chapters on engineering graphics and freehand sketching has been improved with the addition of videos demonstrating freehand sketching techniques and projection theory. Sixteen sketch videos, each about ten minutes in length, are included. They cover perspective (one and two point), oblique (cavalier, cabinet), isometric, multiviews, missing views, partial auxiliary views, section views (full, half, offset, aligned), as well as rotated views. Seven videos describing how perspective, oblique, orthographic, and axonometric projections are generated are also included.

A new chapter on human-centered design and design thinking has been added. This chapter complements an existing chapter on engineering design, as well as a chapter on product dissection and reverse engineering. The new chapter includes a brief history of engineering design, two design paradigms (based on the work of Herbert Simon and Donald Schön), wicked problems, divergent and convergent thinking and questioning, the double-diamond design process, human-centered design, and design thinking, including observation, the importance of extreme users, empathy, ideation, brainstorming, prototyping, testing and iteration, radical collaboration, as well as a sidebar on T-shaped individuals.

The chapter on CAD appearing in the second edition has been split and expanded into two chapters, CAD: Solid Modeling and CAD: NURBS and Freeform Modeling. New material added to the chapter on CAD solids includes cloud-based CAD, CAD topology, assembly joints, and a sidebar on building information modeling. New material on freeform modeling has been combined with second edition material on NURBS surfaces to form the new chapter on CAD surface modeling. The main focus of this new material are T-splines. Freeform materials also include sections on the topological limitations of NURBS surfaces, subdivision modeling, and a sidebar on the Bézier award.

The second edition chapter on reverse engineering tools has also been split and expanded into two chapters, Additive Manufacturing (AM) and 3D Scanning. The chapter on additive manufacturing includes new sections on unique characteristics of AM, AM technologies and their classification, 3D printer file formats, STL repair tools, low cost AM, and design for AM. The chapter on 3D scanning includes new material on mesh terminology, an expanded discussion of the 3D scanner pipeline, expanded coverage of 3D scanning technologies (including structured-light, time-of-flight, reality capture, photogrammetry), improved coverage of the mesh reconstruction process, including typical workflows like scan to mesh, scan to surface, and scan to parametric solid.

Finally, the chapter on engineering design now includes a sidebar on the patterns and behaviors distinguishing beginning versus experienced designers, the chapter on product dissection includes new student projects and e-portfolios, the chapter on product documentation includes a new section on model-based definition and an industry spotlight on Fiskars Group, and the chapter on simulation now includes a section on generative design, including a generative design workflow.

Molly Hathaway Goldstein is co-author for this edition and is now instructor for both of the previously mentioned courses. As a UIUC freshman, she took the first-year course, and later served as a graduate teaching assistant for that course. After working for several years in industry, she earned her PhD in engineering education from Purdue, and returned to Illinois in 2018.

Special thanks to PhD candidate Chengbin Zhang who, among many other things, made the projection videos for this text. There are many student employees who have made significant contributions to engineering design graphics at Illinois, but Zhang stands at the top of the list. Our thanks and appreciation go out to him.

I first met David Ian Weightman in 2009, and we have been collaborating ever since. Everything I know about industrial design and design thinking begins with David. He has made important contributions to both courses, but especially the advanced course where for ten years he served as de facto co-instructor. Thanks to his recruitment efforts, many talented industrial design students have taken the course, thus creating a rich cross-disciplinary environment for the mutual benefit of all involved.

James M. Leake
University of Illinois at Urbana-Champaign
August 2021

CONTENTS

CHAPTER 1

ENGINEERING DESIGN

▋ INTRODUCTION

Design is the central activity of the engineering profession. ***Engineering design*** can be defined as a set of decision-making processes and activities that are used to determine the form of a product, component, system, or process, given the functions desired by the customer.[1] The term ***function*** refers to the behavior of the design, that is, what does the design need to do? ***Form***, on the other hand, has to do with the appearance of the design. A product's form refers to its size, shape, and configuration, as well as the materials and manufacturing processes used to produce it.

Engineering design is a part of the larger ***product realization process***. As we see in Figure 1-1, product realization starts with a customer need and ends with a finished product that satisfies this need. The product realization process consists of design and manufacturing processes that are used to convert information, materials, and energy into a completed product. The stages of the product realization process include sales and marketing, industrial design, engineering design, production design, manufacturing, distribution, service, and disposal. ***Product development*** refers to the first stages of the product realization process up to manufacturing. Product development includes

engineering design, as well as sales/marketing, industrial design, and production design.

▋ ASPECTS OF ENGINEERING DESIGN

Our notion of engineering design encompasses many different aspects. For instance, it is a ***process***, one that prominently involves both ***problem-solving*** and ***decision-making*** activities. Engineering design also employs both ***analysis*** and ***synthesis***. By nature it is ***interdisciplinary*** and ***iterative***. Even prior to the emergence of such modern concepts as ***concurrent engineering*** and the design team, there has always been a strong ***social*** aspect to engineering design. Finally, in keeping with the main topic of this book, engineering design is characterized by strong ***graphical*** elements.

Engineering design is really about solving problems. In fact, a simple definition of engineering design is "a structured problem-solving approach."[2] Concisely described, the design process is no more than identifying a problem, carefully researching and defining the problem in order to better understand it, creatively generating possible alternative solutions to address the problem, evaluating these candidate solutions

[1] Rudolph J. Eggert, *Engineering Design*, Pearson Prentice Hall, 2005.

[2] Arvid Eide et al., *Engineering Fundamentals and Problem Solving*, McGraw-Hill, 1997.

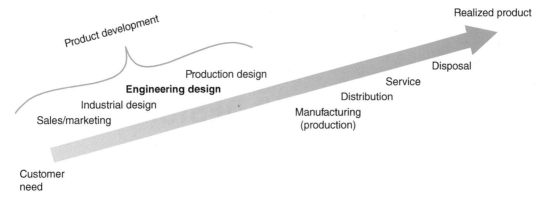

Figure 1-1 Product realization process (Eggert, Rudolph, J., *Engineering Design, 2nd Edition*, © 2010, Page 12. Reprinted by permission of High Peak Press, Meridian, ID)

to ensure their feasibility, making a rational decision, and then implementing it. One of the hidden merits of an engineering education is that this problem-solving framework becomes so ingrained that it can easily be adapted to deal with life's many problems, technical or otherwise.

Although the need to make decisions is apparent throughout the course of a design's evolution, decision making in conjunction with engineering design typically refers to that part of the design process where competing feasible solutions are evaluated and an optimal solution is decided upon. Because of the numerous *tradeoffs* involved, these types of decisions are often difficult to make. Examples of such tradeoffs include strength versus weight, cost versus performance, and towing power versus free-running speed. Optimizing one criterion often means sacrificing the optimum position of the other criterion. Decision making is still something of an art, requiring solid information, good advice, considerable experience, and sound judgment. In recent decades, however, a mathematically based *decision theory* has been developed, and it is commonly used in the development of commercial products.

Synthesis is the process of combining different ideas, influences, or objects into a new unified whole. From this perspective, engineering design can be viewed as a synthesis technique, one used for creating new products based on customer needs. More specifically, synthesis refers to creative approaches used to generate potential solutions to a design problem.

Analysis, on the other hand, is the process of breaking a problem down into distinct components in order to better understand it. From the perspective of engineering design, analysis often refers to the tools used to predict the behavior and performance of potential solutions to a design problem.

Good design requires both divergent thinking (i.e., synthesis), which is used to expand the design space, and convergent thinking (i.e., analysis), which is used to narrow the design space by focusing on finding the best alternatives in order to converge to an optimal solution. Chapter 2 includes a section on divergent and convergent questioning.

As a synthetic process, engineering design is interdisciplinary in nature. Although it relies heavily on basic science, mathematics, and the engineering sciences, engineering design still retains its artistic roots. Eugene Ferguson, in his book *Engineering and the Mind's Eye*, notes that "many of the cumulative decisions that establish a product's design are not based in science." He points out that even though design engineers certainly make decisions based on analytical calculations, many important design decisions are based on engineering intuition, a sense of fitness, and personal preference.[3] Chapter 2 discusses the artistic roots of engineering design in detail.

[3] Eugene Ferguson, *Engineering and the Mind's Eye*, MIT Press, 1997.

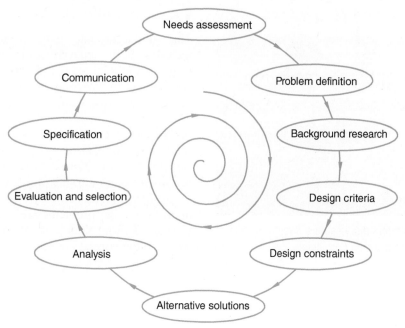

Figure 1-2 Engineering design process

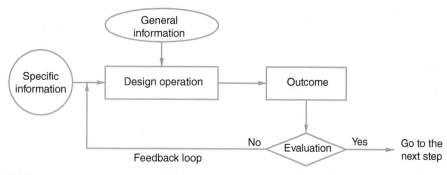

Figure 1-3 Design process feedback loop (Based on [4])

An important element of engineering design is communication. It is often noted that a significant portion of an engineer's time is spent in communication (oral, written, listening) with others. Add to this the fact that modern engineering design draws on decision making, optimization, engineering economy, planning, applied statistics, materials selection and processing, and manufacturing, and the interdisciplinary nature of engineering design should be clear.

Although engineering design is frequently described as a sequential process that moves from one stage to the next, it can also be portrayed as a design spiral, where each stage is visited more than once (see Figure 1-2). In any case, feedback loops, like that shown in Figure 1-3,[4] are built into each step in the design process. Based on the latest information, it is often necessary to iterate the various design process steps.

As we have already mentioned, the actual engineering design process is not nearly as structured as that described in engineering design textbooks. There has always been a

[4] George Dieter, *Engineering Design: A Materials and Processing Approach*, McGraw-Hill, 1991.

Figure 1-4 Engineers discussing a design project (Courtesy of Jensen Maritime Consultants, Inc.)

strong social aspect to the design process, perhaps best captured in the notion of the napkin or back-of-the envelope sketch.[5] In fact, a reasonable definition of the engineering design process is "a social process that identifies a need, defines a problem, and specifies a plan that enables others to manufacture the solutions."[6] This social aspect of engineering design is further reinforced by the emphasis in the recent decades on project design teams. Figure 1-4 shows two engineers holding a discussion over an engineering drawing.

Graphics is used throughout the entire engineering design process. Freehand sketching is useful both for generating ideas and for communicating them. Computer-aided design (CAD) is used not only for traditional documentation purposes but also to evaluate design alternatives. Parametric solid models capture design geometry, but they also serve as digital prototypes used in downstream analysis, prototyping, and manufacturing applications, as well as for marketing and sales. This digital prototype is in fact a 3D CAD relational database representing all aspects of the product's design.

ANALYSIS AND DESIGN

The undergraduate engineering curriculum includes both analysis and design content. Engineering analysis uses science, mathematics, and computational tools to predict the behavior of an engineering design. Analysis is characterized by breaking a complex problem up into manageable pieces. Analysis attempts to simplify the real world by using models. In a typical analysis problem, the input data are given, and there is a single solution to the problem.

Design problems employ synthesis as well as analysis. Design problems are open-ended; that is, there is more than one solution. In a design problem the correct answer is unknown. For this reason, optimization and decision-making techniques are often employed in association with design. Design problems are typically constrained by time, money, and legal issues; they tend to be multidisciplinary and of undefined scope.

PRODUCT ANATOMY

A *product* is a designed object or artifact that is purchased and used as a unit.[7] Products vary in complexity depending on the number, type, and function of their components. A *component* is either a part or a subassembly. A *part* is a piece that requires no assembly, whereas an *assembly* is a collection of two or more parts. A *subassembly* is an assembly that is included in another assembly or subassembly. For example, a bicycle is a product that can also be thought of as an assembly composed of subassemblies and parts. A rear hub is a subassembly on a bicycle, whereas a handlebar is a part on a bicycle. Both the hub and the handlebar are bicycle components.

DESIGN PHASES

Product design projects, as conducted by large manufacturing companies like Boeing or Motorola, pass through several distinct phases. See Figure 1-5. In the *formulation* phase information

---

[5] For example, the original concept for the Seattle Space Needle was sketched on a placemat in a Seattle coffee house.
[6] Karl Smith, *Project Management and Teamwork*, McGraw-Hill, 2000.

[7] John Dixon and Corrado Poli, *Engineering Design and Design for Manufacturing: A Structured Approach*, Field Stone Publishing, 1999.

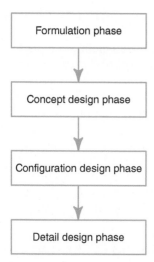

Figure 1-5 Engineering design phases

is gathered in order to fully understand the design problem. Next, functional (customer, company) requirements of the product as well as constraints and evaluation criteria are identified; finally, engineering performance targets are developed. In the *concept design* phase the product is decomposed initially by component (form) and then by function in order to better understand it. Concept alternatives are next generated to meet each product subfunction; these subfunction alternative concepts are then analyzed to determine their feasibility, and finally, they are evaluated on the basis of the evaluation criteria. The concept design phase concludes with the selection of the best concept alternatives for each product subfunction. By the conclusion of the *configuration design* phase, the type and number of components of the product, their arrangement, and their relative dimensions have been determined. In the *detail design* phase, a package of information that includes drawings and specifications sufficient to manufacture the product is prepared.

▌ DESIGN PROCESS OVERVIEW

The design phases describe the actual procedure employed by a manufacturing company to bring new products to market. The engineering design process, on the other hand, is more an instructional framework used to describe the idealized steps taken by an engineering team in developing a design.

The engineering design literature includes many different versions of this process, each with a different number of stages, or steps. The design process template (see Figure 1-2) employed in this work[8] includes (1) needs assessment, where specific needs (problems) are identified, (2) problem definition, where the problem is clarified, (3) background research, where research is conducted to gather additional information about the problem, (4) design criteria, where desirable characteristics of the design are identified, (5) design constraints, where quantitative boundaries that limit the possible solutions to the problem are identified, (6) alternative solutions, where candidate solutions are conceptualized and generated, (7) analysis, where the alternative solutions are analyzed in an effort to determine their feasibility, (8) evaluation and selection, where the remaining solutions are evaluated with respect to a weighted criteria and the best candidate is selected, (9) specification, where drawings and technical specifications documenting the design are created, and (10) communication, where written reports and oral presentations describing the design are developed. Many of the steps (e.g., problem definition) may be revisited multiple times as additional information comes to light. Further, the stages may not necessarily be performed in the order specified here, and some steps may be omitted entirely.

It is worth noting that the pace in an engineering design office is driven by the available billable hours. Engineering design is tightly constrained by the time available to get the job done. If the billable hours are limited (and they always are), then tough decisions must be made, and some of the idealized steps described here are abbreviated— or even eliminated.

▌ NEEDS ASSESSMENT

The engineering design process starts with the recognition of a need that can potentially be satisfied using technology. In many cases, needs are

---

[8] Adapted from Eide.

identified not by engineers but rather by actual users, sometimes with the assistance of experts. See Chapter 2 for additional information on the user's role in the design process. Potential clients may also approach design firms with a specific need. In a large manufacturing company, the sales or marketing department maintains contact with its customer base and identifies many of these needs. In a more systematic approach, product ideas are generated by a product planning group.

Needs arise for a variety of reasons, including (1) a product redesign in order to make it more profitable or effective, (2) the establishment of a new product line, (3) a need to protect public health and safety or to improve quality of life, (4) an invention, often by an individual, that is then commercialized, (5) opportunities created by new technology or scientific advances, and (6) a change in rules, requirements, or the like. The outcome of the needs assessment is a list of needs or requirements, which then becomes a part of the problem definition.

▌ PROBLEM DEFINITION

Once a list of needs has been developed, the next step in the design process is to clearly and carefully formulate the problem to be solved. At first glance this step may seem somewhat trivial, but the true nature of the problem is not always obvious. A poorly defined problem can lead to a solution search that is either misdirected or too limited in scope. A good problem definition should focus on the desired functional behavior of the solution, rather than on a specific solution. For example, if the issue is lawn maintenance, then "Design a better lawn mower" presupposes the solution. If the problem is restated as "Design an effective means of maintaining lawns," then the options are kept open.[9] When the problem is defined as broadly as possible, novel or unconventional solutions are less likely to be overlooked. Another tip when formulating a problem is to focus on the source of the problem, rather than on the symptoms. In any event, a complete definition of the problem should include a formal *problem statement*.

A problem statement attempts to capture the essence of a problem in one or two sentences. A problem is characterized by three components: (1) an undesirable initial state, (2) a desired goal state, and (3) obstacles that impede efforts to go from the undesired to the desired state.[10] A good initial problem statement should consequently describe the nature of the problem, as well as express what the design is intended to accomplish. In further iterations, the obstacles that prevent reaching the desired state may also be included.

▌ BACKGROUND RESEARCH

The problem having been broadly defined, the next step in the engineering design process is research. Background research is conducted in order to obtain a deeper understanding of the problem. Information is sought regarding the target user, the intended operating environment, additional constraints that bound the solution, prior design solutions, and so on. Useful questions that may be posed at this stage include: What has been written on the topic? What must the design do? What features or attributes should the solution have? Is something already on the market that may solve the problem? What is right or wrong with how it is being done? Who markets the current solution, how much does it cost, and how can it be improved upon? Will people pay for a better one if it costs more?

Common sources include existing solutions, the library, the Internet, trade journals, government documents, professional organizations, vendor catalogues, and experts in the field. Regarding existing solutions, *reverse engineering* is the process of physically disassembling an existing product to learn how each component contributes to the product's overall performance. Manufacturing companies use reverse engineering techniques to gain insight into a competitor's design solutions. Applied systematically, *best-in-class* solutions can then be identified for a broad range of

[9] Restating the problem in this way led to the invention of the Weed Eater® lawn trimmer.

[10] G. Pahl and W. Beitz, *Engineering Design: A Systematic Approach*, Springer-Verlag London, 1996.

Table 1-1 **Tugboat design criteria**

Specific	General
Bollard pull (thrust, towing power)	Appearance
Habitability	Cost
Maneuverability	Durability
Seakeeping	Ease of maintenance
Speed (free-running)	Ease of operation
Stability (safety)	Environmental protection
Stable work platform	Reliability
Visibility from the pilothouse	Use of standard components
Workmanship	

Figure 1-6 Tugboat photograph (Courtesy of Jensen Maritime Consultants, Inc.)

common engineering problems. Reverse engineering is discussed further in Chapters 3 and 14.

DESIGN CRITERIA

Design criteria[11] describe desirable characteristics of the solution. They tend to originate from experience, research, marketing studies, customer preferences, and client needs. Design criteria are used at a later stage in the design process to qualitatively judge alternative design solutions. Design criteria may be categorized as either general or specific. General criteria categories applicable to most all design projects include cost, safety, environmental protection, public acceptance, reliability, performance, ease of operation, ease of maintenance, use of standard components, appearance, compatibility, durability, and so on. Criteria categories specific to a certain design might include weight, size, shape, power, physical requirements for use, reaction time required for operation, and noise level. Table 1-1 provides examples of both general and specific criteria for a tugboat design similar to that shown in Figure 1-6.

DESIGN CONSTRAINTS

Design constraints limit the possible number of solutions to a given design problem. Design constraints are quantitative boundaries associated

[11] In engineering design literature, design criteria are frequently expressed as design objectives or as design goals.

with each design objective. They establish maximum, minimum, or permissible ranges for physical or operational properties of the design, environmental conditions affecting the design or impacting the environment, and ergonomic requirements, as well as economic or legal constraints. A feasible design solution must satisfy all of the design constraints.

Design constraint categories include (1) *physical*—space, weight, material, etc., (2) *functional or operational*—vibration limits, operating times, speed, etc., (3) *environmental*—temperature ranges, noise limits, effect on other people, (4) *economic*—cost of existing competitive solutions in the marketplace, (5) *legal*—governmental and other regulations, and (6) *ergonomic* (or human) factors—strength, intelligence, anatomical dimensions, etc. Some design constraints, organized by category, for a tugboat are shown in Table 1-2.

ALTERNATIVE SOLUTIONS

Having thoroughly researched the problem, the next step is to systematically generate as many alternative product solutions as possible for subsequent analysis, evaluation, and selection. The group technique most commonly used for generating ideas is *brainstorming*. The objective of brainstorming is to develop as many ideas as possible in a limited amount of time. Roughly an hour

Table 1-2 Tugboat design constraints (sample)

Physical

ASTM A-36 steel used for hull
Fresh water > 6,000 gallons
Fuel > 30,000 gallons
Heel less than $\frac{1}{4}$ degree
Length Overall (LOA) < 100'
Railing height > 39 inches

Functional/Operational

Accommodations for 6 crew
Bollard pull > 30 tons
Horsepower > 2500 BHP
Running speed > 12 knots

Economic

Cost < $8 million

Environmental

Engine room noise level < 120 decibels

Legal (Regulatory Bodies)
American Bureau of Shipping (ABS)

American Society of Mechanical Engineers (ASME)
American Society for Testing and Materials (ASTM)
Environmental Protection Agency (EPA)
International Maritime Organization (IMO)
Occupational Safety and Health Administration (OSHA)

Ergonomic

Accommodations headroom > 6 feet

---

is typically allotted for a brainstorming session. Emphasis is placed on the quantity, rather than the quality, of the ideas. Free expression is essential; the ideas can be evaluated later. Upon occasion, one group member's seemingly impractical notion serves as the inspiration for a teammate's more viable idea. It is important that all group members participate as equals. A group leader can start with a clear statement of the problem, invite ideas regarding the solution, and set the pace of the session. Using marker pens and large Post-it notes (or rolls of butcher paper), the ideas should be recorded for all of the participants to see. Brainstorming is discussed further in Chapter 2.

As we discussed in the section on product anatomy, a product is composed of different components. Each component has its own form, as does the product. Similarly, a product has a primary function, but it also has secondary functions, or subfunctions. For example, the primary function of a coffee maker is to make coffee, but it also needs to store ground coffee and water, hold a filter, brew coffee, convert electricity to heat, and keep the coffee pot warm. To ensure a robust design, multiple concept alternatives need to be generated for each product subfunction.

A *morphological chart* is a tool that can be used to expand the number of concept alternatives. Product subfunctions are listed versus concept alternatives in a table. Subfunctions are listed vertically, in a column, and the corresponding concept alternatives horizontally, in a row. Table 1-3

Table 1-3 Morphological chart for a portable water filter

Function	Solution 1	Solution 2	Solution 3	Solution 4	Solution 5
draw water	manual	hose	hose w/ float	hose w/ float, prefilter	
move water	hand pump	lever pump	gear pump	dual piston pump	gravity feed
filter water	ceramic	ceramic w/ carbon core	fiberglass	labyrinth	iodine resin
collect water	manual	threaded adaptor base	built-in collection bottle		

shows a morphological chart for a portable water filter. With this technique, a vast number of possible solutions can be identified. Although some of these solutions may be incompatible, morphological analysis is nonetheless useful in identifying a great many combinations, some of which may be unusual or innovative.

In order to compete globally, large manufacturing companies engaged in product realization systematically optimize a product's design at the subfunction level. For first-year student design projects, though, it may be sufficient to generate, evaluate, and then select from product level alternative solutions. For example, each member of the design team may propose an alternative solution to the problem, as shown in Figure 1-7.

Many designs, too, have distinct solutions that have developed over time. Drawbridges, for example, include single and twin span bascule, vertical lift, and rotating solutions (see Figure 1-8), each one serving as an alternative solution to the drawbridge problem.

Students may well come up with their own innovative solutions to these time-honored alternatives, as seen in the student drawbridge concept design solution shown in Figure 1-9.

■ ANALYSIS

The concept alternatives generated in the previous stage must be analyzed and then evaluated in order to determine which concepts are worth pursuing. It is possible that some of the candidate concepts that have been generated thus far are not feasible. Design analysis, which uses mathematical and engineering principles to evaluate the performance of a solution, may be used to verify the functionality and manufacturability of the concept alternatives. The objective is to eliminate any infeasible solutions. Although the candidates are at this point no more than abstract concepts, still it may be possible to eliminate some of them by using back-of-the-envelope calculations, investigating their manufacturability, and the like.

■ EVALUATION AND SELECTION

After analysis, the remaining feasible candidate concepts are evaluated. Design evaluation

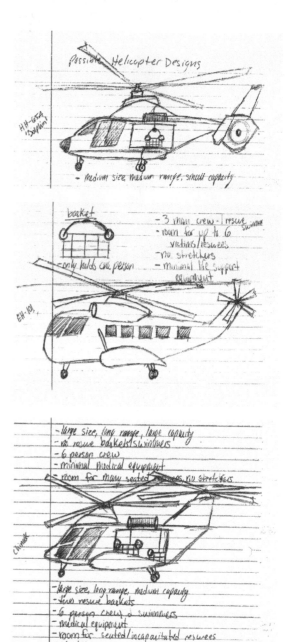

Figure 1-7 Alternative solution sketches (Courtesy of Matthew Patton)

is a stage in the design process when results from analyses are assessed to determine which alternative is best. The alternatives are evaluated with respect to weighted evaluation criteria. The evaluation criteria are derived from the

(400tmax/iStock/Getty Images)

(Lee Foster / Alamy stock photo)

(ntzolov/E+/Getty Images)

(Potifor/Shutterstock.com)

Figure 1-8 Drawbridge alternative solutions

design criteria established earlier in the design process, although these criteria may have been modified and further refined based on knowledge gained in the course of the design's development.

It is the responsibility of the design team to come to a decision regarding the relative importance of the different criteria. The evaluation criteria are then weighted in terms of their assigned importance. It is customary to have the criteria weights add up to 100. Once the weighting has been established, the concept alternatives (or product subfunctions) can be evaluated with respect to the weighted criteria. Table 1-4 provides an example of the evaluation procedure used by a student group in a first-year engineering design graphics course to design a dune buggy. An image of the resulting dune buggy concept design appears in Figure 1-10.

Figure 1-9 Innovative student drawbridge design solution (Courtesy of Yang Cui, Allison Dale, David Shier, Michael Marcinowski)

Table 1-4 Alternative design evaluation process

Weighted Design Criteria

Cost	30%
Safety	10%
Weight and Power	15%
Durability	15%
Ease of Operation	20%
Simple Construction	10%

Alternative Design Rankings (1–10)

Weighted Design Criteria	Brian A's Drawing	Dan's Drawing	John's Drawing	Brian S's Drawing	Nilay's Drawing
Cost	8	10	9	8	8
Safety	7	8	6	9	10
Weight and Power	10	6	9	10	7
Durability	8	7	7	9	8
Ease of Operation	7	10	9	6	9
Simple Construction	9	7	6	8	8

Alternative Design Results (Rankings Multiplied by Percentages)

Weighted Design Criteria	Brian A's Drawing	Dan's Drawing	John's Drawing	Brian S's Drawing	Nilay's Drawing
Cost	2.4	3	2.7	2.4	2.4
Safety	0.7	0.8	0.6	0.9	1
Weight and Power	1.5	0.9	1.35	1.5	1.05
Durability	1.2	1.05	1.05	1.35	1.2
Ease of Operation	1.4	2	1.8	1.2	1.8
Simple Construction	0.9	0.7	0.6	0.8	0.8
Total	**8.1**	**8.45**	**8.1**	**8.15**	**8.25**

Another example is provided in Figure 1-11. In this industry sponsored project, a senior design team was asked to develop a new and improved endcap display to be used in grocery and other stores. In an effort to evaluate different concepts with respect to the design criteria, the group developed the evaluation matrix shown in Figure 1-11. Each column in Figure 1-11 shows a different endcap criterion.

These include modularity, adjustable shelving, mobility, futuristic design, ease of cleaning, beverage advancement, packout, and header. In the cells beneath each column, images illustrating different ways to satisfy each specific criterion are

Figure 1-10 Concept dune buggy design (Courtesy of Dan Fey, Nilay Patel, Jack Streinman, Brian Aggen, Brian Shea)

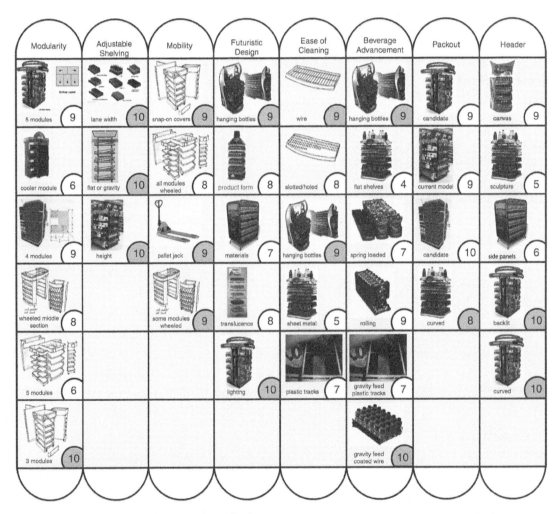

Figure 1-11 Evaluation matrix for an endcap display (Courtesy of Franklin Wire and Display; Jennifer Bessette, Michelle Wentzler, Madison Major, Faye Hellman)

Figure 1-12 Endcap display design (Courtesy of Franklin Wire and Display; Jennifer Bessette, Michelle Wentzler, Madison Major, Faye Hellman)

shown. Also included in each cell is a rating (the circled number in the lower right-hand corner) assigned by the design team. This rating is used to indicate the performance of the particular solution in meeting the criterion. Using this and other techniques, the design team produced the endcap design shown in Figure 1-12.

▌ SPECIFICATION

Having selected the best concept alternatives, or simply the best alternative solution, the different product features, parts, and components are configured and arranged using computer-aided design (CAD). *Configuration design* refers to that part of product development where planners determine the number and type of components, parts, or geometric features; how they are spatially arranged or interconnected; and the approximate relative dimensions of the components, parts, and features.

This is typically accomplished using a general arrangement or layout drawing. Figure 1-13 shows general arrangement and outboard profile drawings for a 96′ tugboat. The tug design employs two rudder propellers mounted at the stern of the vessel. The rudder propellers can be fully rotated through 360 degrees, independently of one another, giving the vessel exceptional maneuverability and propulsion characteristics.

In addition to the general arrangement drawing, a written specifications document is typically produced at this time. A dictionary definition of *specification* is "a detailed description of a particular thing, especially one detailed enough to provide somebody with the information needed to make that thing." The design specification document is consequently a comprehensive description of the product, including the intended uses, physical dimensions, materials employed, operating conditions, and performance targets, as well as the functional, maintenance, testing, and delivery requirements for the designed object. Relevant codes and regulations to be adhered to in the course of the product's design and manufacture are also spelled out in these specifications. Figure 1-14 shows an excerpt taken from the written specifications developed for the tugboat shown in Figure 1-13.

These drawings and specifications may then be used in a formal bidding process called a Request for Proposals (RFP). The drawing package would be sent out to different contractors, in this case shipyards, to be competitively bid upon for the right to construct the vessel. Assuming that the RFP process is successful and an agreement is reached to continue the product's development,[12] the project would then move from the configuration to the

[12] In a manufacturing firm, the decision whether to continue the product's development beyond the configuration design phase would be made internally.

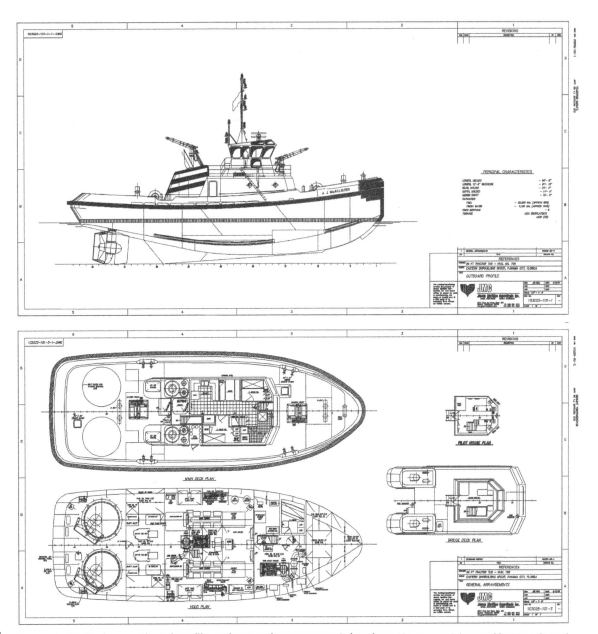

Figure 1-13 Tugboat outboard profile and general arrangement drawings (Courtesy of Jensen Maritime Consultants, Inc.)

detail design phase. The outcome of the detail design phase is a complete package of information, including many more drawings and more detailed written specifications, sufficient to enable the contractor to fabricate the product.

These ***working drawings*** are also called production or detail drawings. Working drawings are discussed in Chapter 12. They typically include part detail drawings, assembly drawings (see Figure 1-15), and a bill of materials (BOM),

GROUP 100—HULL STRUCTURE

SECTION 101—GENERAL INFORMATION

See Section 012 for OWNER furnished equipment.

The CONTRACTOR shall design and construct the principal hull structure as described herein. The hull and deckhouse is constructed of steel. All plating shall be ABS Grade A and/or ASTM A-36. All steel shall be wheel-abrated and pre-primed. Steel Certificates shall be provided to the OWNER, along with location of the certified steel.

All workmanship and welding shall be in accordance with the following:

ABS "Rules for Building and Classing Steel Vessels Under 90 meters (295 feet) in Length" (for Hull)

ABS "Rules for Building and Classing Steel Vessels for Service on Rivers and Intracoastal Waterways," Part 3, Section 7, Passenger Vessels (for Superstructure)

Classification Society certification of the structure is not required, but structural calculations supporting compliance with one of the above listed Rules shall be submitted to the OWNER for approval in areas that the CONTRACTOR proposes any changes. Calculations shall show clear references to rules being used.

Construction details shall be in accordance with ABS' "Guide for Shipbuilding and Repair Quality Standards for Hull Structures During Construction," latest edition. Plating shall be fitted fair and free from buckles or uneven sight edges. All formed plates or shapes shall be formed true to the required alignment, shape or curvature. Where flanges are used for attachments, the faying edges shall be beveled and free from hollows. Shims shall not be used to correct improper fit. Members shall be in alignment before welding is undertaken. No fairing compounds shall be used. Warpage or distortion that prevents the installation of the final welded assembly into the boat is not acceptable.

"Panting" or "oil-canning" of any panel in shell, deckhouse or decks is not permitted. Filling compound shall not be used to compensate for unfairness in the boat structure. Every effort shall be exercised to construct a vessel with fair and undistorted surfaces. This shall include diligence for careful fit-up, proper weld sequences and utilizing minimum weldments to achieve required structural strength.

The maximum unfairness between hull, deck and house stiffeners shall be t/2, where "t" is the plate thickness.

Penetrations of deck, bulkhead and shell plating shall be reinforced as required to maintain structural integrity.

All cuts shall be neatly and accurately made with edges cleaned for welding. All sharp edges exposed to personnel or equipment shall be dressed or ground to avoid injury to operating or maintenance personnel or damage to equipment. Internal corners shall be filleted and external corners shall be rounded off. Ragged edges or sharp projections shall be removed.

All burning of notches and holes in the structure shall be carefully laid out and shall be regular in outline with jagged edges removed and with rounded corners. Semicircular "rat holes" shall be provided in framing members in way of passing welded seams or butts in the attached plating. Notches and rat holes shall be machine cut.

Where it is necessary to provide holes for passage of wiring, piping or ductwork through vessel structure, special penetration details shall be used to maintain structural and fire insulation integrity and oil/water/air tightness as appropriate. All penetrations of fire boundaries shall be collared and insulated on both sides of the fire boundary and as otherwise necessary to meet regulatory requirements. In general, penetrations through bulkheads shall utilize Nelson Firestop $4'' \times 6''$ transit or equal. All details of such penetrations shall be submitted to the OWNER for approval.

All edges or corners of the bulwarks which could have a fire hose dragged over or around them shall be radiused using $1\text{-}1/2$ to 2 inch diameter pipe (or equivalent flanged radius).

The CONTRACTOR shall ensure that precautions are taken to prevent rusting of exposed plate and new welds. The OWNER is extremely concerned that excessive rust bloom will compromise the ability of the paint system to obtain a secure hold on the sub-surface, especially in way of the welds. Painting shall be conducted within a controlled environment, protected from the elements, through use of either permanent or portable structures. See SECTION 631 for a description of surface preparation and other coating requirements.

Figure 1-14 Excerpts from written specifications (Courtesy of Jensen Maritime Consultants, Inc.)

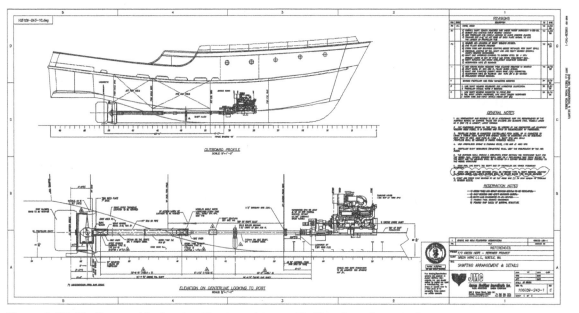

Figure 1-15 Shaft assembly drawing (Courtesy of Jensen Maritime Consultants, Inc.)

or parts list. An assembly drawing with a bill of materials is shown in Figure 1-16.

Note that the term *specification* is also used in engineering design to refer to a target value, performance measure, or benchmark employed to evaluate the success of a design. Taken in this sense, specifications begin to emerge early in the design process, as the designer clarifies the client's or company's understanding of the problem, and the potential user's needs. Examples of performance specifications taken from the written tugboat specifications document cited in Figure 1-14 include:

- *Propulsion system shall consist of a pair of EPA Tier 1 certified, 4-cycle diesel engines with a minimum continuous rating of approximately 2480 BHP.*

- *The boat shall undergo a two-hour endurance trial run at the maximum engine rating. During this trial it shall be demonstrated that all mechanical parts of the propulsion unit and all auxiliaries are in satisfactory operating condition, and that propulsion system steady-state conditions are within engine manufacturer's tolerances. Inspections shall be carried out for leaks in all piping systems and any*

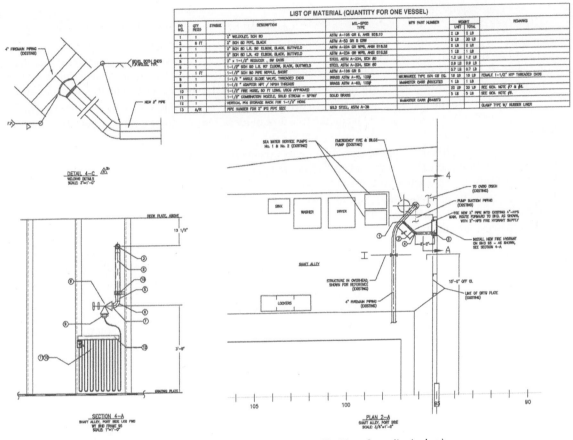

Figure 1-16 Assembly drawing with BOM (Courtesy of Jensen Maritime Consultants, Inc.)

structural defects. The readings of all installed gauges and meters shall be recorded at 15-minute intervals.

▌ COMMUNICATION

At this point, the product's development transitions from design to manufacture. The final step in the engineering design process is to communicate the design using written reports and oral presentations.

Written Reports

Although they are better known for their math skills than for their verbal skills, most experienced engineers are in fact good writers. Written communication consumes a large portion of an engineer's work week. Written documents composed by engineers include letters, emails, memoranda, technical reports, technical papers, and proposals. The audience for these technical documents includes other engineers, designers, supervisors, vendors, regulatory agencies, and customers. Regardless of the audience and type of report, technical writing requires a clear, concise expression of what can be very complex issues, concepts, and arguments.

A typical outline for a technical report might include the following sections:

Introduction	Discussion (of results)
Background	Conclusions
Procedure	References
Results	Appendices

When writing a technical report or paper, it is recommended that you break the writing process up into several distinct sessions. Try to start each session when you are at your best (e.g., early morning, late afternoon). Once the session begins, try to ensure that you won't be interrupted for at least a few hours. Finally, wait a day before starting the next session. Shown below is a series of session steps that can be used to write papers and technical reports.

1. *Brainstorm ideas*—In the first session, write down everything that you wish to say, making no effort to structure the material. The emphasis should be on the free flow of ideas. This session is probably best done using a pen and paper.

2. *Preliminary outline*—In the second session, attempt to structure the brainstorming output in bullet form. What are the topic headings and subheadings? How are the topics ordered? What material falls under which heading? At this stage, transition to a word processor, if you have not already done so.

3. *Detailed outline*—Read the initial outline, revising and amplifying as needed. Move things around, and add and delete material. Do the topic headings and bullet items easily transition and flow? Is there sufficient content to make the argument? If not, do more research to fill in the gaps. Amplify on the bullet items, adding detail as needed. Move from phrases to complete sentences.

4. *Start writing*—Normally with the introduction, but not necessarily. This is where the true struggle takes place. Progress can be painfully slow; concentration, determination, and grit are essential. Aim for coherence and continuity. After a significant struggle, you may have to settle for less. As long as an honest effort is made, a fresh start will almost always help. A good day's work is a couple of pages.

5. *Carefully and critically edit* what has already been written for coherence, continuity, grammar. Revise, rewrite, rearrange.

6. Critically reread the document from start to finish, making final editorial changes as necessary.

Oral Presentations

As an engineer, you will occasionally be called upon to present the results of your work. Presentations are made to deliver progress and summary reports, to pitch an idea or proposal, and also to present a paper at a conference or technical society. Whatever the purpose, it is important to carefully prepare for, and then practice, the delivery of the presentation. When designing the presentation, it is important to know your audience so that you can anticipate and fulfill its needs. Good visual aids and graphics are important, as is a well-organized, logical development of the main ideas.

Some oral presentation guidelines include[13]:

1. Speak clearly so that you can be heard at the back of the room.
2. Vary your pitch or tone of voice occasionally.
3. Add enthusiasm to the delivery.
4. Expect to be somewhat nervous. Take a deep breath and relax a moment before beginning.
5. Start on time, stick to the schedule, and finish on time; do not go over.
6. Maintain eye contact with your audience. Pick a few individuals in the room and look at them as you present.
7. Use a pointer when appropriate.
8. Introduce all team members and the project title.
9. Show an overview slide describing the structure of the presentation.
10. Make your slides clear and crisp. Make sure all slides have sufficient contrast to show the text clearly.
11. Make bullets brief, typically with no more than five words.
12. Each bullet slide should require your explanation in order to be well understood by the

[13] Adapted from the Senior Project Design Manual used in GE 494, Senior Project Design, at the University of Illinois Urbana-Champaign.

audience. The audience will rely on you to bring the presentation together.

13. Photos should be large and clear. Use labels, arrows, etc. to define and point out important features.

14. When showing a graph, briefly define the axes and then tell the purpose of the graph and what it is intended to show.

15. Use PowerPoint® transitions sparingly.

16. Be prepared; know the purpose of each slide. Do not include any information on a slide that you cannot explain.

17. In developing your ideas, move from the general to the specific.

18. Distribute any handouts before the presentation begins.

19. Be clear about your conclusions and what they mean. Show how your conclusions and final recommendations satisfy the problem statement.

20. Have a *war chest* of extra slides for the Q&A period. Anticipate questions and prepare extra slides that will help you answer them.

21. In your practice sessions, notice how often you say "Um," "you know," and so on. Try to avoid this nervous habit. Silence is best when you have nothing to say.

Beginning versus informed designer patterns

Beginning designers tend to approach design situations differently than more advanced, "informed" designers. The nine engineering design strategies and associated patterns in Table 1-5 contrast beginning versus informed design behaviors.

Table 1-5 **Beginning versus Informed Designer Patterns**[14]

Engineering design strategy	Beginning designers	Informed designers
Understand the challenge: Problem solving vs. Problem framing	Do not grasp the basics of a design task, or treat it as a well-defined, straightforward problem that they prematurely try to solve	Understand basics of the design problem, and then delay making design decisions in order to explore, comprehend, and frame the problem better
Build knowledge: Skipping vs. Doing research	Skip doing research and instead build solutions immediately	Conduct investigations and research to learn about the problem, relevant cases, and how the system works
Generate ideas: Idea scarcity vs. Idea fluency	Work with just a few or just one idea, which they can get fixated or stuck on, and may not want to discard, add to, or revise	Practice idea fluency in order to work with lots of ideas by doing divergent thinking, brainstorming, etc.
Represent ideas: Surface vs. Deep drawing & modeling	Propose superficial idea that do not support deep inquiry of a system, and that would not work if built	Use multiple representation to explore and investigate design ideas and support deeper inquiry into how systems work
Weigh options and make decisions: Ignore vs. Balance benefits & tradeoffs	Make design decisions without weighing all options, or attend only to pros of favored ideas, and cons of lesser approaches	Use words and graphics to displace and weight both benefits and tradeoffs of all ideas before picking a design

(continued)

[14] David P. Crismond and Robin S. Adams, The Informed Design Teaching & Learning Matrix, *Journal of Engineering Education, 101*, no. 4 (2012): 738.

Table 1-5 Continued

Engineering design strategy	Beginning designers	Informed designers
Conduct experiments: Confounded vs. Valid tests & experiments	Do few or no experiments on prototypes, or run confounded tests by changing multiple variables in a single test	Conduct valid experiments to learn about materials, key design variables, and the system work
Troubleshoot: Unfocused vs Diagnostic troubleshooting	Use an unfocused, nonanalytics way of viewing prototypes during testing and troubleshooting ideas	Focus attention on problematic areas and subsystems when troubleshooting devices and proposing ways to fix them
Revise/Iterate: Haphazard or linear vs. Managed & iterative designing	Design in haphazard ways where little learning gets done, or do design steps once in linear order	Do design in a managed way, where ideas are improved iteratively via feedback. Strategies get used as many times as needed, in any order
Reflect on Process: Tacit vs. Reflective design thinking	Do tacit designing with little self-monitoring while working or reflecting on process	Practice reflective thinking by keeping tabs on design strategies and thinking

(Modified from [14])

■ CONCURRENT ENGINEERING

In the face of ever-increasing global competition, it has become necessary for successful manufacturing companies engaged in product development to continuously (1) shorten product development times, (2) improve product quality and performance, and (3) reduce product cost. A modern approach used to accomplish these daunting tasks is called concurrent engineering.

In the years after the Second World War, product development as conducted in large manufacturing companies was essentially a serial process, as shown in Figure 1-17a. The specialized functions of the company, represented as distinct departments, worked in comparative isolation. The free exchange of information across functional boundaries was not encouraged. The most commonly cited example of this behavior occurred at the interface between design and manufacturing, where the design group developed a design in isolation, and then *tossed it over the wall* to manufacturing. Manufacturing was then left to modify the design in order to meet manufacturing process, material, and equipment constraints. Although these changes were costly, a lack of competition made them bearable.

With the emergence, in recent decades, of a formidable global competitiveness, these inefficiencies can no longer be tolerated. ***Concurrent engineering*** is a team approach to product design in which team members, representing critical business functions, work together under the coordination of a senior manager. The cross-functional team is typically composed of members from such areas as sales/marketing, industrial design, design engineering, industrial engineering, manufacturing engineering, purchasing, and production. See Figures 1-17a and b for a graphical comparison of the concurrent and traditional engineering models.

One of the benefits of concurrent engineering's team approach is the open exchange of product information. Concurrent engineering is sometimes represented in association with the single most important source of product information, the CAD database, as shown in Figure 1-18.

Although concurrent engineering strives to incorporate all aspects of a product's life cycle (i.e., design, manufacture, distribution, service, and disposal) into the product's development, the central motivation for concurrent engineering is to ensure that manufacturing considerations are taken into account early and throughout

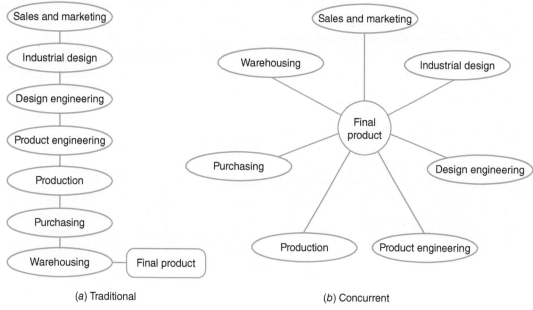

Figure 1-17 Traditional versus concurrent engineering

(a) Traditional

(b) Concurrent

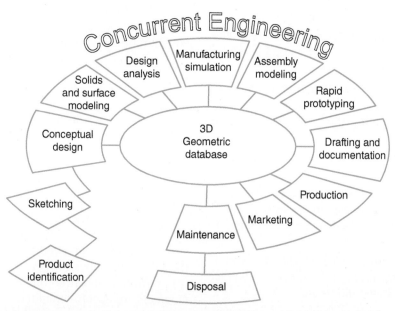

Figure 1-18 Concurrent engineering and the CAD database (Courtesy of Barr, Kreuger, and Juricic)

the entire course of the design process. In the following section, on design for manufacture and assembly, this point is addressed more fully. The chapter then concludes with a discussion of another one of concurrent engineering's most important attributes, teamwork.

Design for Manufacture and Assembly

Design changes occurring late in the design process are especially detrimental to a product's cost, quality, and time to market. Design for Manufacture (DFM) and its cousin Design for

Assembly (DFA) are relatively new areas that strive to minimize these late-occurring design changes by formalizing the relationship between design and manufacture. Whereas Design for Manufacture aims to improve the fabrication of individual parts, Design for Assembly strives to reduce the time and cost required to assemble a product. Taken together, DFM and DFA practices help ensure that the product and the manufacturing processes used to produce it are designed together. Recently a new term, Design for X (DFX), has become popular. DFX describes any of the various design methods that focus on specific product development concerns. These include Design for the Environment, Design for Reliability, Design for Safety, Design for Quality, and so on.

Here are some of the more important DFMA guidelines:

- *Minimize the total number of parts*—The fewer the number of parts, the easier and faster the product's assembly.

- *Minimize part variations*—This can be accomplished, for example, by using fasteners of the same size throughout.

- *Design parts to be multifunctional*—A single part can be designed to serve more than one purpose, eliminating the need for additional parts.

- *Design parts for multiuse*—Use the same part in different products.

- *Design parts for ease of fabrication*—Near-netshape manufacturing processes (e.g., injection molding) are best; avoid machining if possible.

- *Design parts with self-fastening features*—Use snap and press-fits whenever possible; avoid screws, bolts, and so on.

- *Use modular design*—Design self-contained elements, blocks, or chunks that can be fabricated under ideal conditions and then connected to the main assembly.

- *Minimize assembly direction*—Design products to be assembled along a single, preferably vertical axis (this is also called z-axis loading).

- *Minimize handling requirements during assembly*—Positioning parts is costly; design features can be used to simplify positioning; avoid parts that tangle.

- *Maximize compliance in assembly*—Use generous tapers, chamfers, radii; guiding features; use one component as a base.

■ TEAMWORK

As we have seen, the adoption of concurrent engineering is credited with enhancing productivity in design and manufacture, enabling companies engaged in product development to maintain and even increase market share, despite the recent upsurge in global competition. Teamwork, as a result, is the single most important distinguishing factor of the concurrent engineering philosophy. Teams are increasingly common on projects and in engineering firms. And even apart from this recent trend, engineering design has always been a social activity relying on teamwork skills like communication, collaboration, and cooperation.

A *team* is a group of people with complementary skills and knowledge who work together toward common goals and hold one another mutually accountable. *Teamwork* consists of a demonstrated collaborative attitude and the ability to accomplish team goals.

There is of course no assurance that a team will outperform a collection of individuals. Research suggests that there are four identifiable levels of team performance. These include teams that: (1) perform below the level of the average member, (2) don't quite get going but struggle along at, or slightly above, the level of the average member, (3) perform quite well, and (4) perform at an extraordinary level, where the members are deeply committed to one another's personal growth and success.[15] Research also suggests that sufficient time is necessary for a team to reach the higher performance levels.[16]

Successful teams are characterized by certain traits or skills. These include the use of group

[15] Smith, 2000.
[16] Eggert, 2005.

Table 1-6 Code of Cooperation, Boeing Airplane group, training manual for team members

Every member is responsible for the team's progress and success.

1. Attend all meetings and be on time.
2. Come prepared.
3. Carry out assignments on schedule.
4. Listen and show respect for the contributions of other members; be an active listener.
5. Constructively criticize ideas, not persons.
6. Resolve conflicts constructively.
7. Pay attention; avoid disruptive behavior.
8. Avoid disruptive side conversations.
9. Only one person speaks at a time.
10. Everyone participates; no one dominates.
11. Be succinct; avoid long anecdotes and examples.
12. No rank in the room.
13. Respect those not present.
14. Ask questions when you do not understand.
15. Attend to your personal comfort needs at any time, but minimize team disruption.
16. Have fun.

norms, as well as communication, leadership, decision making, and conflict management. *Group norms* are guidelines or standards of behavior that are agreed upon by the group. By establishing and adhering to these guidelines, groups can avoid many disruptive conflicts. Table 1-6[17] shows an example of group norms.

Group decision making is considerably more difficult than a decision taken individually. There are a number of ways to make a group decision, ranging from decision by authority to consensus. A decision based on authority takes the least amount of time, whereas decisions based on consensus require more time but generally result in the best decision. A consensus decision is one in which the team thoughtfully examines all the issues and agrees upon a course of action that does not compromise the strong convictions of any team member.

Conflicts inevitably arise between team members. A conflict is a situation in which the action of one person interferes with the actions of another person. Conflicts can be either constructive or

[17] Taken from Smith, 2000.

destructive. In a constructive conflict, the source of the disagreement is based on ideas or values, whereas in a destructive conflict, the dispute is based on personality. In order to function properly, it is essential for teams to develop rules that prohibit destructive conflict.

There are a number of strategies for dealing with conflict. These include:

1. *Avoidance*—simply ignore the situation, and hope that the problem resolves itself over time.
2. *Smoothing*—let the other side have his or her way.
3. *Forcing*—one side imposes a solution.
4. *Compromise*—meet the other side halfway. Though sometimes successful, this approach does not get to the underlying cause of the disagreement.
5. *Constructive engagement*—the two parties strive to get to the heart of the matter and then work to resolve the situation through negotiation. Only this approach offers the possibility of reaching a satisfactory resolution to a serious conflict.

TEAMWORK

∎ QUESTIONS

TRUE OR FALSE

1. Engineering design involves both analysis and synthesis.

2. A component can be either a part or a subassembly.

3. Design criteria serve to establish maximum, minimum, or permissible ranges for physical or operational properties of a design.

4. Design for Manufacture and Assembly (DFMA) guidelines include the minimization of the number of parts in a product.

MULTIPLE CHOICE

5. Which of the following is not a stage of the product development process?
   a. Sales/marketing
   b. Industrial design
   c. Engineering design
   d. Product design
   e. Manufacturing

`SS` 6. A good problem statement includes:
   a. An undesired initial state
   b. A desired goal state
   c. Obstacles that prevent going from the undesired to the desired state
   d. All of the above

7. Design _____ are used to qualitatively judge alternative design solutions.
   a. Specifications
   b. Criteria
   c. Constraints
   d. Phases

8. Informed designers often differ from `SS` beginning designers in that *informed designers*:
   a. Treat a design task as a well-defined, straightforward problem
   b. Use gestures, words, and artifacts to explore and communicate their design plans
   c. Make drawings, construct physical proto-types, and create virtual models that help them develop deeper understandings of how their designs function
   d. Propose and sketch ideas that superficially resemble viable solutions
   e. All of the above
   f. a and b
   g. b and c
   h. c and d

CHAPTER

2

HUMAN-CENTERED DESIGN AND DESIGN THINKING

■ INTRODUCTION

Engineering design is both an art and a science. Chapter 1 describes engineering design as a science, while the focus of this chapter is engineering design as an art. See Figure 2-1. This chapter begins with some history, the development of engineering as a discipline, the evolution of engineering design, and the advance of engineering education, especially in the twentieth century. It is helpful to know this historical

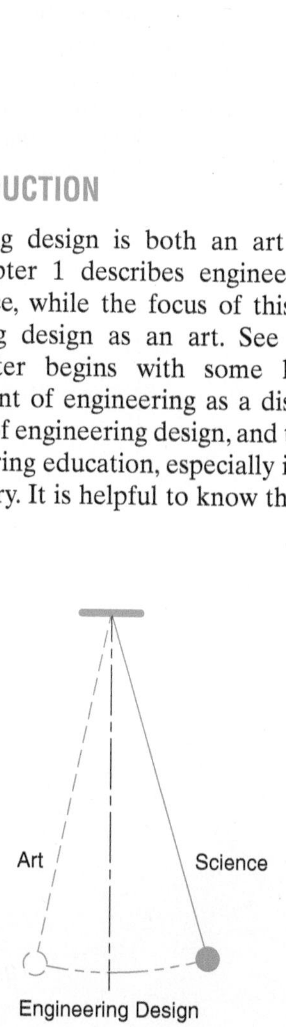

Figure 2-1 Engineering design: art and science

development in order to appreciate engineering design as both art and science.

Next comes a discussion of two design paradigms, design as rational problem solving, the "science of design," and design as reflective practice. The latter provides a foundation for a discussion of human-centered design and design thinking, the main topics of this chapter.

But why two designs? In the mid-twentieth century, that is, World War II and after, manufacturers needed to get products to market quickly and at reasonable cost in order to be competitive. In the later part of the twentieth century, that is, the late 70s through the 90s, a third ingredient, quality, became essential for success. Best-in-class companies now needed to deliver products to market quickly, at reasonable cost, and these products needed to provide quality, reliability, and customer satisfaction. In the early twenty-first century, good design has become the key differentiator; products now need to meet four criteria: time, money, quality, and design. Whereas design used to be viewed as a styling add-on, it is now of strategic importance. More than anything else, human-centered design and design thinking have been critical in allowing leading companies to set themselves apart from the competition.

ENGINEERING DESIGN: ART AND SCIENCE

Introduction

The Accreditation Board for Engineering and Technology (ABET) defines engineering design as follows:

> Engineering design is a process of devising a system, component, or process to meet desired needs and specifications within constraints. It is an iterative, creative, decision-making process in which the **basic sciences, mathematics, and engineering sciences are applied** to convert resources into solutions.

This definition can be traced back to Nobel laureate Herbert Simon's 1968 lecture, "The Science of Design: Creating the Artificial." Based on this definition, as well as Simon's use of the term "science of design," we would have to say that engineering design is more a science than an art.

Design: A Fundamental Human Activity

However, in that same lecture, Simon also says that "Everybody designs who devises courses of action aimed at changing existing situations into preferred ones." This quote makes it clear that *design is a fundamental human activity*. We are all designers, in that we all plan and act in order to improve our circumstances (i.e., devise courses of action aimed at changing existing situations into preferred ones). Figure 2-2 provides examples of human beings devising seats out of discarded materials.

This idea, that design is a fundamental human activity, is empowering. It helps to democratize design, making it an activity that is available to everyone. The concept of design is a little intimidating for most people. The recognition that we all do it makes it more accessible, and we can better see our creative potential. It is comforting to realize that, when making the decision to become an engineer or a designer, for example, you already know a thing or two about it. Knowledge gained through learning design can not only be applied professionally, but even to design one's own life.[1] In the end, design is about solving problems, and life is certainly filled with these.

Engineering Design from 1400 to 1900

The names of the first engineers are largely unknown to us. Think, for example, of the master builders of the Romanesque (see Figure 2-3) and,

[1] Bill Burnett and Dave Evans, *Designing Your Life: How to Build a Well-Lived, Joyful Life*, Knopf Publishing, 2016.

Figure 2-2 Bastard chairs (Michael Wolf Estate/laif/Redux)

Figure 2-3 Romanesque cathedral example (Church of St. Mary at Mount Naranco, 848) (Alberto Imedio/Wikimedia Commons/CC BY-SA 4.0)

later, Gothic (see Figure 2-4) cathedrals of medieval Europe. The identities of these individuals are mostly lost to us.

It is not until the early Italian Renaissance, in the 1400s, that the first recognizable "artist-engineers," beginning with the Florentine, Filippo Brunelleschi (see Figure 2-5) and culminating with Leonardo (see Figure 2-6), emerge

Figure 2-4 Gothic cathedral example (Lincoln Cathedral, 1092) (Richard Croft/East front/Wikimedia Commons/CC BY-SA 2.0)

Figure 2-5 Filippo Brunelleschi (Bildagentur-online/Universal Images Group/Getty Images)

in historical time. Both of these individuals are known first and foremost as artists,[2] and secondarily as engineers. At the time of the Renaissance, engineering was just beginning to emerge as a recognized discipline; so little distinction is made between artists and engineers, they are one and the same. Brunelleschi and Leonardo both began their careers by serving apprenticeships in the workshops of established master artisans.[3] Artisanal workshops like these have existed for centuries, and provide the framework necessary to preserve, protect, develop, and refine technical knowledge, and then pass it along to posterity.

Engineering as a profession did not appear until around the time of the industrial revolution. The first schools of engineering were founded in France in the mid- to late-eighteenth century.[4]

[2]Brunelleschi is best known as an architect, but he is also credited as the first person to describe how *perspective* works. Leonardo is best known as a painter, but he is equally famous for his engineering notebooks.

[3]Brunelleschi first apprenticed in a goldsmith's workshop; Leonardo apprenticed in Andrea del Verrocchio's workshop.

[4]The École Royale du Génie de Mézières (Royal Engineering School of Mézières) was a military engineering school in what is now Charleville-Mezieres, France. It was founded in 1749. The Ecole des Ponts et Chaussées (School of Bridges and Roads), established in Paris in 1747/1775, is the world's first civil engineering school. The École Polytechnique was established in Paris in 1794 by the mathematician Gaspard Monge, the inventor of descriptive geometry.

Figure 2-6 Leonardo da Vinci (Ivan Burmistrov/iStockphoto)

Figure 2-7 Engineer's notebook example—Francesco di Giorgio, column lifter (Science History Images/Alamy Stock Photo)

In the period from 1400 to 1900, the information developed and made available for use in educating engineers is overwhelmingly visual, nonverbal, and intuitive. Examples include engineer's notebooks (Figure 2-7) authored by two Sienese engineers,[5] Leonardo and their descendents, illustrated machine books called "theaters of machines" (Figure 2-8), working models (Figure 2-9), and mechanism collections featuring gears, linkages, and so on (Figure 2-10).

The tools of visualization developed and used by the earliest engineers included pictorial perspective, orthographic projection, engineering drawings, and study models.[6] Throughout this 500-year period, ending at the dawn of the twentieth century, engineering design was clearly more art than science.

Engineering Education after 1900

By 1900, the distinction between art and engineering was well established. At this time, most of an engineer's deep understanding continued to rely upon visual, nonverbal thinking and engineering intuition. This kind of knowledge is an intrinsic and inseparable part of engineering.

By the turn of the century, the curriculum in engineering schools was hands-on, with a practical emphasis. Students took classes in machine shop, surveying, and drawing. Technical drawing courses include mechanical drawing and descriptive geometry. See Figure 2-11. Mechanical drawing instruction used physical models coordinated with shop practice; machines designed in the drafting room were later built in the shop by students.[7]

[5]Mariano di Jacopo, known as Taccola (1382 to c. 1453), and Francesco di Giorgio Martini (1439–1501).

[6]Eugene S. Ferguson, *Engineering and the Mind's Eye*, MIT Press, 1997.

[7]Jerry S. Dobrovolny, *A History of the Department of General Engineering, 1868–1995*, University of Illinois at Urbana-Champaign.

Figure 2-8 Agostino Ramelli, Diverse and Ingenious Machines, 1588 (Agostino Ramelli/Library of Congress/Public Domain)

In 1921, the author's own Department[8] required 17 hours of practice-based instruction—in graphics (eight hours), machine shop (six hours), and surveying (three hours). By 1994, that same Department required four hours of graphics instruction,[9] while shop and surveying courses had been eliminated.

As the twentieth century progressed, engineering programs everywhere were gradually purged of vocationalism. Hands-on shop, surveying, and drafting courses were eliminated, and replaced with more math, science, and engineering science courses. These courses emphasized analysis, mathematical modeling, and theory-based approaches.

Several factors contributed to this move away from hands-on practical knowledge in engineering education over the course of the twentieth century. These factors include the influence of positivism, World War II, the American Society of Engineering Education (ASEE)'s Grinter Report, and the success of the Soviet *Sputnik* space mission in 1957.

[8]Industrial and Enterprise Systems Engineering, originally the Department of General Engineering, University of Illinois at Urbana-Champaign.

[9]This has since been reduced to three credit hours

Figure 2-9 Models submitted to move Vatican obelisk (The New York Public Library)

Positivism is a powerful philosophical perspective that tracks the rise of science and technology in the nineteenth century. Positivism was institutionalized in the modern university in the United States in the later part of that century. One of the central tenets of positivism is that empirical science is the only source of positive knowledge in the world. Only statements based on experience and observation have value. This led to the idea that the problems of practice (i.e., design problems) can be solved by the application of scientific theory. Science-based, technical practice began to replace craft and artistry. University-based scientists and scholars create fundamental theory, which professionals and technicians apply to practice. This led to the familiar split between

Figure 2-10 Reuleaux Kinematic Mechanisms Collection example

hands-on, practical knowledge. Vannevar Bush, an electrical engineer and MIT Dean of Engineering, was the Director of the Office of Scientific Research and Development (OSRD) during the war. Bush tended to value basic scientific research at the expense of applied engineering. Bush's book, *Science, the Endless Frontier* (published in 1946), stresses the importance of basic research by scientists. Bush is considered to be the architect of the National Science Foundation (NSF), founded in 1950. The NSF is to this day the leading funder of research done at elite engineering schools.

Just after the war, the ASEE conducted an internal review that led to the publication of the Grinter Report in 1955. The report recommended the elimination of courses having high vocational and skill content, as well as those primarily attempting to convey engineering art and practice. In addition, the report recommended teaching six engineering sciences—mechanics of solids, fluid mechanics, thermodynamics, heat and mass transfer, electrical theory, and nature and properties of materials.

In 1957, the Soviet Union successfully launched Sputnik, the world's first artificial satellite. This successful space mission caused considerable soul searching in the United States, where the feeling was that the United States was falling behind the Soviet Union in science and technology. This resulted in additional emphasis on science and technological research, as well as reforms in many areas, including education.

research and practice, with those who create new theory of higher status than those who apply it.

World War II provided additional impetus for the elevation of engineering science over

Figure 2-11 Turn-of-century engineering drawing class at the University of Illinois at Urbana-Champaign. Note the kinematic devices in the cabinet along the wall

▌ TWO DESIGN PARADIGMS

A *paradigm* is a philosophical and theoretical framework of a scientific school or discipline within which theories, laws, and generalizations and the experiments performed in support of them are formulated. Two design paradigms emerged in second half of the twentieth century: (1) design as a rational problem solving process and (2) design as a process of reflection, reflection-in-action, or reflective practice. They represent two fundamentally different ways of looking at the world, one based on positivism and the other on constructivism,[10] or even, one based on science, the other on art or humanism. The impact of these two paradigms is with us today, respectively as engineering design and human-centered design thinking.

Design as Rational Problem Solving

Herbert A. Simon (1916–2001) was an American economist, political scientist, and cognitive psychologist. He was a faculty member at Carnegie Mellon University for more than 50 years. In 1975 he received the Turing Award,[11] and in 1978 the Nobel Prize in Economics. Simon's "science of design" was outlined in his 1968 lecture, "The Science of Design: Creating the Artificial," and subsequently published as a chapter in Simon's influential book, *The Sciences of the Artificial* (1996, 3rd ed.).

For Simon, "artificial" means made by humans. He wrote that "the task of the science disciplines is to teach about *natural* things: how they are and how they work," whereas "the task of engineering schools is to teach about *artificial* things: how to make artifacts that have desired properties and how to design."

Simon's formulation of a science of design was motivated by the feeling that the discipline of design at that time was ". . . intellectually soft, intuitive, informal, and cookbooky," that "academic respectability calls for subject matter that is intellectually tough, analytic, formalizable, and teachable," but also to counter the fact that engineering schools, owing to positivism's influence, had gradually became schools of physics and mathematics.

Simon's science of design is defined as an objective, value-neutral, quantifiable, and mathematical field of research centered on problem solving. Here design is framed as a logical search for satisfactory[12] criteria that fulfill a specific goal. Simon attempted to scientize design, using classical sciences like physics as a model for a new science of design.

Simon aspired to make design into a technology-driven, rational problem-solving process. Design is viewed as a search process, where it is necessary to define and explore a design space, and then seek out optimal (or at least satisficing) solutions. Simon's view of design became the dominant methodology within engineering. Much of the material in Chapter 1 is influenced by Simon.

Simon's work on design, based on rational problem solving, has provided a stable basis for conducting design research, and informs much of our knowledge about design.

Despite the success of Simon's approach, there are limits to using mathematical logic in solving design problems. The rational problem-solving model of the design process is often juxtaposed with Donald Schön's reflective practice, a model that allows for professional expertise, intuition, and artistry.

Design as Reflective Practice

Donald A. Schön (1930–1997) was a philosopher and professor of urban planning at the Massachusetts Institute of Technology (MIT) for almost 30 years. Schön's book, *The Reflective Practitioner: How Professionals Think in Action*, was published in 1983. The book is written in reaction to positivism's influence on academia in the twentieth century, as well as to Simon's science of design. In the book, Schön argues

---

[10]Constructivism is the belief that reality is socially constructed, that there is no single reality, and that reality is a subjective creation.

[11]Annual award given by the Association for Computing Machinery (ACM), and generally recognized as the highest distinction in computer science.

[12]Simon introduced the term "satisficing" to describe a decision-making strategy that involves searching through available alternatives until a satisfactory/sufficient solution is found. He used this term to explain the behavior of decision makers under circumstances in which an optimal solution cannot be determined.

that universities ignore practical competence and professional artistry, and that schools of engineering have become schools of engineering science. He assails the view that professional activity should consist of problem solving made rigorous by the application of scientific theory and technique. He then attempts to answer the question "How do professional practitioners think in action?"

Much of the book consists of detailed descriptions of professionals/practitioners engaging in what Schön calls *reflective conversation*, including:

- A professor of architecture working with a student
- A supervisor of psychoanalysis working with a resident therapist
- A student mechanical engineering design team working on an industry sponsored project
- A brief historical development of the transistor
- The career of a systems engineer
- A town planner meeting with a developer and an architect
- Various situations involving technical and factory-floor management

Schön proposes that engineering design can be understood as "a *reflective conversation* with the materials of the situation."[13] The book includes two figures: (1) Stages of Engineering Design, Conceived as a Reflective Conversation, and (2) Stages of Development of the Transistor, reproduced here as Figures 2-12 and 2-13. In both

Stages of Engineering Design, Conceived as a Reflective Conversation	
Phenomenon to be explained or remedied	Inquiry
The traditional process.	Thermodynamic studies of the oxide layer. Experiments to duplicate results of the old process.
Why do both new and old processes work?	Experiments with dissolved oxygen.
Blistering, uncontrolled variables.	New experiments, coupled with design of production furnace.
Heat-sagging. Acceptable color and hardness on one side only.	Modified furnace design. Vertical furnace.

Figure 2-12 Stages of Engineering Design, Conceived as a Reflective Conversation

[13]For example, we can say that a designer has a reflective conversation with a sketch.

Stages of Development of the Transistor	
Theories	Experiment, phenomena, invention
Pre-World War II model of semiconductor.	Initial attempts to make semiconductor based on analogy with vacuum tube.
Shockley's theory of the electric field.	Field-effect experiments. Unexpected failure of predictions.
Bardeen's theory of surface states.	New experiments. Discovery of point contact transistor effect.
Theory of minority carriers.	Design of junction transistor.

Figure 2-13 Stages of Development of the Transistor

figures, actions are taken, leading to frequently puzzling outcomes. The outcome is then evaluated, leading to another action, and the process is repeated until a solution is converged upon.

Schön sees design as a process of **reflection-in-action**, by which he means to reflect on your actions in the moment, as you are doing them.[14] In the reflection-in-action paradigm, designers alternate between framing the problem, making design moves, and evaluating those moves. Framing refers to problem conceptualization, that is, defining goals and objectives. A move is a design decision. Evaluation leads to further moves in the design. See Figure 2-14.

Reflective practice, then, is the ability to reflect on one's actions in order to engage in a process of continuous learning. Reflective practice tries to explain how professionals meet the challenges of work with a kind of improvisation that is improved through practice. In the 1990's

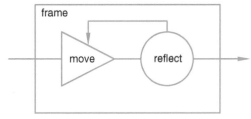

Figure 2-14 To frame (provide context), to move (action), to evaluate (reflection)

[14]Schön also identified ***reflection-on-action***, which he defined as reflecting on action after the action has been completed.

Schön's reflective practice was referred to as "design thinking research."[15]

The design as rational problem-solving paradigm is well-suited for situations where the problem is fairly straightforward, or well-structured, for example in the detail design phase. The design as reflective practice paradigm works well for ill-structured, or messy problems, typically in the conceptual stage of the design process. An early criticism of Simon's rational problem-solving approach is that it is unable to handle *wicked* problems.

◼ WICKED PROBLEMS

Many problems dealt with in science and engineering are well-structured, or "tame." These problems can be clearly defined, and solutions based on technical knowledge are available.

In contrast, **wicked problems** are a "class of social system problems which are ill-formulated, where the information is confusing, where there are many clients and decision makers with conflicting values, and where the ramifications in the whole system are thoroughly confusing."[16] Complex, ill-defined, ambiguous, open-ended problems are wicked, as are problems embedded within larger problems, and problems for which there is no obviously correct solution. In an open society, planning and policy problems (involving political, economic, and environmental issues) are inevitably wicked. Examples of these kinds of wicked problems include climate change, rural poverty, and obesity in America.

Design problems are often wicked because they tend to be poorly defined, they involve stakeholders with differing perspectives, and they have no obviously correct, let alone optimal, solution. These wicked design problems cannot be solved using rational problem-solving methods; they demand creative solutions.

[15]Nigel Cross, *Designerly Ways of Knowing*, 2001: "This (Schön's) approach particularly has been developed in a series of conferences and publications throughout the 1990s in 'design thinking research.'"

[16]Horst Rittel, design theorist and university professor, first at the Ulm School of Design in Germany, then at the University of California Berkeley, Professor of the Science of Design.

◼ DIVERGENT AND CONVERGENT QUESTIONING

A designer needs to know what the client wants. To learn from the client, a designer must be able to ask good questions. Effective inquiry involves two kinds of questioning, **convergent** and **divergent**. Convergent questions aim to get at the truth, to converge on the facts of the situation. Divergent questions expand the design space, provide options, and increase the number of alternatives. Figure 2-15,[17] shows that divergent questioning is used to create choices, while convergent questioning is used to make choices.

Convergent questioning traces back to Aristotle, who asserted that *knowledge resides in the questions to be asked and the answers to be provided*.[18] In Aristotle's systematic questioning procedure, the questioner asks low-level questions first, followed by deep reasoning questions. Low-level questions ask about the existence, essence, and attributes of a phenomenon, while deep reasoning questions ask about the phenomenon itself. The aim is to converge on the verifiable facts.

Convergent thinking is a practical way to decide among existing alternatives. A funnel serves as a useful metaphor for convergent questioning, with the flared opening at the top representing a broad range of initial possibilities and the small spout at the bottom representing a narrowly converged

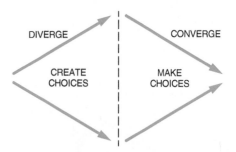

Figure 2-15 Diverge—create choices, converge—make choices (based on figure in *Change by Design*, p. 72)

[17]Adapted from Tim Brown's *Change by Design*, p. 72.
[18]Dym et al., "Engineering Design Thinking, Teaching and Learning," *Journal of Engineering Education*, 2005.

solution. The convergent, analysis phase of problem solving drives us toward solutions.

Divergent questions move from facts to possibilities. These questions have no truth value, and they cannot be verified. The aim of divergent thinking is to query the future, multiply options, and create new alternatives. Hypothetical questions are divergent. Here is an example from a student design group:

"Suppose that the different color filters (needed for the microscope) could be selected with a rotating wheel, like the one used on a hose nozzle?" This hypothetical question resulted in the multifilter housing design for an existing USB microscope, as shown in Figure 2-16.

Metaphorical thinking can generate lots of powerful ideas, both in engineering design and in scientific investigation. Schön in his book *The Reflective Practitioner* recounts a story about a group of product development researchers working to improve the performance of a new paintbrush. A breakthrough came when one of the researchers observed "You know, a paintbrush is a kind of pump!" This insight led the group

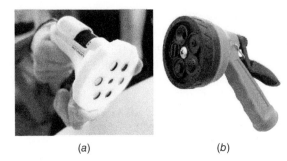

(a)   (b)

Figure 2-16 (*a*) Multifilter housing design for an existing USB microscope; (*b*) hose nozzle

to a variety of inventions. Schön called insights like these, paintbrush as pump, generative metaphors. Think too of **biomimetics**, the imitation of biological systems for solving human problems, an entire creative discipline based on simile.

For example, birds are able to engage in a multitude of complex flight maneuvers by changing the shape of their wings during flight. These abilities are not shared with today's small unmanned aerial vehicles (UAVs). Figure 2-17*a* shows the primary degrees of freedom of bird

Primary degrees of freedom of bird wingtips. Dihedral angle (left), incidence angle (center), gap space (right)

(a)

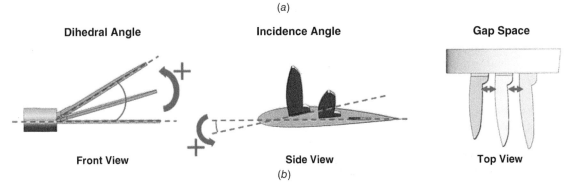

(b)

Figure 2-17 Primary degrees of freedom of bird wingtips (*a*) and their simulation (*b*) (courtesy of Dr. Aimy Wissa)

wingtips. Current research at UIUC aims to model these features in order to improve the maneuverability of UAVs (Figure 17*b*).

Convergent questions are closely associated with the problem definition phase of the design process and with analysis activities. Divergent questioning is associated with the ideation (e.g., brainstorming) phase of the design process, as well as with synthesis activities. This back-and-forth interplay between divergent and convergent questioning is central to design thinking.

▌ DOUBLE DIAMOND DESIGN PROCESS

According to Don Norman,[19] there are two main components of design: (1) find the right problem and (2) meet human needs and capabilities. These in turn give rise to the two phases of the design process, phase one: find the right problem and phase two: find the right solution. This led the British Design Council in 2005 to describe the design process as a double-diamond, as seen in Figure 2-18. The figure shows two diamonds, connected end to end. On the left, find the problem; on the right, find the solution. To achieve these objectives, an iterative combination of divergent and convergent thinking is used in each phase.

Norman observed that you start with the problem that you are asked to solve, but this is unlikely to be the real, fundamental, or root problem. It is only a symptom of the underlying problem. Think of the original problem as a suggestion, not as a final statement.[20]

Technical problem solving only works for well-formed problems, and well-formed problems are rarely given, they must be constructed. To convert a problematic situation into a problem that can be solved, the designer must first make sense of an uncertain situation. Finding the right problem to solve, also called *problem setting*, is difficult. To do so the relevant design issues must be identified, and a proper context for defining the problem must be provided. These activities are called *naming* and *framing*, respectively. Reflection, information gathering (research, interviews, observation), and iteration are all used to help find the root problem. Once we are satisfied that we have identified the correct problem, we can proceed to solve it. This second phase, find the right solution, introduces the next topic, Human-Centered Design (HCD).

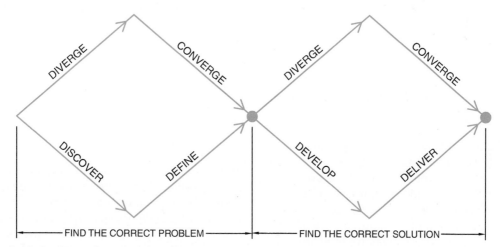

Figure 2-18 Double-diamond design process

---

[19]Don Norman, *The Design of Everyday Things*, 2013, p. 220.

[20]See the section on Problem Definition in Chapter 1 for an example of a problem statement that needed to be re-framed before the correct problem could be solved.

▊ HUMAN-CENTERED DESIGN

Recall that design is a fundamental human activity, and that "everybody designs who devises courses of action aimed at changing existing situations into preferred ones." Design comes naturally to humans; we are both designers and consumers of design. Humans design for and with other humans. Engineering design has great impact on society, and on the world we live in. Just think of the products, tools, artifacts, and devices we use daily, and the built environment that we live in, all designed by humanity.

There is, however, a lot of bad design, much of it because it fails to account for human needs.[21] *Human-centered design* first appeared in computer software, to help eliminate poor design. The influence of human-centered approaches quickly spread to other areas in design, including product design. In fact, it has been argued that there is a paradigm shift occurring in design from "technology-centered design" to "human-centered design,"[22] where technology-centered design is described as a process in which the designers or their clients make design decisions that are imposed on the intended users.

Human-centered design, by contrast, starts with the people being designed for. Throughout the process the intended user is the central focus of HCD. A designer sets out to understand user needs, capabilities, and behaviors, and to match them to the requirements of the intended design. This deep understanding comes primarily through observation, since people are often unaware, or unable to articulate what is needed.

HCD is both a philosophy and a process. Seen as a design philosophy, HCD aims to involve users throughout the design and development process, seeking to better understand user needs, practices, and preferences. It considers the whole user experience and looks to understand users holistically.

Don Norman[23] uses a cyclical representation to emphasize the iterative nature of the human-centered design process, as shown in Figure 2-19.

[21]Don Norman, *The Design of Everyday Things*, 2013.

[22]Krippendorff, *The Semantic Turn: A New Foundation for Design*, 2006.

[23]Norman, p. 222.

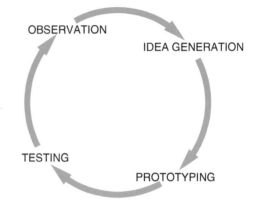

Figure 2-19 Iterative cycle of human-centered design: observation idea generation prototyping testing

Norman's iterative cycle takes place within the double diamond design process, and consists of four activities: observation, idea generation, prototyping, and testing.

To summarize, Don Norman defines human-centered design as follows:

- A process of ensuring that people's needs are met (human-centered focus)
- The resulting product is understandable and usable (focus on usability)
- It accomplishes the desired tasks (focus on functionality)
- The experience of use is positive and enjoyable (focus on meaning)

▊ DESIGN THINKING

Introduction

"Design thinking is a **human-centered** approach to **innovation** that draws from the designer's toolkit to integrate the needs of people, the possibilities of technology, and the requirements for business success."[24] Figure 2-20 illustrates this definition, showing that for design thinking to succeed three conditions—the desires of users,

[24]Tim Brown, Chair of IDEO.

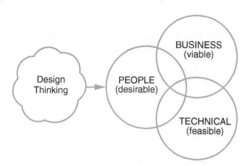

Figure 2-20 Venn diagram of showing the conditions for design thinking

the feasibility of the technology, and the viability of the business plan—must all be met.[25]

Three Spaces of Innovation

There are many different ways to represent the design process, Norman's iterative cycle of human-centered design (Figure 2-19 above) among them. Another example is Bernard Roth's[26] (and many others) five principles of design thinking:

1. Empathize
2. Define the problem
3. Ideate
4. Prototype
5. Test and get feedback

Tim Brown, in his book *Change by Design*, suggests that the design process is best thought of as a system of overlapping innovation spaces, rather than as a sequence of orderly steps. The three spaces of innovation are inspiration, ideation, and implementation.

In the inspiration space, opportunities motivating a search for solutions are identified, while insights are gathered in the field by observing the

user. Tasks or action words associated with the inspiration space include observe, inquire, empathize, finding stories, and insight.

In the ideation space, these insights are used to generate, develop, and test ideas and concepts. Brainstorming, sketching, and prototyping are all activities closely associated with the ideation.

In the implementation space the best ideas are developed into a concrete, fully conceived plan of action. Prototype, experiment, test, telling stories, and iterate are all associated with the implementation space. Figure 2-21 summarizes these design activities.

In the following sections, the different design process activities are discussed.

Inspiration

The inspiration space includes observation and empathy. These related activities can lead to insight into the unmet needs of users. *Insight* is the capacity to gain an accurate and deep intuitive understanding of a person or thing.

Observation

Observation refers to the act of going out into the field to observe and interact with people in their natural setting, wherever the product or service is

[25]Morris Asimow's *Introduction to Design* (1962) describes engineering design as a "synthesis of technical, human, and economic factors: and it requires the consideration of social, political, and other factors whenever they are relevant" (p. 2).

[26]Bernard Roth, *The Achievement Habit: Stop Wishing, Start Doing, and Take Command of Your Life*, 2015.

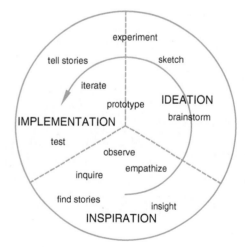

Figure 2-21 Three spaces of innovation, and some associated activities

likely to be used. This technique, borrowed from anthropology, is called *applied ethnography*.[27]

Imagine, for example, that you are asked to identify and develop potential products or services for bicycle messengers, or perhaps for people who commute to work or school using a bike. In addition to creating a survey and conducting interviews, a designer may decide to follow a bicycle messenger as they go about their daily routine. While the survey or interview will certainly provide useful information, observation is more likely to provide actual insight into the needs of the bike messenger or bicycle commuter community. Observation can lead to insight, and these hard-earned insights may produce inspired products and services that will improve lives.

Extreme users are especially likely to yield insights into potential products, services, and experiences that have the power to change lives. Typical bike commuters, those at the center of the bell curve, may confirm what we already know, and perhaps be able to verify that some project idea is valid. A novice cyclist at one extreme and the most experienced at the other are more likely to provide insights that trigger future offerings. Figure 2-22 helps illustrate these ideas on extreme users.

Empathy

Empathy is the capacity to see, experience, and understand the world through the eyes of others, to relate to their feelings and emotions. Empathy is to walk a mile in another person's shoes.

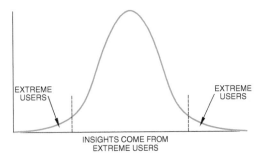

Figure 2-22 Bell curve showing extreme users

[27]See, for example, Tom Kelley's *Ten Faces of Innovation*.

Observation and empathy are closely related. To learn what the critical issues are and to build upon the ideas of others, we need to care about and connect with the people we are observing. While empathy is at the core of human-centered design, it plays no role in technology-driven design.

A prime example of empathy in design is the desktop metaphor for personal computers. Before 1980 computers were only used by scientists and engineers. The analogy of a computer being like a desktop with documents, folders, a trash bin, and so on, helped regular people to visualize why a computer might be useful in a home or office environment.

Ideation

Linus Pauling, a recipient of two Nobel Prizes,[28] is quoted as saying something like "If you want to have good ideas you must have many ideas." So clearly, to ideate a lot of thinking is required, but how are ideas generated? Techniques like sketching, meditation,[29] taking a walk, thinking about a problem before going to sleep, and even psychedelics[30] are all associated with idea generation.

The ideation process can be viewed as a back-and-forth alternation between divergent and convergent thinking. Convergent thinking is characterized by the collection and analysis of data, leading to convergence upon a single solution. Convergent thinking provides a practical way of deciding between alternatives and driving toward solutions. The metaphor of a funnel is useful in describing convergent thinking, with lots of input and a single output.

In divergent thinking, on the other hand, the objective is to multiply options in order to create choices. Divergent thinking is inherently creative,

[28]Chemistry in 1954, Peace Prize in 1962.

[29]https://medium.com/the-book-mechanic/how-vipassana-helps-yuval-noah-harari-write-his-book-73a1b1c9e978 https://www.vox.com/2017/2/28/14745596/yuval-harari-sapiens-interview-meditation-ezra-klein.

[30]Michael Pollan, *How to Change Your Mind: What the New Science of Psychedelics Teaches Us about Consciousness, Dying, Depression, and Transcendence*, 2018.

and capable of sparking innovation. Ideation can be described by this rhythmic exchange between divergent thinking—where new options emerge—and convergent thinking, where options are eliminated, and choices are made.

In the ideation space, insights obtained from the inspiration phase are used to generate, develop, and test ideas and concepts. Brainstorming and prototyping are activities that are closely associated with the ideation space.

Brainstorming

Brainstorming, also discussed in Chapter 1, is an important ideation technique with well-established guidelines. Tom Kelley writes in *The Art of Innovation*, "Brainstorming is practically a religion at IDEO, one we practice nearly every day." Kelley points out that because many people are already familiar with the concept of brainstorming, the practice tends to be trivialized. When used correctly though, collaborative brainstorming can yield great benefits. Brainstorming can be used at the start of a project or to solve an urgent problem that crops up. Some brainstorming guidelines include the following:

- Begin with a clear statement of the problem, one that is open-ended and not too broad.
- Alternatively, the Stanford Design Project Guide suggests using "How might we . . .?" (HMW) questions as a good way to launch a brainstorming session.
- Rules include go for quantity, encourage wild ideas, be visual, and defer judgment.
- The presence of senior executives is counterproductive.
- Stay focused on the topic; stick to one conversation at a time.
- Build on the ideas of others.
- Sixty minutes is a good length.
- Use colorful marker pens and Post-its to capture ideas. These ideas can then be grouped, sorted, eliminated, ranked, and so on.
- Use blocks, foam core, tubing, duct tape, hot-melt glue guns, and other prototyping basics.

- Number the ideas—it provides a goal (100 ideas per session is reasonable), helps gauge the fluency of the session, and provides markers.
- Change gears when you plateau; it helps to have a good facilitator.
- Use the space—cover the walls and tables with the ideas.
- Do stretching and warm-up exercises before you begin.

Prototyping

A **prototype** is a simple experimental model of a proposed solution for products, services, or experiences. A prototype is anything tangible[31]—typically a physical model, or a mock-up, but it might also be a sketch, a film, or a simulation. **Prototyping** is "thinking with your hands."[32] Giving form to our ideas and making them perceptible helps us to think more deeply about them.[33] Just as a designer has a reflective conversation with a sketch, a design team has a reflective conversation with a prototype.

Prototyping takes place across all three innovation spaces—inspiration, ideation, and implementation. Designers use prototypes in a great many ways, including:

1. For communication, visualization, to give form to, to reduce cognitive load (when dealing with complexity)
2. For exploration, experimentation, to learn and think clearly about
3. For evaluation, testing, validation, refinement, or to eliminate
4. To identify new directions, to advance, to move forward

with design ideas, concepts, assumptions, and with the project as a whole. Figure 2-23 shows a

[31]Capable of being precisely identified or realized by the mind.

[32]David Kelley, IDEO co-founder.

[33]"Design thinking is inherently a prototyping process. Once you spot a promising idea, you build it. In a sense, we build to think." Tim Brown, IDEO.

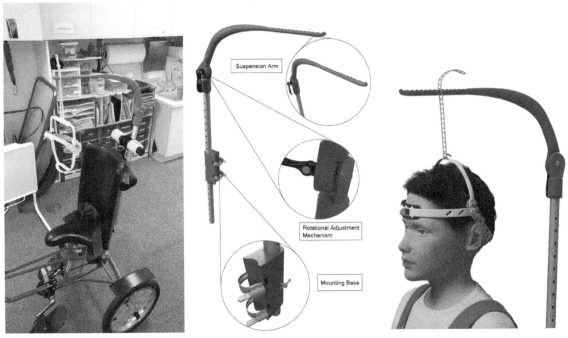

Figure 2-23 Working prototype of stabilizing arm designed by a student team. Device allows a person with cerebral palsy to ride a bike (Courtesy of Gerald Eilers, Matt Tenhagen, James Alrich, and Michael Baker)

working prototype, and Figure 2-24 compares an early prototype with the final product.

Prototypes benefit the design process by increasing the flow of ideas, helping to identify design flaws, generating feedback, and by speeding the process of innovation. Research suggests that building and testing physical models helps identify flawed features and reduces *design fixation*.[34] Products designed using physical prototypes are better and have fewer fixations.

Early prototypes should be simple, cheap, and easy to build. All prototypes should only take the time, effort, and expense required to generate useful feedback and move the project forward.

Rough prototyping materials include foam core, cardboard, balsa wood, acrylic sheets,

[34]Design fixation refers to a situation where a designer experiences an example of an existing design, then creates a new product with features similar to the prior example. Design fixation can occur unconsciously. Recently activated concepts are more likely to be retrieved. A student's tendency to commit to design ideas they think of first are likely to contain elements of fixation.

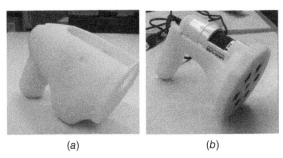

(a)   (b)

Figure 2-24 Comparing an (a) early prototype with the (b) final product

modeling clay, and even pipe cleaners. Cutting tools include X-acto knives and extra blades, box cutters, scissors, and a large cutting mat. Adhesives include five minute epoxy glue, permanent spray adhesive, white glue, and a hot glue gun and glue sticks. Scotch, double-sided, and masking tapes are all useful, as are different grades of sandpaper. Figure 2-25 shows a variety of prototyping materials and tools.

For more refined prototyping, laser cutters and 3D printers are especially useful. Laser

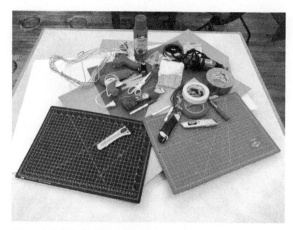

Figure 2-25 Prototyping materials and tools

cutters are used to cut two-dimensional sheets in acrylic, plexiglass, wood, cardboard, and leather. Figure 2-26 shows a typical laser cutter.

Implementation

The implementation space may include prototyping, testing, and iteration, as well as downstream marketing activities. In this phase, the best ideas are developed into a concrete, actionable plan for taking the product or service to market. The plan must be convincingly tested, and then communicated with sufficient clarity in order to gain acceptance.

Figure 2-26 Laser cutter

Test and Iterate

Testing begins with the recruitment of users, small groups of people who closely match the intended audience. Have the users interact with the prototypes while the design team observes and asks questions. The design team may choose to video record the sessions. Does the prototype meet the design requirements? Based on this feedback, the team develops more prototypes, and does more testing. See Figure 2-27.

Testing is done in both the problem specification phase to ensure that the problem is well understood, and in the problem solution phase to ensure that the design meets the needs and abilities of those who will use the product.

Iteration is a cyclical process in which the repetition of a sequence of operations yields results successively closer to a desired result. The back-and-forth interplay between rapid prototyping and testing, including feedback from the user, is at the core of design iteration and allows for a continual refinement and enhancement of the project goals and requirements. In cycling through the design phases, the tests are more targeted and efficient. With each passing cycle, the ideas are clearer, the specifications are better defined, and the prototypes move closer to the actual product. The design process converges upon a solution.

Failure is a valuable part of the design process and is to be encouraged. In the moment, it seems that failure is a complete waste of time, amounting to nothing. This is not the case however, unless we fail to learn from our mistakes. Failure is an inevitable part of all learning experience, one that builds character and resilience. Throughout the process there is a focus on action, on experimentation, on doing. Reflect-in-action but avoid overthinking. Move fast and break things.

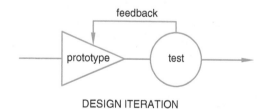

Figure 2-27 Design iteration: prototype and test

The term *T-shaped* is used to describe people having skills and knowledge that are both deep and broad. T-shaped individuals have the capacity and the disposition for collaboration across disciplines. They share two characteristics, empathy and an enthusiasm for other disciplines. Multidisciplinary people – architects who have studied psychology, artists with MBAs, engineers with art history degrees – frequently fit this T-shaped description.

As seen in Figure 2-28, the vertical axis represents expertise in a specific field of endeavor. T-shaped people possess a depth of skill in their chosen profession that allows them to make important contributions to a project. Whether as an engineer, industrial designer, anthropologist, or behavioral scientist, possession of such a specialized expertise is an essential characteristic of being T-shaped. Think, for example of Malcolm Gladwell's ten-thousand-hour rule.[35] Throughout their careers, these individuals have demonstrated both talent and hard work in order to achieve mastery in their discipline

The horizontal, or breadth, axis is where the design thinker is made. These right-brained thinkers are adept at dealing with complex, messy, wicked problems. T-shaped individuals show empathy for people and for disciplines beyond their own. They learn through doing and experimentation. They are open to all things new, curious, and optimistic.

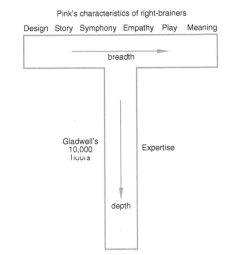

Figure 2-28 T-shaped individuals

Daniel Pink's six characteristics of right-brain thinkers[36]—design, story, symphony, empathy, play, and meaning—provide an excellent accounting of the strengths of these multifaceted individuals. T-shaped individuals are problem solvers, storytellers, directors (symphony), empathizers, playful, and driven by a search for significance.

Radical Collaboration

Collaboration means working together to accomplish a goal. When the collaborators have varied backgrounds and diverse viewpoints, and their insights and innovative solutions stem from this diversity, then this may be called *radical collaboration*. Stanford's d.school[37] describes radical collaboration as follows:

> To inspire creative thinking, we bring together students, faculty, and practitioners from all disciplines, perspectives, and backgrounds. Different points of view are key in pushing students to advance their own design practice. Our methods become a shared language for groups to navigate the ups and downs of messy challenges.

For radical collaboration to work, it is important to have a diverse group of people involved in the process. Multidisciplinary, T-shaped people embrace diversity. They also tend to make good communicators and collaborators. Diversity can lead to more creativity and better decision making.

Involving clients in the design process can radically improve collaboration and promote innovation. One example comes from Proctor &

[35]See *Outliers*, 2008 or *Complexity and the Ten-Thousand-Hour Rule*, New Yorker, 2013.

[36]Daniel Pink, *A Whole New Mind: Why Right-brainers Will Rule the Future*, 2005.

[37]https://dschool.stanford.edu/about.

Gamble (P&G), and their CEO, Alan G. Lafley. Through Lafley's leadership, P&G has become one of the world's most admired and successful companies, with products like Tide, Crest, Pampers, Swiffer, and Febreze. Working with the design consultancy IDEO, in 2002 P&G began an initiative to use design as a source of innovation and growth. In innovation workshops, P&G employees learn principles of design thinking. Business schools today use P&G's success as a twenty-first century case study in disruptive innovation.

Collaborative teams composed of T-shaped participants can lead to truly ***interdisciplinary teams***. Multidisciplinary teams are composed of people from different disciplines working together, but with each person focused on their own expertise. Projects tend to become negotiations, resulting in some form of compromise. Interdisciplinary teamwork, on the other hand, is characterized by a melting pot of ideas, a synthesis of approaches, and a fusion of knowledge and techniques derived from the respective disciplines. There is a collective ownership of ideas, and everyone takes responsibility for the project.

■ QUESTIONS

SHORT ANSWER

1. What does it mean to say that design is a reflective practice?
2. Explain the relevance of observation in human-centered design.
3. List three to five brainstorming techniques.
4. What is iteration, and how might iteration lead to better design ideas?

TRUE OR FALSE

5. Divergent questions aim to get at a truth and ultimately lead to one best solution.
6. A human-centered design approach requires understanding the experience of use.
7. Prototypes should always be made with a CAD program.

MULTIPLE CHOICE

8. Who said the following: "Everyone designs who devises courses of action aimed at changing existing situations into preferred ones." SS
   a. Herbert Simon
   b. Vannevar Bush
   c. Donald Schön
   d. Russell Ackoff
   e. Horst Rittel
9. Which of the following is **not** an important contributor to the 20th century science of design trend in engineering education in the United States?
   a. Positivism
   b. Wicked problems
   c. World War II
   d. Grinter Report, 1955
   e. Sputnik, 1957

CHAPTER

3

PRODUCT DISSECTION

■ INTRODUCTION

Chapter 1 began with the observation that design is the central activity of the engineering profession. The chapter then attempted to describe the nature of engineering design: what it is, how it is done, and so on. Engineering design is truly about finding solutions to *real-world* problems. Good design calls for verbal (common sense, reasoning), graphical (sketching, CAD modeling), and social (communication, collaboration, teamwork) skills, as well as analytical (mathematical, scientific) skills focused on finding innovative solutions to open-ended problems.

Engineering design is difficult to master. There is no *cookbook approach* to design. Fortunately, the design truism "All design is redesign" provides another avenue for learning design, other than "Just do it." Young engineers are frequently surprised (and perhaps disappointed) to find how rare original design is. Very few design projects start with a blank sheet of paper. Most design efforts begin with an existing design or product that serves as the basis for the new project. Given today's global market, this reality has become more evident. In order to remain competitive, modern manufacturing companies must roll out new products (e.g., digital cameras, computers, cell phones) with astonishing frequency.

The subject of this chapter is product dissection. The techniques of product dissection and reverse engineering (see Chapter 14) are based on the notion that by taking apart and studying mechanical or electrical devices, we can better understand not only the specific product, but also the design process that produced it. After all, one of the biggest obstacles to learning design is that we are forced to begin with a blank sheet of paper. Starting with an existing product that is disassembled, analyzed, sketched, modeled, and finally improved upon, our understanding of design can only grow.

Product dissection is an approach to learning about engineering concepts and design principles by exploring engineered products. In a product dissection project, the relationship between the form and function of a device is investigated. Also called mechanical dissection, product dissection enables us to see how others have successfully solved a particular design problem.

In a typical dissection exercise, student teams disassemble, study, and then reassemble commercial products such as bicycles, power tools, and coffee makers. Frequently cited benefits of this hands-on activity include (1) improved mechanical aptitude, (2) basic knowledge of manufacturing processes, materials and material selection, and the product decomposition hierarchy, (3) exposure to the design process, (4) appreciation for how ergonomics influences product design, and (5) awareness of how customer needs are translated into product functions that are then converted into commercial products.

PRODUCT SUITABILITY

The following guidelines are helpful when one is selecting a product or mechanism suitable for disassembly.

- Assemblies with 10 to 15 parts are ideal, but a range from 5 to 30 parts is workable.
- Try to select products with moving parts so that you can use CAD software animation and motion analysis capabilities.
- Look for products that can easily be disassembled and reassembled, either by hand or with simple tools; avoid products with welded housings and glued joints.
- Try to select an inexpensive product; if it must be broken to be disassembled, a second one can be affordably purchased.
- The device should still be in working order, so that it is possible to see how the external functionality is achieved internally.
- The artifact should be easily portable; it should not be large or heavy, nor should it be too small.
- The product should not require complicated assembly instructions.
- You should *want* to reverse-engineer the product.
- Ideally, the internal functioning of the product should not be readily apparent.
- The product should offer the potential for improvement.

Figure 3-1 shows several commercial products that meet these requirements.

PRODUCT DISSECTION PROCEDURE

In addition to imparting knowledge of mechanisms and product design, product dissection experiences also offer many opportunities to develop graphical and visualization skills. Product

Figure 3-1 Commercial products used for product dissection projects: bicycle front hub, kitchen item, toy

dissection activities can be expanded to include freehand and digital sketching, instrument drawing, and CAD modeling. Assuming that a parametric assembly model of the product has been created, the virtual prototype can be documented (working drawings, rendered views, physical prototypes), and analyzed (weight estimate, tolerancing, motion analysis, stress analysis). In an effort to continuously improve their products, manufacturing companies routinely employ reverse engineering practices. In an academic setting, a product improvement component can likewise be incorporated into the dissection experience. Finally, as in other design projects, it is important that the dissection team communicate its findings. Thus a product dissection project typically incorporates the following features:

- Pre-dissection analysis
- Dissection
- Product documentation
- Product analysis
- Product improvement
- Product reassembly
- Communication

In the course of the subsequent discussion, a pair of locking pliers made by Craftsman Tools (see Figure 3-2) will be used to illustrate the dissection process. Locking pliers are used by trade professionals, as well as around the house, for clamping, tightening, twisting, and turning.

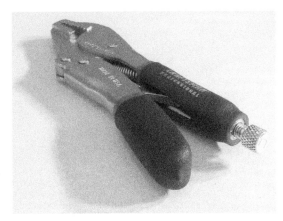

Figure 3-2 Craftsman Professional locking pliers (Courtesy of Craftsman Tools, Inc.)

▌ PRE-DISSECTION ANALYSIS

Before taking the product apart, it is worthwhile to evaluate the product's functionality and market segment. With regard to functionality, the device should be observed in operation. What does the device do? Make some predictions regarding how the device works. How many parts does the product contain? What sort of mechanisms allow the device to work as it does? What scientific principles were used in the design of the mechanism?

Locking pliers like the Craftsman Professional 7-inch locking straight-jaw version are popular for welding, metalwork, and in other trades. They are also found in home tool boxes. This multipurpose hand tool functions as an adjustable wrench, but it can also be used as a clamp or vise capable of holding a workpiece in a fixed position. The primary mechanism employed in locking pliers is a *four-bar linkage*, although the principle of *leverage* is also used. The locking pliers appear to contain about ten different parts.

Next, create a black-box diagram that represents the overall intended function of the product. Inputs and outputs can be categorized as material, energy, or signals (information). A black-box representation showing the functional structure of the locking pliers is shown in Figure 3-3.

A list of customer needs should be identified to establish the market segment. Who are the target customers? Who are the principal competitors? Identify some features of competing products.

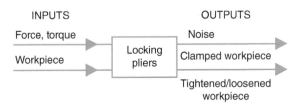

Figure 3-3 Black-box representation of locking pliers' functionality

What are the advantages and disadvantages of this product versus those of the competition?

A list of the qualities that are important to customers who seek locking pliers might include the following:

- Tough, durable
- Reliable
- Easy to adjust
- Easy to use
- Can be locked in any position
- Easy-to-release locking mechanism
- Comfortable grip
- Affordable

Although the primary customers for locking pliers are welders and other metal workers, the basic design has been modified to expand the target user base to include other crafts. The Craftsman locking pliers appear to have one principal competitor, along with several smaller competitors who typically produce less expensive versions of the locking pliers concept. Most of the designs include a trigger component that is used to release the locking mechanism, but the Craftsman tool has a patented design that permits an easier release with no need for the trigger component. The Craftsman tool also includes a nonslip, cushion grip for increased reliability and comfort.

▌ DISSECTION

In dissecting the product, the principal goal is to understand how the product works and is assembled. Another goal is to identify potential

As mentioned earlier, potential candidates for product dissection should be easy to take apart, either by hand or with simple tools. Useful tools include screwdrivers, wrenches, and pliers. In difficult situations, a Dremel with cutting wheels, an electric drill, shear cutters, a hammer, or a hacksaw may have to be used. The Craftsman locking pliers proved difficult to disassemble, owing to its rugged design. Steel rivets, used to connect the locking pliers' components, had to be removed.

Good lighting is certainly important when dissecting products. A workbench equipped with a vise or clamps is also helpful. A camera is useful in documenting the dissection process, especially when it is used to show the order and orientation of parts as assembled (i.e., exploded views). Figures 3-4 through 3-12 provide a visual record of the dissection process for the locking pliers.

As the product is disassembled, it is a good idea to keep a written record of the disassembly sequence. This product disassembly plan can later be used to help create an exploded view and an assembly animation file. If at all possible, try to avoid breaking parts. If breakage is unavoidable and the product is inexpensive, a second product may be purchased so that a working version will remain available.

Craftsman locking pliers disassembly steps

The following steps provide a disassembly sequence for the locking pliers shown in Figure 3-2.

1. Unscrew the adjusting screw from the fixed handle. Note that as the adjusting screw is removed, the jaws of the locking pliers open (Figure 3-4).

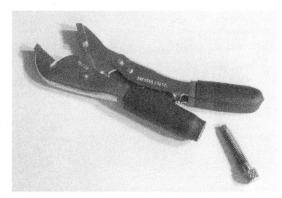

Figure 3-4 Adjusting screw removed

2. Free the end of the toggle link from the fixed handle channel by pushing down on the toggle link hump feature (thus closing the jaws) and then rotating the link end out of the channel (Figure 3-5).

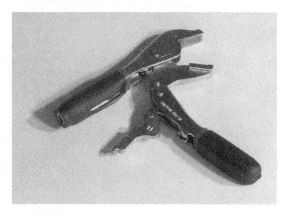

Figure 3-5 Toggle link end released from fixed handle channel

(*Continued*)

DISSECTION

3. Using a small screwdriver or other tool, remove the spring that spans between the fixed handle channel and the jaw. Note that the spring serves to resist the closing of the opposable jaws.

4. Using a flat screwdriver or other tool to stretch and loosen them, slide the cushion grips from both the fixed handle and the movable handle. Figure 3-6 shows the cushion grips and the spring.

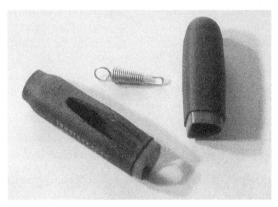

Figure 3-6 Cushion grips and spring

5. After clamping the fixed handle in a bench vise, use an electric drill, a rivet removal tool (Figure 3-7), and regular drill bits to remove the head of the rivet connecting the fixed handle to the jaw (Figure 3-8). Once the rivet head has been removed, use a hole punch and hammer to completely remove the rivet. Figure 3-9 shows the fixed handle subassembly components.

Figure 3-7 Rivet removal tool kit

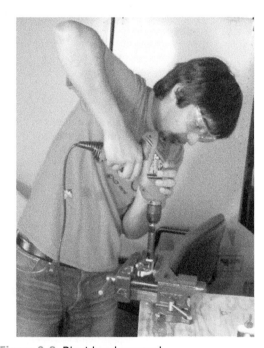

Figure 3-8 Rivet head removal

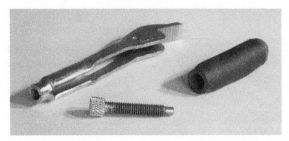

Figure 3-9 Fixed handle subassembly components

6. Use the process described in step 5 above to remove the rivet connecting the jaw to the movable handle subassembly.

7. Repeat the process described in step 5 above to remove the rivet connecting the movable handle to the compound toggle link subassembly. Figure 3-10 shows the movable handle subassembly components.

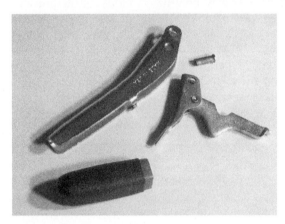

Figure 3-10 Movable handle subassembly components

8. Figure 3-11 shows the compound toggle link subassembly. The flat rivet connecting the compound link to the toggle link should be removed using the technique described in step 5. In practice, however, this proved to be too difficult. But it was not difficult to obtain the dimensions of both parts while the subassembly was still assembled (see Figure 3-16, which includes a tracing of the subassembly profile). Figure 3-12 shows the disassembled components of the locking pliers.

Figure 3-11 Compound toggle link subassembly

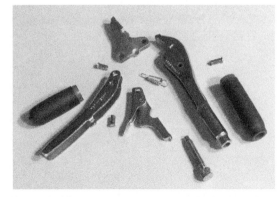

Figure 3-12 Disassembled parts for the locking pliers

▪ PRODUCT DOCUMENTATION

Product documentation can include a number of elements, among them a product decomposition diagram, freehand sketches of the parts and assembly, and a parametric assembly model of the product. If the product is modeled, then the list of documentation items can be expanded to include working drawings, 3D prints, and rendered views.

A product component decomposition diagram, as shown in Figure 3-13, can be used to capture the relationships among the product's parts and subassemblies. Note that in order to construct the diagram, it is necessary to identify all parts with the appropriate names—another useful activity.

Product dissection exercises provide an opportunity to use the sketching techniques discussed in earlier chapters on these open-ended problems. Depending on the dissected product, any number of sketches may be appropriate,

including multiviews, pictorial views (isometric, oblique, perspective), section views (full, offset, half, broken-out, revolved, removed), and auxiliary views. Assembly sketches (e.g., section, broken-out) can also be executed, although exploded views are best done within CAD software. Figure 3-14, for example, shows several section views of a pump head part, made in the course of dissecting a portable water filter product.

Although the tools needed to execute the part sketches are normally limited to pencil and paper, digital calipers (see Figure 3-15) are also necessary if the sketches are to include the dimensions needed to create an accurate CAD model of the product. This offers an opportunity to put previously acquired dimensioning skills—and even tolerancing skills—into practice.

Figure 3-16 displays traced sketches of three different locking pliers parts to be used for dimensional take-off. Note that a circle template and triangles have been used in the execution of these sketches. Also note that hole features are

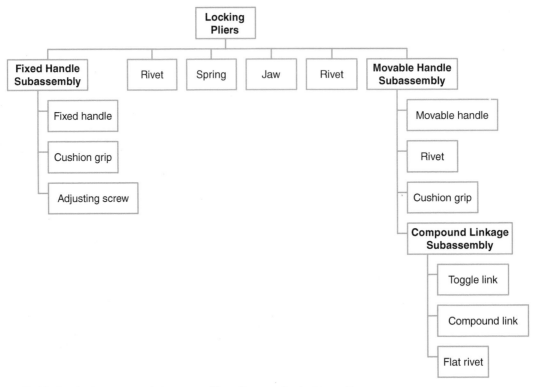

Figure 3-13 Product component decomposition diagram for locking pliers

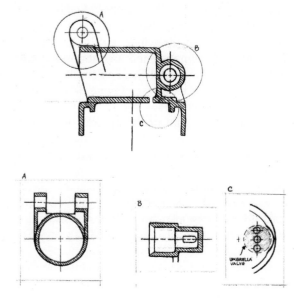

Figure 3-15 Digital calipers

Figure 3-14 Section view sketches of portable water filter pump head (Courtesy of Andrew Block)

used as the origin in two of these part sketches. Compare the sketch of the jaw part with the parametric sketch of the same part that appears

in Figure 3-17. One of the critical hole features is located at the sketch origin; the second hole is on the y axis.

Once the parts are modeled, subassemblies and a final product assembly model may also be created, assuming that parametric solid modeling software is being used. In the event that some parts have complex, sculpted surfaces, it may be possible to use a portable coordinate-measuring machine (see Figure 14-2 in Chapter 14) or 3D scanner (Figure 14-6 in Chapter 14) to capture the part's surface features. Figure 3-18 shows several CAD models of different parts from the locking pliers.

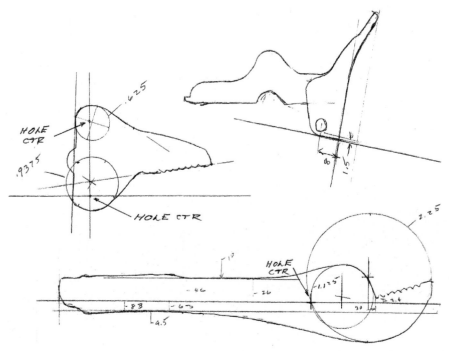

Figure 3-16 Locking pliers part sketches

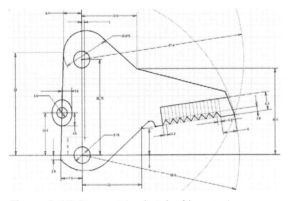

Figure 3-17 Parametric sketch of jaw part

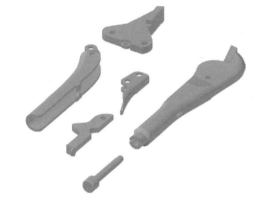

Figure 3-18 CAD models of locking pliers parts

Assuming that the dissected product has been modeled (parts, subassemblies, final product assembly), working drawings can also be derived. In addition to dimensioned and annotated part drawings (see Figure 3-19), an assembly drawing including section views (see Figure 3-20) as well as an exploded view (see Figure 3-21) can all be developed. Note that in the process of creating the exploded view most parametric modeling software programs also produce an animation file. This animation file documents the product's assembly, since the disassembly steps are played in reverse order. A parts list or bill of materials (BOM) can also be created in association with the exploded or other assembly views. See Figure 3-22.

If a 3D printer is available, additional product documentation can include a physical prototype of individual parts, of a critical subassembly, or even of the entire product. Figure 3-23 shows a 3D print of the assembled locking pliers parts. Assuming that a rendering module is either included within the CAD system or available separately, a photorealistic rendering of the product may also be created. Figure 3-24 shows a rendered view of the locking pliers.

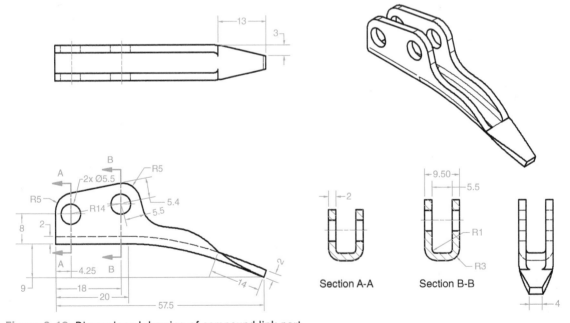

Figure 3-19 Dimensioned drawing of compound link part

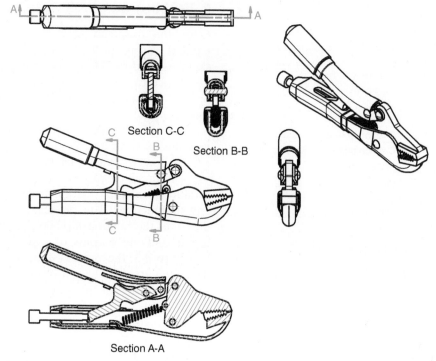

Section C-C

Section B-B

Section A-A

Figure 3-20 Assembly drawing views of the locking pliers

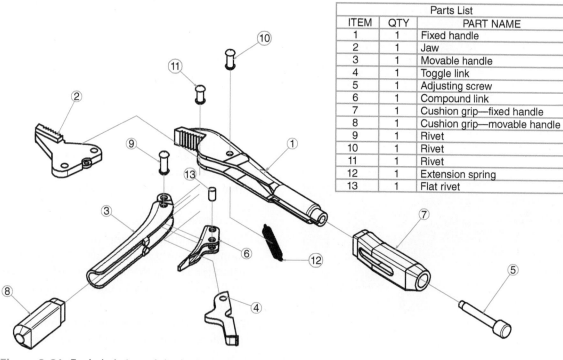

Parts List		
ITEM	QTY	PART NAME
1	1	Fixed handle
2	1	Jaw
3	1	Movable handle
4	1	Toggle link
5	1	Adjusting screw
6	1	Compound link
7	1	Cushion grip—fixed handle
8	1	Cushion grip—movable handle
9	1	Rivet
10	1	Rivet
11	1	Rivet
12	1	Extension spring
13	1	Flat rivet

Figure 3-21 Exploded view of the locking pliers

Parts List		
ITEM	QTY	PART NAME
1	1	Fixed handle
2	1	Jaw
3	1	Movable handle
4	1	Toggle link
5	1	Adjusting screw
6	1	Compound link
7	1	Cushion grip—fixed handle
8	1	Cushion grip—movable handle
9	1	Rivet
10	1	Rivet
11	1	Rivet
12	1	Extension spring
13	1	Flat rivet

Figure 3-22 BOM for the locking pliers

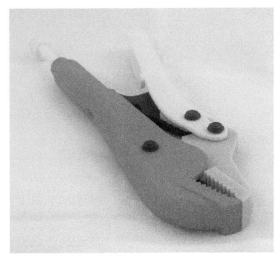

Figure 3-23 3D print of the assembled locking pliers parts

Figure 3-24 Rendered view of the locking pliers

▌ PRODUCT ANALYSIS

Once the dissected product has been documented (using part decomposition diagrams, sketches, models, 3D prints, and rendered views), the device can also be analyzed. Some analysis options include functional decomposition analysis, material analysis and weight estimation, manufacturing process analysis, kinematic analysis, and stress-strain analysis.

The goal of functional decomposition analysis is to break down, or decompose, the primary product function hierarchically into different subfunctions. The product's basic function describes the relationship between the available inputs and the outputs of a product. As seen in Figure 3-3, the product functionality can be described graphically as a black box. Inputs are shown entering on the left, and outputs exiting on the right. Verbally, product functions take the form of a combination of an action verb and a noun. For example, the overall product function of the locking pliers is *clamp workpiece* or *tighten (loosen) workpiece*.

In order to successfully realize the basic product function, a number of secondary functions, or **subfunctions**, must be satisfied. Like the basic product function, these subfunctions can be expressed as action verb + noun combinations. Careful observation of the functioning of the locking pliers, for example, reveals the following subfunctions:

1. Adjust jaws—the adjusting screw is first used to set the jaw size (i.e., distance between jaws) to slightly less than that of the workpiece to be clamped.
2. Open jaws—the locking pliers are then opened by pushing the movable handle away from the fixed handle.
3. Clamp workpiece—after placing the jaws around the workpiece, the jaws are closed by pushing the movable handle toward the fixed handle, effectively clamping the workpiece in place.
4. Tighten/loosen workpiece—if required, the workpiece can now be tightened, loosened, twisted, or turned.

5. Open jaws—to release the workpiece, the movable handle is once again pushed away from the fixed handle, releasing the workpiece.

In the function structure of the locking pliers shown in Figure 3-25, these **critical-path subfunctions** appear in sequence across the middle of the box, connected by arrows.

In addition to these critical-path subfunctions, a number of other secondary functions must also be addressed. Some of these secondary functions support the various critical-path functions. For example, in order for the jaws of the locking pliers to open, the range of rotation of the compound link must be limited inside the movable handle channel, at which point the toggle link disengages from the compound link, allowing the jaws to open.

Other subfunctions are shown at the top of the box in Figure 3-25. These secondary functions are sometimes referred to as all-time functions because they must be met at all stages in the operation of the device. In the case of the locking pliers, the principal components, which together form a four-bar linkage, must be permanently connected and yet free to rotate with respect to one another.

Note that there are a number of different ways to graphically represent the functional structure of a product, Figure 3-25 being one example. The goal of functional decomposition is to gain insight into how a basic product function is realized. Decomposing an overall function into several subfunctions makes the design problem more manageable.

Note also that there tends to be a mapping between subfunctions and components. As an alternative to (or in conjunction with) a functional decomposition diagram, it is useful to identify the function of the different components. For example, in the locking pliers, solid rivets are used to connect the various components; a spring connecting the movable jaw to the fixed handle tends to keep the jaws separated; and handles are provided so that the tool can be grasped.

The main components of the locking pliers are **forged**, using high-grade, heat-treated alloy steel. The threads of the adjusting screw have been **roll threaded**. The cushion grips are made of rubber

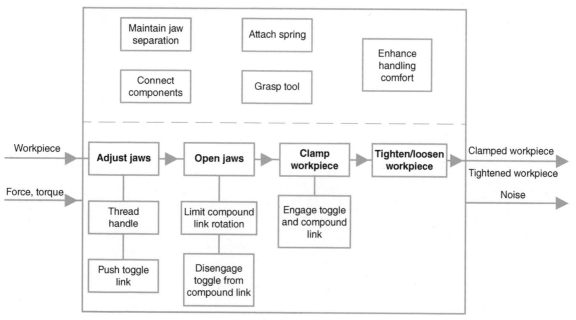

Figure 3-25 Functional decomposition of the locking pliers

Table 3-1 **Weight estimate of the locking pliers**

Item	Qty	Part name	Material	Mass (grams)
1	1	Fixed handle	Steel, High Strength Low Alloy	156.06
2	1	Jaw	Steel, High Strength Low Alloy	79.32
3	1	Movable handle	Steel, High Strength Low Alloy	77.70
4	1	Toggle link	Steel, High Strength Low Alloy	27.48
5	1	Adjusting screw	Steel, Mild	20.20
6	1	Compound link	Steel, High Strength Low Alloy	16.70
7	1	Cushion grip—fixed handle	Rubber	9.82
8	1	Cushion grip—movable handle	Rubber	6.42
9	1	Rivet	Steel, High Strength Low Alloy	4.22
10	1	Rivet	Steel, High Strength Low Alloy	3.61
11	1	Rivet	Steel, High Strength Low Alloy	3.27
12	1	Extension spring	Steel	1.80
13	1	Flat rivet	Steel, High Strength Low Alloy	1.77
			Estimated weight	**408.35 grams**

and have been manufactured using ***injection molding***.

Using the solid assembly model, it is possible to estimate the virtual product's weight and then compare your estimate with the actual product weight. A weight estimate based on the reverse-engineered CAD assembly model of the locking pliers appears in Table 3-1. The estimated weight of 408 grams is reasonably close to the actual weight of 372 grams, roughly a 9% difference. Probable reasons for the difference between the estimated and actual weights include

inaccuracies in modeling and in material density. The weight estimate shown in Table 3-1 was created by exporting the CAD generated parts list to a spreadsheet, where modifications were made.

Assuming that a virtual assembly model has been created and that the assembly constraints have been properly defined, animations showing the motion of any moving parts within the product assembly can be displayed. The range of motion of these moving parts can also be determined. Figure 3-26 shows the locking pliers with

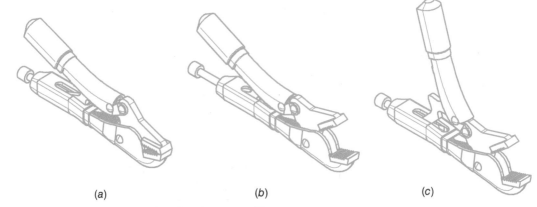

(a)　　　　　　　　　(b)　　　　　　　　　(c)

Figure 3-26 Locking pliers in closed and open positions

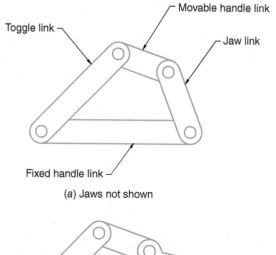

(a) Jaws not shown

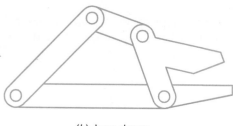

(b) Jaws shown

Figure 3-27 Simple four-bar linkage model of locking pliers

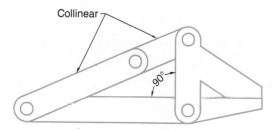

Figure 3-28 Simple four-bar linkage model of locking pliers with jaws closed

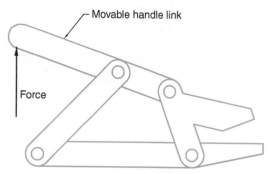

Figure 3-29 Simple four-bar linkage model of locking pliers with jaws open—movable handle pushed outward

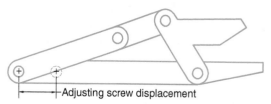

Figure 3-30 Simple four-bar linkage model of locking pliers with jaws open—turn adjusting screw

(*a*) the jaws closed, (*b*) the jaws open using the adjusting screw, and (*c*) the jaws open by pushing the movable handle outward.

In its simplest form, a pair of locking pliers employs a one-DOF four-bar mechanical linkage, similar to that shown in Figure 3-27. As shown in Figure 3-28, when the jaws are closed, the fixed handle and jaw links form a right angle, and the movable handle and toggle links are collinear. The jaws can be opened either by moving the movable handle link outward (see Figure 3-29) or by using the adjusting screw, which in effect changes the length of the fixed handle link (see Figure 3-30).

In fact, the Craftsman locking pliers are more complicated, incorporating as it does a fifth compound link between the movable handle and the toggle link. This compound link, which serves to make the pliers easier to unlock once they are clamped, is at the heart of a patent held for this particular design.

In the event that the CAD software used to create the product's virtual model includes a finite element analysis module, it is possible to estimate the level of stress, strain, or deflection found in a single part, or even in an assembly. To do so, it is necessary to model the forces acting on the part, as well as the way in which the part is supported. Figure 3-31*a* shows the loading environment (loads, supports) on a portable water filter handle, assuming that the water filter

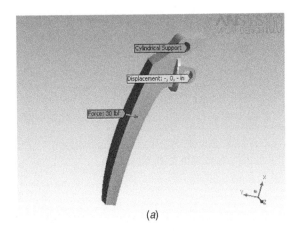

(a)

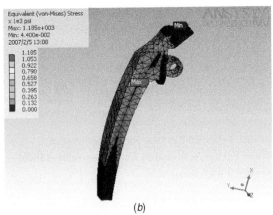

(b)

Figure 3-31 Loading and stress pattern on portable water filter handle part

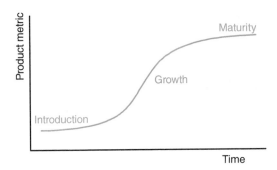

Figure 3-32 S-curve showing product innovation over time

becomes clogged. Figure 3-31*b* shows the resulting stress distribution for this loading, based on the von Mises stress criterion. See the discussion of finite element analysis in Chapter 15 for additional information.

■ PRODUCT IMPROVEMENT

The ultimate goal of reverse engineering is product improvement. In the course of the product dissection process, as the product is used, dissected, documented, and analyzed in various ways in order to understand how it works, ideas for improving on the existing product often come to light.

Generating improvement ideas is easier for some products than for others. This depends largely on the level of maturity of the product and of the technology on which it is based. Technology and product innovations tend to follow an "S-curve." That is, when an important measure or metric of the technology is plotted against time, the curve tends to have an S shape, as seen in Figure 3-32. Early in the development of a new type of product, innovations are slow to occur, because the market is relatively new. As demand for the product grows, so does the amount of innovation. Finally, as the product reaches maturity, the number of innovations begins to fall off. It is consequently easier to identify product improvement ideas for newer products and technologies.

Having been commercially produced since at least 1924, locking pliers are an extremely mature product. Literally dozens of U.S. patents related to locking pliers have been awarded, most recently in 2003. The Craftsman design described in this chapter is covered by a U.S. patent that was awarded in 1991.

For this reason, rather than attempting to generate additional product improvement ideas, let's look at some innovations that have taken place over the years in the design of locking pliers. One of the common innovations associated with locking pliers is to vary the jaw configuration in order to create new, more specialized tools. These variations have included curved-jaw and straight-jaw designs, long noses and bent noses, wrench designs, and C-clamp designs, as well as dedicated sheet metal and welding tools. The incorporation of a wirecutting tool was another popular innovation. Providing cushioned grips improved the product's ergonomics and its resistance to becoming slippery. Many of the patents associated with locking pliers are related to the development of improved locking and release mechanisms.

Benchmarking is the systematic process of measuring products and services against the toughest competitors. Competitive benchmarking is routinely employed in industry to help identify *best practices*. With respect to the locking pliers, the evaluation of competing products can help identify desirable features for adoption.

Finally, the Design for Manufacture and Assembly (DFMA) guidelines, briefly discussed in Chapter 1, offer many potential product improvement ideas. Probably the most important DFMA guideline is to minimize the number of parts. In recent decades, the manufacturability and ease of assembly of literally thousands of products have been improved upon through a careful application of these guidelines.

▍REASSEMBLY

Assuming that all components have survived the dissection procedure intact, the product should be reassembled and tested. This next-to-last step serves to complete the project loop, adding further insight into and appreciation for the product's design. The product disassembly plan, exploded view, and assembly animation file can all be used to assist in the product's reassembly.

▍COMMUNICATION

Like any other well-planned team design project, the product dissection experience offers excellent opportunities for developing both oral and written communication skills. For additional information, see the sections on written reports and oral presentations in Chapter 1.

▍QUESTIONS

TRUE OR FALSE

1. Engineering design follows a well-documented *cookbook approach*.
2. Recognizing principal competitors is not important when identifying the market segment of a product.
3. Benchmarking is a comparative analysis of a product against its competitors.
4. Product innovation is associated with mature products.

MULTIPLE CHOICE

5. DFMA stands for:
   a. Decomposition for mechanical assemblies
   b. Design for machine analysis
   c. Dissection for mechanical analysis
   d. Design for manufacture and assembly

Gopher grabber
Caitlin Connolly, Travis Constantine,
Matt Kirby, Nolan Lock, John Zhao

Unicycle
Kevin Dineen, Jeremy Bertoni,
Rachel Marshall, Taylor Colclasure

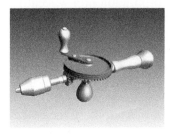

Hand drill
Peter Ensinger, Richard Li,
Kevin Lill, Ryan Moser

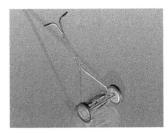

Lawnmower
Matthew Davis, Nigam Shah, Tony
Cohen

Apple peeler
Danielle Malone, Henry Wolf,
Qiongyu Lou, Usman Bhatti

Rope pump
Alex Wiss-Wolferding, Leo Xing,
Donnie Manhard

Rifle sight
Kendall Rak, Dayton Brenner, Joshua
Simmons, Josh Ellis, Peter Zich

Leonardo's catapult
Derek Wagner, Nimit Bhatia, Rahul
Prabhakran, Rui Li

Nerf gun
Brenden O'Donnell, Erik Babcock,
Levente Taganyi, Jordan Bohmbach

Door knob
Max Affrunti, Kevin Chua,
Ben Chui, Zak Bertolino

Human treadmill crane
Awnye Taylor, Christine Rhoades,
Katherine Mathews, Silas Tappendorf

Catapult
Adam Fabianski, Eric Mason,
Alyssa Cast, Haotian Wu

Eggbeater
Mike Kuo, Daniel Levitus, Monica Mingione, Brian Spencer, Janice Yoshimura

Percussion drill
Yi Duan, Jeff Joutras, Xingan Kan, Hong Kim

Plastic shredder
Marian Domanski, Nigel Li, Zelong Qiu, Meng Wang

Grain grinder
Adam Cornell, Michael Fischer, Kshitij Malik, Bo Xu

Apple cider press
Loren Anliker, Matthew Murphy, Josh Jochem

Treadle pump
Tyler Hill, Esther Kim

Bingo ball dispenser
Cody Suba, Lauren Shaw, Nick Henry, Allison Falkin

Fishing reel
Dave Bartalone, Drew Bishop, Mingyo Jung, Joe Martinez, Matt Wright

Lifestraw
Jackson Kontny, Allen Chang, Siyi Tu

Metronome
Kevin Anderson, Kathryn Beyer, Tim Corcoran, Matthew Gill, Cameron Sarfarazi

Windmill
Tori Ammon, Siddhant Anand, Augusto Canario, Dan Malsom, Matthew McClone

Crown cork machine
Justin Cruce, Scott Lunardini, Matt Magill, Clare Roman, Derek Vann

Toy drone
Chengin Zhang, Jingxin Duan,
Jennifer Marten

Toy microscope
Emily Hettinger, Lucas Kinsey,
Matthew Hanley, Wanxing, Shi

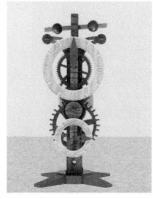

Da Vinci clock
Ramya Gandhi, Can Balkose, Kanat
Colak, Ayush Tiwary

PRODUCT DISSECTION PROJECT IDEAS

Household		
Air freshener, Air Wick	Fan, desktop, portable	Processor fan
Alarm clock	Faucet, kitchen	Push lawnmower
Analog clock	Fire alarm bell	Quick painter
Apple peeler	Fire extinguisher	Ratchet strap
Beer tap	Flashlight	Rotary telephone
Bidematic	Floor lamp	Salad spinner
Blender	Flour sifter	Screwdriver, bit shooter
Can crusher	Flush converter	Seed spreader
Can opener	Fly swatter gun	Slap chop
Car jack, scissor	Food processor	Spray nozzle
Card shuffler	Fruit peeler	Sprinkler, pulsating
Cargo strap	Hair dryer	Strainer
CD player	Hair trimmer	Sump-pump
Coffee maker	Hand drill (manual)	Swiffer wetjet
Coin sorter	Hose nozzle	Swiss army knife
Combination lock	Shake light	Timer, kitchen
Cookie scooper	Ice cream scooper	Trashcan, step
Corkscrew	Lawnmower crankshaft	T.V. antennae
Crank flashlight	Locking pliers	Toaster

Deadbolt lock	Makeup mirror	Toilet fill valve
Desk lamp	Melon baller	Toothbrush
Door jam	Misting fan	Umbrella
Doorknob	Mouse trap	Utility knife
Drain assembly	Padlock	Utility lighter
Egg beater, manual	Parking meter	Waffle maker
Electric razor cleaner	Pepper grinder	Wall clock
Electric screwdriver	Pickup tool	Water chill system
Eyelash curler	Pitcher pump	Water clock
Fabric shaver	Pliers	Water sprinkler
Entertainment/Outdoors		
ABslide	Hockey table	Scooter
Action figure	Honda mower starter	Skateboard
Air pump	Hungry Hungry Hippos	Solar car
Airsoft gun	Leatherman multi tool	Space shuttle
Beacon light	Lego set	Squirt gun
Bike, BMZ, fixed gear	Longboard	Super soaker
Bike gears	Maverick nerf gun	Tickle-Me-Ernie
Bike pump	Metronome	Tonka truck
Bingo ball dispenser	Model airplane	Toy bubble launcher
Binoculars	Model rocket	Toy car
Butane lighter	Model train	Toy cap gun
Camera	Mountain board	Toy cash register
Camping stove	Mouse trap board game	Toy corn popper
Compass	Music box	Toy gun
Dart gun	Nerf gun	Toy helicopter
Derailleur	Paintball gun	Toy pinball machine
Dominos dispenser	Pogo stick	Toy wagon
Drum	Polaroid camera	Transformer toy
Etch-A-Sketch	Popzooka gun	Video game controller
Fishing reel	Ratchet bar clamp	Vise

(*Continued*)

Frame pump	Razor scooter	Voice changer
Fused motor assembly	Remote control car	Unicycle
Garden watering assembly	Rifle sight	Water gun
Gri-gri	Rock 'em sock 'em Robots	Weight bar clamp
Guitar	Rubik's cube	Xbox controller
Guitar hero remote control	Scissor jack	Zippo lighter
Gumball machine		
Office/School/Lab		
Electric pencil sharpener	One-hole punch	Stapler
Computer mouse	Packing tape dispenser	Swivel office chair
Correction tape dispenser	Pencil sharpener	Three-hole punch
Chemical dispenser	Revolving date stamper	Vacuum pump, laboratory
Mechanical pencil	Staple-less stapler	
Appropriate Technology		
Apple cider press	Nutsheller, Universal	Solar cooker
Crank flashlight	Percussion drill	Treadle pump
Grain grinder	Plastic shredder	Water pump, windmill
Hand water pump	Rope pump	Water pump, solar powered
Hydraulic ram pump	Screenless hammermill	Windmill
LifeStraw		
History of Technology		
Catapult	Cross bow, Leonardo da Vinci	Treadwheel crane, human
Catapult, Leonardo da Vinci	Trebuchet	

WHERE TO LOOK—STORES WHERE APPROPRIATE PRODUCTS, DEVICES, AND MECHANISMS CAN BE FOUND

Arts and crafts	Hardware	Office supplies
Automotive	Hobby	Outdoor recreation (camping, hiking, climbing, hunting, fishing)
Boating	Home electronics	Sporting goods
Cycling	Household, kitchen	Toy
Department	Lawn and garden	

CHAPTER 4

FREEHAND SKETCHING

■ INTRODUCTION

Prior to the introduction of the personal computer in the early 1980s and the accompanying introduction of computer-aided design (CAD) software packages such as AutoCAD® shortly thereafter, most engineering drawings were executed manually using equipment like drafting machines, T-squares, triangles, and compasses. Today, however, nearly all engineering drawings are executed using a CAD system. This sea change in the way technical drawings are produced has had a major impact on the engineering graphics curriculum. Instrument drawing (with T-squares, triangles, and so on), for example, has largely been replaced with *freehand sketching*.

Engineering is a creative endeavor with roots that can be traced back to great Italian Renaissance artists like Leonardo, Michelangelo, Raffaello, and Donatello. Although it is certainly true that engineering is technology driven, it is important not to lose sight of the discipline's rich graphical and creative traditions. Great engineering is evidenced as much through the ability to communicate ideas via freehand sketching as it is by manipulating differential equations, by making a computer "sing," or for that matter, by presenting carefully reasoned, well-crafted prose.

While some of us are blessed with the natural ability to draw, most are not. All of us, though, can improve in our ability to communicate, document, and visualize using freehand sketches. All that is required is practice.

Freehand sketching has a number of important uses. Engineers often need to make sketches out in the field. These sketches are subsequently converted to CAD back at the office, in order to document modifications that are to be made to an existing structure. Sketching is also a just-in-time communication tool used by engineers, designers, and craftspeople, as well as with clients and supervisors. Freehand sketching is used creatively for brainstorming ideas, inventing, and exploring alternatives. Figure 4-1, for example, shows sketches done by engineering graphics students as part of their concept design projects. Either sketching or instrument drawing is essential in order to practice the language of engineering graphics. In her work, Sorby[1] has amply demonstrated that freehand sketching is an excellent way to improve spatial visualization skills.

Engineers use technical sketches to create rough, preliminary drawings that represent the main features of a product or structure. Freehand sketches should not be sloppy. Above all, careful attention should be paid to the proportions of the sketch. Although not drawn to a specific scale, freehand sketches should appear to be proportionally accurate to the eye.

[1] S. Sorby, Developing 3-D Spatial Visualization Skills, *Engineering Design Graphics Journal*, 63, no. 2 (Spring 1999).

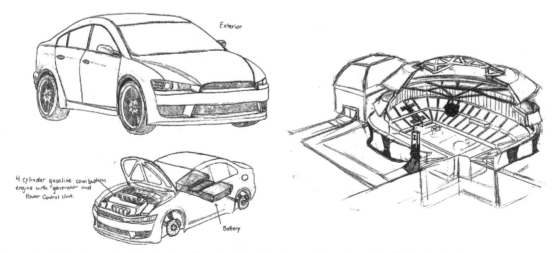

Figure 4-1 Design project brainstorming sketches (Courtesy of Jonathan Schmid, Donjin Lee)

■ SKETCHING TOOLS AND MATERIALS

All that is really needed for sketching is paper, pencils, and an eraser. Most any kind of paper will do for sketching; the reverse side of already used paper works well. Rectangular grid paper is frequently used for sketching. Engineering firms often use custom-ordered A (8½″ × 11″) or A4 metric size tablets of rectangular grid paper. The grid is actually on the back side of the paper, but it is dark enough to be visible on the front side. The paper also includes a custom title block.

This paper is useful for combining text, calculations, and sketches.

Isometric grid paper, shown in Figure 4-2, is helpful when one is first learning to create isometric sketches. This type of grid paper is not typically encountered in engineering design firms.

Most engineers use fine-line mechanical pencils, like those shown in Figure 4-3, both for sketching and for calculations. Each mechanical pencil is designed to hold a specific lead size. Common lead sizes include 0.3 mm, 0.5 mm, 0.7 mm, and 0.9 mm. The lead size is the diameter of the lead. The best size for general usage is 0.5 mm, whereas 0.7 mm is good for bolder stokes. An important advantage of mechanical over wooden pencils is that mechanical pencils do not require sharpening.

Pencil lead is available in different hardness grades, as shown in Table 4-1. The harder the

Figure 4-2 Isometric grid paper

Figure 4-3 Fine-line mechanical pencils

Table 4-1	**Pencil lead grades**	
Range	Grades	Purpose
Hard	9H, 8H, 7H, 6H, 5H, 4H	Accuracy, precision
Medium	3H, 2H, H, F, **HB,** B	General purpose
Soft	2B, 3B, 4B, 5B, 6B, 7B	Artistic rendering
	harder → softer	

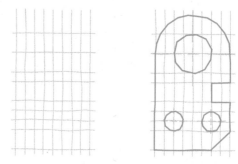

Figure 4-5 Construction lines with bold lines drawn over

Figure 4-4 Eraser (Adha Ghazali/EyeEm/Getty Images)

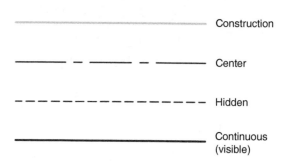

Figure 4-6 Common types of sketching lines

lead, the lighter, sharper, and crisper the resulting line. Soft lead, on the other hand, results in darker lines that tend to smear easily. Medium-grade leads (3H, 2H, H, F, HB, and B) are good for general-purpose drafting, and HB is the most commonly used lead grade.

Finally, a good eraser that does not tend to smudge is an important sketching tool. See Figure 4-4.

▊ SKETCHING TECHNIQUES

Line Techniques

A freehand sketch should begin with proportionally laid out *construction lines*. Construction lines are thin and drawn lightly; they serve to guide the lines to follow. All other lines should be dark, crisp, and of uniform thickness; they can be sketched directly over the construction lines.

See Figure 4-5, where construction lines are first used to lay out the proportions of the sketch, and then the features are indicated using bold lines. If drawn correctly, construction lines need not be erased. It is a common mistake to draw construction lines that are too dark, making them hard to distinguish from other lines.

In addition to the relative lightness or darkness of a line, two line widths, thick and thin, are used in engineering sketches and drawings. Continuous lines indicating visible object edges are thick, while hidden, center, and construction lines are thin. See Figure 4-6 for a brief summary of the characteristics of the most commonly used lines.

A more thorough discussion of the line styles employed in CAD and manual engineering drawings is provided at the end of the chapter.

Sketching Straight Lines

To sketch a straight line, start by marking both end points. Next, place the pencil point at one of these end points. Keeping your eye fixed on the

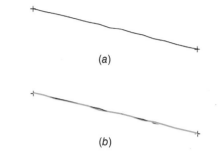

(a)

(b)

Figure 4-7 Sketching a straight line

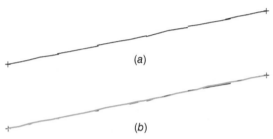

(a)

(b)

Figure 4-8 Sketching a long straight line

Figure 4-9 Wiggles and gaps in sketched lines

Figure 4-10 Sketched line tailing off due to overly firm grip

Keep eye on end point

Figure 4-11 Sketching horizontal lines

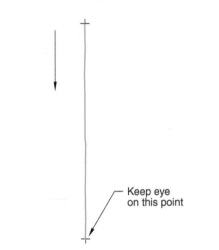

Keep eye on this point

Figure 4-12 Sketching vertical lines

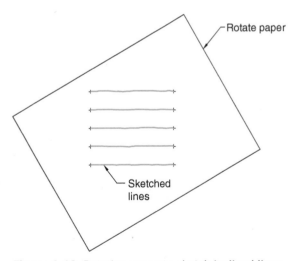

Rotate paper

Sketched lines

Figure 4-13 Rotating paper to sketch inclined lines

point to which the line is being drawn, sketch a light line with a single stroke. Finally, darken the line. See Figure 4-7.

To sketch an especially long line, use several short overlapping strokes, and then darken the line, as shown in Figure 4-8. Slight wiggles are okay as long as the resulting line is straight. Occasional gaps are also acceptable. See Figure 4-9. If the resulting line tails off and is not straight, the pencil is probably being gripped too tightly. See Figure 4-10 for an example.

For right-handers, horizontal line strokes are typically made from left to right, as shown in Figure 4-11.

Vertical lines are usually drawn from top to bottom, as seen in Figure 4-12. Inclined lines at certain orientations can be difficult to draw, in which case the paper can be rotated to a more comfortable position and then drawn, as shown in Figure 4-13. In fact, for lines drawn at any orientation, it is a good idea to rotate the paper to a comfortable position and then sketch the line.

Sketching Circles

A number of different techniques can be used to sketch a circle. In the trammel method, a piece of scrap paper is used to locate points on the circumference of the circle. On a straight edge of the paper, mark two points at a distance equal to the radius of the circle. With one point at the center, rotate the paper and mark as many points on the circumference as desired; then sketch a circle passing through the points. Figure 4-14 illustrates the trammel method.

In the square method, the enclosing square of the circle is first sketched. Next the midpoints of the sides of the square are marked. The midpoints are used as the quadrants of the circle. A circle passing through the quadrant points and tangent to the sides of the square is then sketched. See Figure 4-15 for an example of a circle constructed using the square method. Note that by constructing the diagonals of the square and marking the radius along the diagonals, you can add four more points on the circumference of the circle.

Either of these methods can be used to sketch a circle or arc feature in a *pictorial sketch* (see Chapter 5), as long as the face on which the circle or arc appears is parallel to the view plane. Figure 4-16 shows an example of this for an *oblique* pictorial sketch.

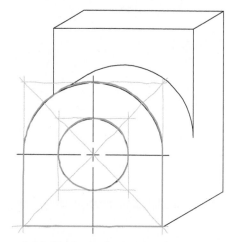

Figure 4-16 Circles and arcs sketched on an oblique pictorial

Sketching Ellipses

The rectangle method can be used to sketch an ellipse. First construct the enclosing rectangle, and then locate the midpoints along the sides of the rectangle. Sketch the ellipse passing the midpoints and tangent to the sides of bounding rectangle. See Figure 4-17 for an example of this method.

In a pictorial sketch a circle will appear as an ellipse, unless the circle is parallel to the view plane. To sketch such an ellipse, first construct

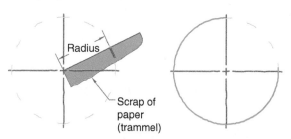

Figure 4-14 Trammel method used to sketch circles

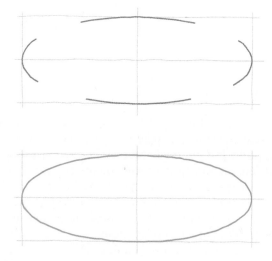

Figure 4-17 Rectangular method used to sketch ellipses

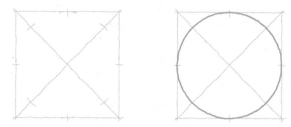

Figure 4-15 Square method used to sketch circles

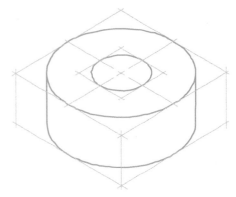

Figure 4-18 Parallelogram method used to sketch ellipses

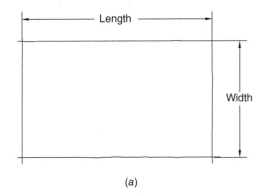

(a)

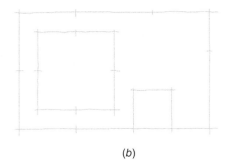

(b)

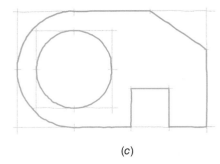

(c)

Figure 4-19 Proportioning with principal dimensions

the enclosing parallelogram. The ellipse will pass through the midpoints of the sides of the parallelogram and will be tangent to the sides of the parallelogram. Figure 4-18 shows an example of the construction of an ellipse on an *isometric* pictorial sketch.

▌ PROPORTIONING

Although freehand sketches are not drawn to scale, it is important to maintain the relative proportions between the principal dimensions of the object. To accomplish this, first estimate the proportional relationship between the object's principal dimensions and lightly block them in, as shown in Figure 4-19a. The estimated dimensions of each feature should then be proportioned with respect to established dimensions, as seen in Figure 4-19b. Work at developing the ability to divide a line in half by eye; the halves can then be further divided into fourths. Finally, use bold lines to fully define the object, as in Figure 4-19c.

While it is assumed that scales are not used when sketching, a *trammel* may be used to improve proportional accuracy. Decide on a unit of length, and then transfer it to the trammel, or straight-edged scrap paper. Additional unit-length graduations can be added to the trammel, which can then be used as a scale to layout the proportions of each object feature. See Figure 4-20 for an example of this technique. Note also that the trammel can be folded to obtain half and quarter lengths.

Estimating Dimensions of Actual Objects

It is sometimes necessary to make a sketch of an actual object. To make such a sketch, hold the pencil at arm's length, with the pencil between the eye and the object. Using the pencil as a sight, establish a proportional relationship between an object edge and the length of the pencil. Do this by aligning the end of the pencil with one end of the object edge. Move your thumb along the

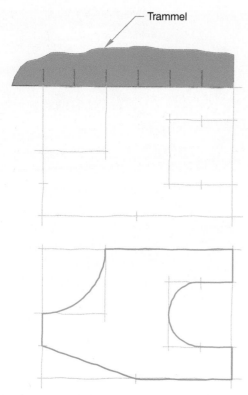

Figure 4-20 Using a trammel to improve proportions

Figure 4-21 Using a pencil to estimate proportional lengths

pencil until it coincides with the other end of the edge. Now use this proportion to estimate the lengths of other object features. See Figure 4-21 for an example of this technique. A similar

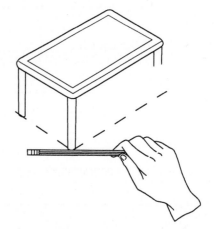

Figure 4-22 Using a pencil to estimate angles

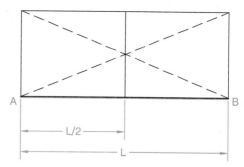

Figure 4-23 Method for partitioning lines

technique can be used to estimate angles, as shown in Figure 4-22.

Partitioning Lines

The following procedure may be used to divide a line into fractional parts. To subdivide the line AB shown in Figure 4-23, construct a rectangle on the line AB. Next draw both of the diagonals of the rectangle. Pass a line perpendicular to AB that passes through the intersection of the diagonals. This line divides AB in half.

To partition AB into thirds, sketch another diagonal from one corner of the original rectangle to the midline on the opposite side. Now sketch a line perpendicular to AB that passes through the point where the two diagonals (full diagonal and the half diagonal) intersect. This determines a one-third length of AB, as seen in Figure 4-24. Repeating this same process will

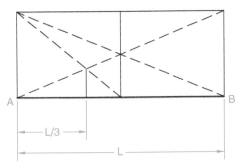

Figure 4-24 Method for partitioning lines into thirds

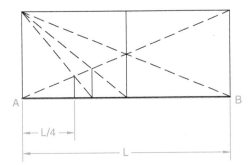

Figure 4-25 Method for partitioning lines into quarters

further divide the line into fourths, fifths, and so on. Figure 4-25 shows this.

■ INSTRUMENT USAGE—TRIANGLES

Although drafting machines and T-squares have fallen into disuse, the construction of parallel and perpendicular lines using triangles remains a useful skill. Every engineer should possess both a 45° triangle and a 30°–60° triangle. These triangles are shown in Figure 4-26.

Parallel Lines

In order to draw a line parallel to another line using two triangles, align the hypotenuse of one triangle with the original line. Next place the hypotenuse of the second triangle along the leg of the first triangle. Now slide the first triangle along the fixed second triangle until the location of the parallel line is reached, and construct the line. Figure 4-27 demonstrates this technique.

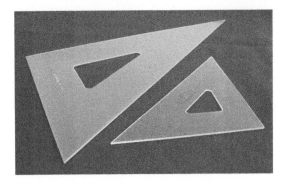

Figure 4-26 30°–60° and 45° triangles

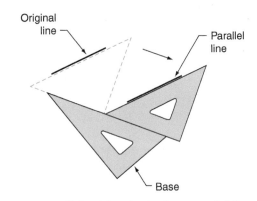

Figure 4-27 Using triangles to draw parallel lines

Perpendicular Lines

Triangles are also useful for drawing one line perpendicular to another. There are a couple of ways to accomplish this. In the first method, shown in Figure 4-28, one leg of a triangle is aligned with the original line, while a second triangle is used

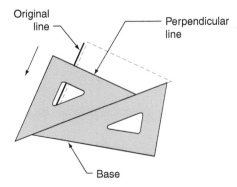

Figure 4-28 Using triangles to draw perpendicular lines: method 1

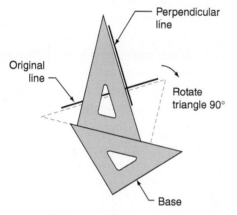

Perpendicular line

Original line

Rotate triangle 90°

Base

Figure 4-29 Using triangles to draw perpendicular lines: method 2

as a base. Sliding the first triangle along the base triangle, you can use the other leg of the first triangle to draw the perpendicular line. In the second method, shown in Figure 4-29, the first

triangle is rotated 90°, rather than slid, prior to drawing the perpendicular line.

■ LINE STYLES

Line styles, sometimes referred to as the alphabet of lines, describe the size, spacing, construction, and application of the various lines used in CAD and manual engineering drawings. These conventions are established in ASME Y14.2M-1992, Line Conventions and Lettering.

Two line widths are recommended on engineering drawings: thick and thin. The ratio of these line widths should be approximately 2:1. The thin line width is recommended to be a minimum of 0.3 mm, with a recommended minimum thick line width of 0.6 mm.

Figure 4-30 shows the various types of lines and their widths. *Visible lines* are used to show the visible edges and contours of objects. *Hidden*

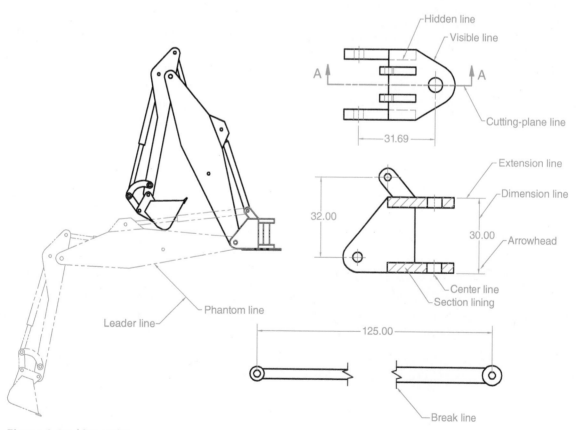

Figure 4-30 Line styles

lines are used to show the hidden edges and contours of objects. *Center lines* are used to represent axes or center planes of symmetrical parts and features, as well as bolt circles and paths of motion. *Phantom lines* are most commonly used to show alternative positions of moving parts. *Cutting-plane lines* are used to indicate the location of cutting planes for section views. The ends of the lines are at 90 degrees and are terminated by arrowheads to indicate the viewing direction. *Section lining* is used to indicate the cut surfaces of an object in a section view. *Break lines* are used when complete views are not needed.

Dimension lines are used to indicate the extent and direction of a dimension and are terminated with uniform arrowheads. *Extension lines*, used in combination with dimension lines, indicate the point or line on the drawing to which the dimension applies. *Leader lines* are used to direct notes, dimensions, symbols, and part numbers on a drawing. A leader line is a straight inclined line, not vertical or horizontal, except for a short horizontal portion extending to the note.

Arrowheads are used to terminate dimension, leader, and cutting-plane lines. Arrowhead length and width should be in a ratio of approximately 3:1. A single style of arrowhead should be used throughout the drawing.

▮ QUESTIONS

TRUE OR FALSE

1. When drawn correctly, it is not necessary to erase construction lines.
2. When sketching a straight line, it is important to keep your eye on the pencil as it moves.

MULTIPLE CHOICE

3. The most popular lead grade is:
   a. 2H
   b. H
   c. F
   d. HB
   e. B
4. Which line type is drawn more thickly than SS the others?
   a. Construction
   b. Continuous
   c. Hidden
   d. Center
5. Trammels can be used for which of the following purposes?
   a. Circle construction
   b. Proportioning the features in a sketch
   c. Transferring dimensions
   d. All of the above
   e. None of the above

A Using the cues provided, sketch straight lines.

B Using the cues provided, sketch the nested shapes.

C Using a trammel, sketch the missing symmetrical half of the object.

D Using a trammel, sketch the missing symmetrical half of the object.

| Drawing 4-1 | Name | Date |

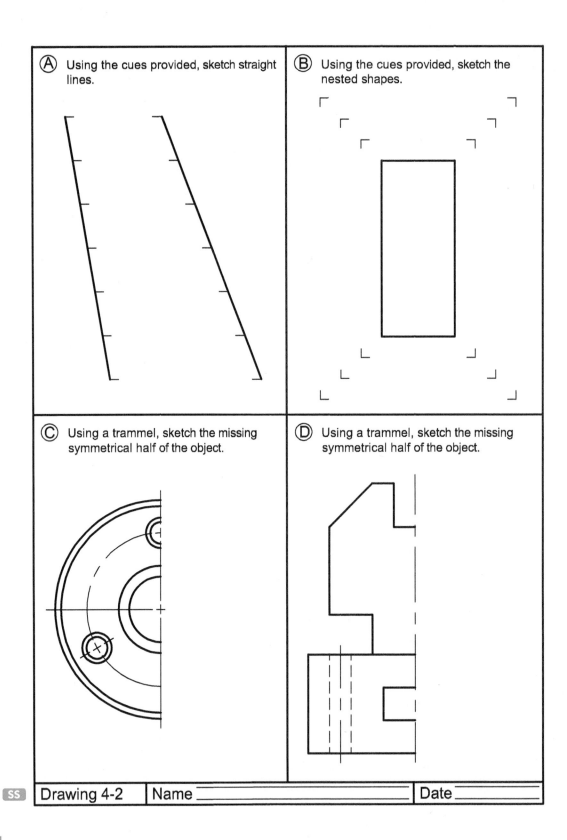

A. Using the cues provided, sketch straight lines.

B. Using the cues provided, sketch the nested shapes.

C. Using a trammel, sketch the missing symmetrical half of the object.

D. Using a trammel, sketch the missing symmetrical half of the object.

| Drawing 4-2 | Name _____ | Date _____ |

Ⓐ Using a trammel, sketch the missing symmetrical half of the object.

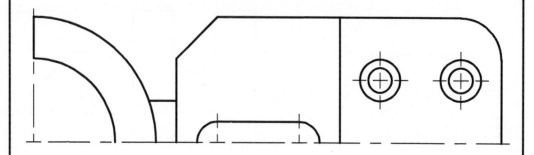

Ⓑ Using a trammel and the cue provided, sketch a replica of the object.

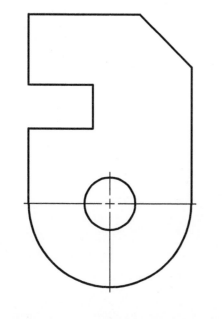

| Drawing 4-3 | Name _____ | Date _____ |

Ⓐ Using a trammel, sketch the missing symmetrical half of the object.

Ⓑ Using a trammel and the cue provided, sketch a replica of the object.

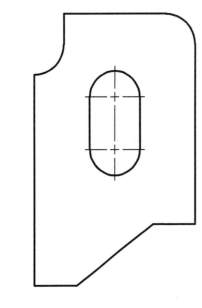

└

| Drawing 4-4 | Name _____ | Date _____ |

CHAPTER

5 PLANAR PROJECTIONS AND PICTORIAL VIEWS

■ PLANAR PROJECTIONS

Introduction

Projection is the process of reproducing a spatial object on a plane, curved surface, or line by projecting its points. Common examples of projection include photography, where a 3D scene is projected onto a 2D medium, and map projection, where the earth is projected onto a cylinder, a cone, or a plane in order to create a map. ***Planar projection*** figures prominently in both engineering and computer graphics. For our purposes, a projection is a mapping of a three-dimensional (3D) space onto a two-dimensional subspace (i.e., a plane). The word *projection* also refers to the two-dimensional (2D) image resulting from such a mapping.

Every planar projection includes the following elements:

- The 3D object (or scene) to be projected
- Sight lines (called projectors) passing through each point on the object
- A 2D projection plane[1]
- The projected 2D image that is formed on the projection plane

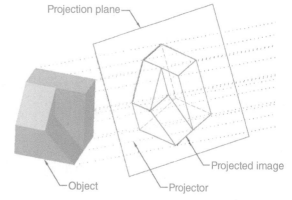

Figure 5-1 Elements of a planar projection

These elements are indicated in Figure 5-1. The projection is formed by plotting piercing points created by the intersection of the projectors with the projection plane. By mapping these points onto the projection plane, the 2D image is formed. In effect, three-dimensional information is collapsed onto a plane.

The Albrecht Dürer drawing shown in Figure 5-2 further illustrates these projection elements. The lute lying on the table is the ***object***, the piece of string is a (movable) ***projector***, and the frame through which the artist looks is the ***projection plane***. Two movable wires are mounted on the frame, allowing the artist to identify a single piercing point. A piece of paper hinged to the frame serves as the basis for the ***projected***

---

[1] Although commonly represented as a bounded rectangle, the projection plane is infinite in extent.

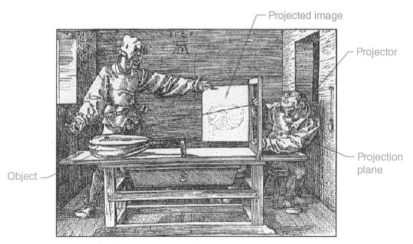

Figure 5-2 Albrecht Dürer, *Artist Drawing a Lute*, 1525 (Albrecht Dürer/Metropolitan Museum of Art/Wikimedia Commons/Public Domain).

image. Once the projected point is transferred to the paper, the artist's assistant moves the string to another point on the lute, and the process is repeated.

Classification of Planar Projections: Projector Characteristics

Planar projections are initially classified according to the characteristics of their projectors. In a *perspective projection*, the projectors converge to a single viewpoint called the center of projection (CP). The *center of projection* represents the position of the observer of the scene and is positioned at a finite distance from the object. Dürer's *Artist Drawing a Lute* (Figure 5-2) illustrates perspective projection, with the eyebolt mounted on the wall serving as the center of projection. If the center of projection is infinitely far from the object, the projectors will be parallel to one another. In this case a *parallel projection* results. Figure 5-3 compares the

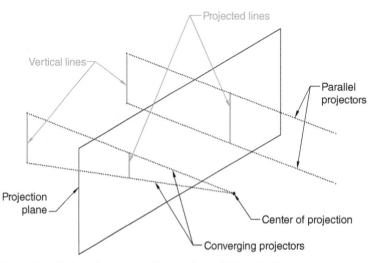

Figure 5-3 Comparison of projectors for perspective and parallel projection

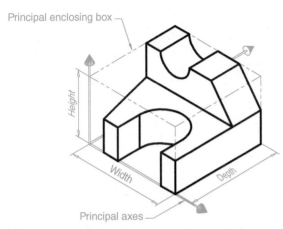

Figure 5-4 PEB, principal dimensions, and principal axes

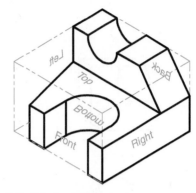

Figure 5-5 Principal planes

Preliminary Definitions

Before moving on to a more detailed discussion of the different kinds of planar projections, it is useful to introduce some additional terminology. Figure 5-4 shows a cut block, typical to the objects used throughout the course of this work. An object's *principal enclosing box (PEB)* just contains the object, and consequently its dimensions are the maximum width, depth, and height of the object. These are referred to as the *principal dimensions* of the object. Mutually perpendicular axes corresponding to the edges of the PEB are referred to as the *principal axes* of the object. The PEB is also referred to as a *bounding box*; these terms can be used interchangeably.

The faces of the PEB are referred to as the *principal planes* or *faces* of the object. As shown in Figure 5-5, these planes are categorized as being either frontal (front, back), horizontal (top, bottom), or profile (right, left), for a total of six.

Foreshortening is an important graphical concept related to projection theory. A dictionary definition of *foreshorten* is to shorten by proportionately contracting in the direction of depth so that an illusion of projection or extension in space is obtained. To demonstrate foreshortening, close

projectors in both a perspective and a parallel projection, where in this case the projected object is simply a vertical line.

one eye and hold your hand in front of you, with your palm perpendicular to your line of sight. Now curl your fingers toward you until they are parallel to your line of sight. The image that you see of your fingers is foreshortened.

The term *pictorial* is used to indicate a kind of projection, view, drawing, or sketch that includes all three dimensions and thus creates the illusion of depth. The principal types of pictorial views are perspective, oblique, and axonometric. Figure 5-6 shows four different pictorial views of the same object. The two on the left (oblique and isometric) are parallel projections, while the two on the right (one-point and two-point) are perspective projections.

Classification of Planar Projections: Orientation of Object with Respect to Projection Plane

Beyond the characteristics of their projectors, planar geometric projections can be further categorized as shown in Figure 5-8. As we will see, these planar projection subclasses are in large part based on the orientation of the object with respect to the projection plane.

In Figure 5-9, three bounding boxes are shown in different orientations with respect to a vertical projection plane. Figure 5-10 shows these same elements when viewed from above. As a result, the projection plane is now seen on edge. Note that the principal axes of each bounding box are also shown. For box A note on the left, two principal axes are parallel to the projection

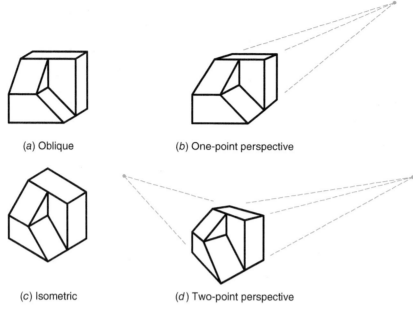

(a) Oblique (b) One-point perspective

(c) Isometric (d) Two-point perspective

Figure 5-6 Pictorial projections

Block coefficient

In ship design, the ratio of a ship's displaced volume to its length, width, and depth up to the waterline (i.e., its PEB) is called the *block coefficient* (see Figure 5-7). The block coefficient can be used to compare the relative fineness between different hull forms. For example, a speedboat or destroyer might have a block coefficient of about 0.38, while an oil tanker has a block coefficient of about 0.80.

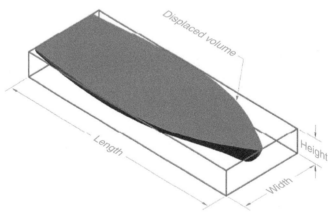

Figure 5-7 Block coefficient

plane, while the third axis is perpendicular to the projection plane. In the case of box B in the center, one principal axis (i.e., vertical) is parallel to the projection plane, while the other two are inclined (i.e., neither parallel nor perpendicular) to the projection plane. In the last case, box C on the right, all three axes are inclined to the projection plane. These possible orientations, together

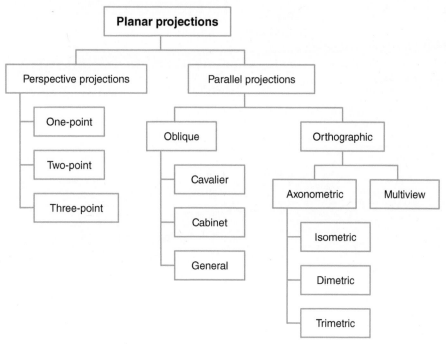

Figure 5-8 Planar geometric projection classes

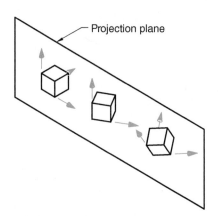

Figure 5-9 Orientation of object with respect to projection plane

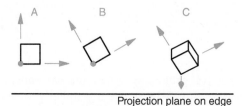

Figure 5-10 Orientation of object with respect to projection plane as seen from above

with the type of projection, either perspective or parallel, determine most of the planar projection categories. Figure 5-11 is based on Figure 5-8 but includes images depicting the orientation of the object with respect to the projection plane.

Further Distinctions Between Parallel and Perspective Projections

In a parallel projection, the center of projection is infinitely far from the object being projected. This means that the projectors are now parallel to one another. While perspective projection is useful in creating a more realistic depiction of an object or scene, a parallel projection is typically used when it is important to preserve the dimensional properties of the object. As an example, compare the projections employing parallel projectors in Figure 5-12 with the converging projectors used in Figure 5-13. In both cases, the projected object face is parallel to the projection plane. Note that parallel projection preserves both the size and the shape of the object face, whereas perspective projection preserves only

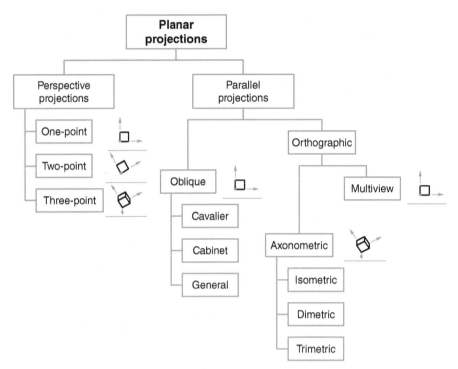

Figure 5-11 Projection classes with orientation shown

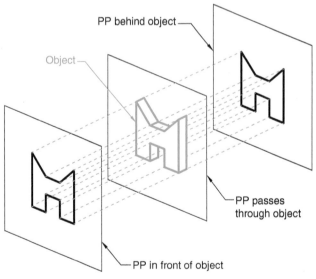

Figure 5-12 Projection plane location in a parallel projection

the shape of the object face (and that the shape may even be inverted!), depending on the location of the projection plane.

In a parallel projection, parallel object edges remain parallel when projected. Unlike perspective projection, there are no **vanishing points**. Figure 5-14 shows a parallel projection and a perspective projection of the same object. Note that whereas the receding edges of the parallel projection on the left remain parallel, the

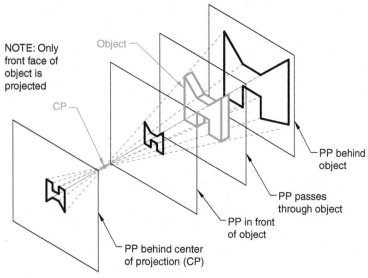

Figure 5-13 Projection plane location in a perspective projection

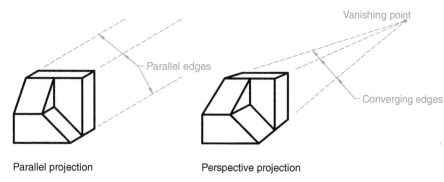

Figure 5-14 Comparison of receding axis edges in a parallel and a perspective projection

corresponding edges on the perspective projection on the right converge to a vanishing point. Although the perspective projection provides a more realistic depiction of the object, the parallel projection, in preserving parallelism, is easier to scale and draw.

Though extremely important in computer graphics, perspective projection is not commonly employed in engineering. For this reason, perspective projection and perspective sketching are treated separately in Chapter 6.

Classes of Parallel Projections

Figure 5-15 shows a breakdown of the different kinds of parallel projection techniques. Oblique and axonometric projections will be discussed in

the remainder of this chapter. Multiview sketching is the subject of the following chapter.

▌OBLIQUE PROJECTIONS

Oblique drawings are traditionally employed when one object face is significantly more complicated than the other faces of the object.

Oblique Projection Geometry

Figure 5-16 shows the geometric arrangement of an oblique projection. As seen in Figure 5-16a, parallel projectors intersect the projection plane at an oblique angle. In addition, one principal face of the object is parallel to the projection plane. In an oblique projection, the object face

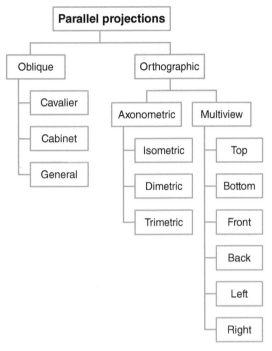

Figure 5-15 Parallel projection classes

that is parallel to the projection plane is projected true size.

Oblique Projection Angle

Two angles can be used to describe the intersection of an oblique projector with a projection plane, as shown in Figure 5-17. An in-plane angle β measures the angle of rotation of the projector about the projection plane normal. The out-of-plane angle *a* is called the ***oblique projection angle***. The oblique projection angle determines the type of oblique projection: cavalier, cabinet, or general.

Classes of Oblique Projections

Figure 5-18 shows an oblique projection setup, where two identical cubes (shown in the top half of the figure) are projected onto a projection plane. The resulting projection is shown in the bottom half of the figure. The front face of

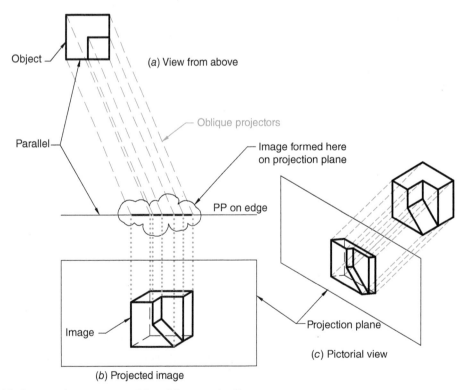

Figure 5-16 Geometric arrangement for oblique projection

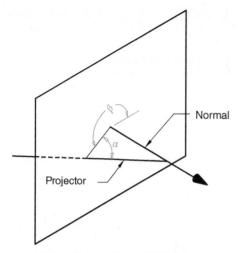

Figure 5-17 In-plane and out-of-plane angles defining an oblique projection

both cubes is projected true size. The oblique projection angle (*a*) used for the cube on the left is 45 degrees, and the resulting projection is called a ***cavalier oblique***. Note that the projection of this cube appears to be elongated in the receding (depth) axis direction. In fact, the measure of all of the edges of both the cube and its cavalier projection are identical. In a cavalier projection, the receding axis is not

foreshortened; it is scaled the same as the other (horizontal, vertical) principal axes.

On the right half of Figure 5-18, the oblique projection angle (*a*) is approximately 63.43 degrees, resulting in a ***cabinet oblique*** projection. In comparing the projected lengths of this cube, you will find that the projected receding edge length is one-half that of the other (horizontal, vertical) edge lengths. The receding axis of a cabinet oblique is foreshortened to exactly one-half that of the other principal axes. Note also that the projection that results from a cabinet oblique appears to the eye to be more proportionally correct than the cavalier oblique. This is because, visually, we expect some foreshortening to occur along a receding axis.

If an oblique projection angle between 45 and 63.43 degrees is used, the result is called a ***general oblique*** projection. On a general oblique, the receding axis is scaled between −1 and 1. This would normally be done to improve the appearance of the projected image. Table 5-1 summarizes the different kinds of oblique projections.

Receding Axis Angle

In an oblique drawing, one axis is horizontal and another is vertical. The third, receding (or depth) axis can be inclined at any angle, but it

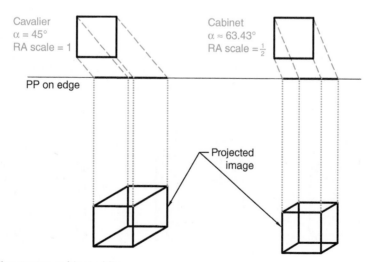

Figure 5-18 Cavalier versus cabinet oblique

By collapsing the 3D oblique projection geometry into 2D, it is easy to understand why a 45 degree projection angle results in no foreshortening (i.e., 1:1 scale), while a ~63.43 degree angle scales a projected length to one-half of the original. Figure 5-19 shows a projection plane on edge and the object, a line of length L. The line is perpendicular to the projection plane, much like the cube edges (not parallel to the projection plane) in Figure 5-18 are perpendicular to the projection plane. (*Note:* Tan 45° = 1, Cot 63.43° = $\frac{1}{2}$.)

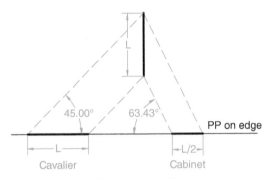

Figure 5-19 In-plane and out-of-plane angles defining an oblique projection

Table 5-1 **Classes of oblique projections**

Type of Oblique	Oblique Projection Angle (α)	Receding Axis Scale
Cavalier	45°	1
Cabinet	63.43°	$\frac{1}{2}$
General	45° < angle < 63.43°	$\frac{1}{2}$ < scale < 1

is normally chosen to be 30, 45, or 60 degrees. This angle determines the relative emphasis of the receding planes on the projection, as seen in Figure 5-20. Note that the receding axis angle should not be confused with the oblique projection angle.

The receding axis angle is related to the in-plane projector angle β discussed earlier and shown in Figure 5-17. As shown in Figure 5-21, the receding axis angle is equal to β – 180 degrees, where β is the in-plane angle of rotation of the oblique projector about the projection plane normal.

ORTHOGRAPHIC PROJECTIONS

Orthographic projection is the most commonly used projection technique employed by engineers. CAD systems typically employ orthographic projection techniques, although the user is often provided with the option of changing to perspective projection.

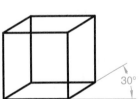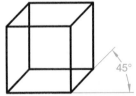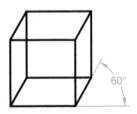

Figure 5-20 Receding axis angle

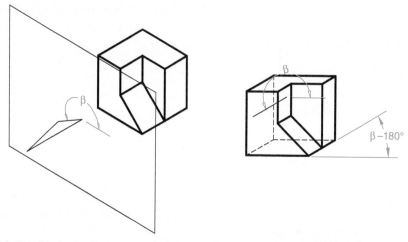

Figure 5-21 Relationship between the receding axis angle and in-plane projector angle β

Orthographic Projection Geometry

Orthographic projection is also a parallel projection technique, but it differs from oblique projection in that the parallel projectors are perpendicular (normal) to the projection plane. Figure 5-22 shows the parallel projectors and their relationship to the projection plane.

Orthographic Projection Categories

Orthographic projections are subdivided according to the orientation of the object with respect to the projection plane. In an *axonometric projection*, all three principal axes are inclined to the projection plane. Figure 5-23

shows this orientation, both as viewed from the front and as viewed from above. No axis is parallel (or perpendicular) to the projection plane. When projected, three principal faces of the object are visible. Axonometric projection results in an orthographic pictorial view.

In a *multiview projection*, one object face and two principal axes are parallel to the projection

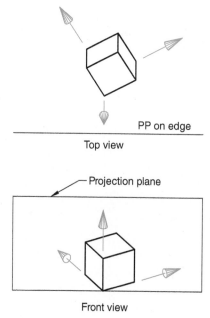

Figure 5-23 Object position in axonometric projection

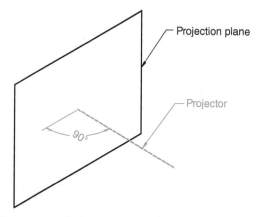

Figure 5-22 Orthographic projection geometry

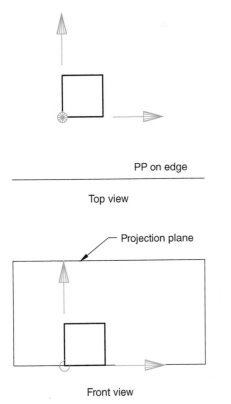

PP on edge

Top view

Projection plane

Front view

Figure 5-24 Object position in multiview projection

plane. When projected, only one object face is visible. In terms of the orientation of the object with respect to the projection plane, multiview projection is identical to oblique projection, as well as to one-point perspective projection. See Figure 5-24.

▋ AXONOMETRIC PROJECTIONS

Axonometric projections are classified according to the angles made by the principal axes when projected onto the projection plane. The upper half of Figure 5-25 shows three different axonometric projections of a cube, along with an attached principal axis triad. In the lower portion of the figure, only the projected axes and the angles between them are shown.

In a *trimetric projection*, depicted on the left in Figure 5-25, none of the angles between the projected principal axes are equal. The middle projection of Figure 5-25 is called a *dimetric projection*, where two of the three angles are equal. In an *isometric projection*, shown on the right in Figure 5-25, all three angles are equal.

Also note that, because all of the equal-length axes are inclined to the projection plane, all of their projections are foreshortened, as depicted in Figure 5-25. The amount of foreshortening is related to the projected angle. In the trimetric projection (on the left) all three axes are foreshortened by different amounts, and all three projected angles are different. Two of the three

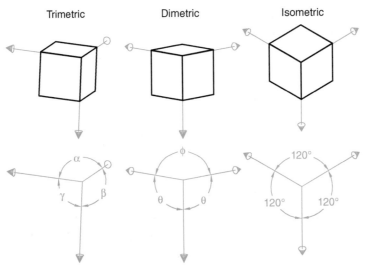

Figure 5-25 Axonometric projection classes

projected axes in the dimetric projection are foreshortened by the same amount, and the angles opposite these axes are also equal. In the isometric projection on the right, there is an equal amount of foreshortening along all three axes, and all of the angles are equal.

ISOMETRIC PROJECTIONS

An isometric projection is foreshortened, or scaled, equally along all three principal axis directions. This fact makes isometric projections particularly useful in engineering. As a pictorial, an isometric projection is relatively easy to visualize, and it is also good at preserving the dimensional properties of the object.

To understand how an object must be oriented in order to obtain an isometric projection, imagine the projection of a cube. Figure 5-26 on the left shows a trimetric view of the cube, on which a cube diagonal has been drawn. If the cube is rotated so that we look down the diagonal, as in Figure 5-26 on the right, we get an idea of how an isometric view is generated.

In most CAD systems an isometric view can be generated automatically. Figure 5-27 shows an example of a CAD viewing tool. Clicking on any diagonal arrow generates one of eight possible isometric views.

More specifically, an isometric view is generated as shown in Figure 5-28:

1. Start with one principal face of the object parallel to the projection plane.

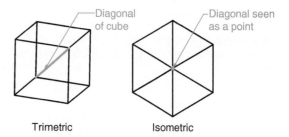

Trimetric   Isometric

Figure 5-26 Looking down the diagonal of a cube

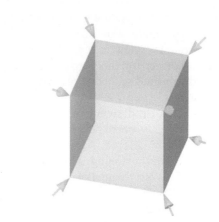

Figure 5-27 CAD system viewing tool

2. Rotate the object about a vertical axis $(45 \pm 90n)$ degrees, where n is an integer.
3. Rotate the object out of the horizontal plane by approximately ± 35.26 degrees.[2]

---

[2]For a derivation of this value, see Ibrahim Zeid, *Mastering CAD/CAM*, McGraw-Hill, 2005, pp. 496–497.

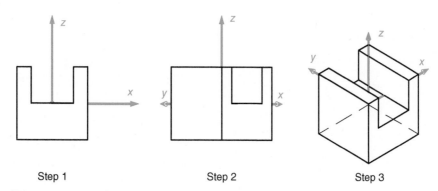

Step 1   Step 2   Step 3

Figure 5-28 Object orientation for an isometric view

Isometric Drawings

As a result of the particular object orientation in an isometric view, all three principal axes are foreshortened by exactly the same amount. This means that if a *solid model* is created in a CAD system, and an isometric view of the object is then printed out at a 1:1 scale, the printed (i.e., *projected*) edge lengths will all be equal to one another, but less than the actual edge lengths. It turns out that in an isometric projection, each principal axis is foreshortened to approximately 82% of its true length. For this reason, when plotting an isometric view in a CAD system, it is possible to correct for this foreshortening by multiplying the desired scale by the reciprocal of 0.82 (i.e., ~1.22).

When a sketch or drawing of an isometric view is made directly on paper, foreshortening effects are typically ignored. An isometric projection without foreshortening is referred to as an ***isometric drawing***. Note that, as shown in Figure 5-29, an isometric drawing is larger than a true isometric projection.

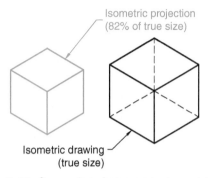

Figure 5-29 Comparison between an isometric projection and an isometric drawing

Multiview Projections

An arrangement for multiview projection is characterized by the following elements (see Figure 5-30).

1. The projectors are parallel to one another.
2. The projectors are normal to the projection plane.
3. The object is positioned with one principal face parallel to the projection plane.

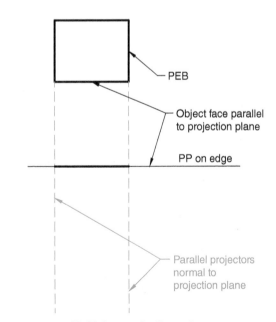

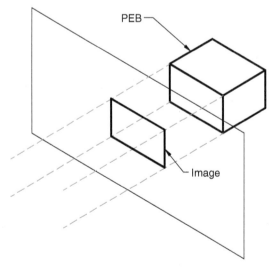

Figure 5-30 Multiview projection setup

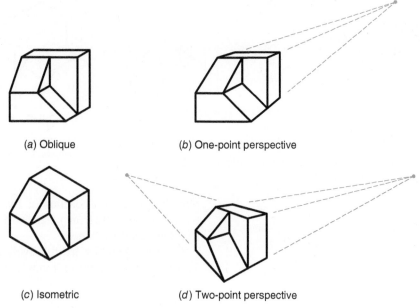

(a) Oblique   (b) One-point perspective

(c) Isometric   (d) Two-point perspective

Figure 5-6 (Repeated) Pictorial projections

As a direct result of these characteristics, multiview projections are good at preserving the object's dimensional information, but more than one view is necessary to fully describe the object.

In the following sections, different methods and techniques for constructing oblique and isometric pictorial sketches will be discussed. Multiview sketching is the subject of Chapter 7.

▮ INTRODUCTION TO PICTORIAL SKETCHING

In a pictorial view, three principal faces of an object are visible. A *pictorial sketch* shows an object's height, width, and depth in a single view. Unlike the multiview orthographic sketches discussed in the following chapter, a pictorial sketch conveys the object's three-dimensional shape in a single view. In this chapter, parallel projection (oblique and isometric) pictorials are discussed. Chapter 6 includes a discussion of perspective (both one-point and two-point) pictorials. Generally speaking, parallel projections preserve the object's dimensional properties, while perspective projections convey a strong sense

of realism. Figure 5-6, which is reprinted above, shows how the same object would appear using these different pictorial projection techniques.

Regardless of the particular pictorial sketch being executed, the same general technique can be employed:

1. Using light construction lines, sketch a properly proportioned bounding box.
2. Continuing with construction lines, add feature details.
3. Starting with curved features, go bold.
4. Compete the sketch using bold lines.

This process is illustrated in Figure 5-31 for an isometric pictorial.

Polyhedral shapes are frequently used for sketching. A *polyhedron* is a 3D solid bounded by a connected set of polygons, where every edge of a polygon belongs to just one other polygon. Polyhedral geometry consists of faces, edges, and vertices, as shown in Figure 5-32.

A *face* is a bounding surface on an object, whereas an *edge* serves as the intersection between two faces. Finally, a *vertex* is the end point of an edge.

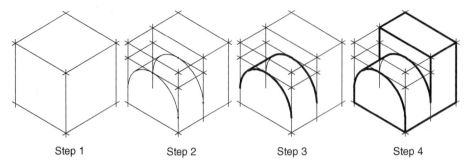

| Step 1 | Step 2 | Step 3 | Step 4 |

Figure 5-31 General procedure for pictorial (isometric) sketch

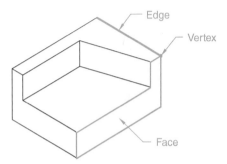

Edge

Vertex

Face

Figure 5-32 Cut block with vertex, edge, and face labeled

▮ OBLIQUE SKETCHES

Introduction

An oblique sketch shows the true size and shape of one principal face of an object. Two receding principal faces are also depicted in order to complete the pictorial view. Traditionally the object is positioned such that its most complex face (i.e., irregular, curved edges, etc.) is shown true size. Relating back to the discussion of oblique projections earlier in this chapter, the true-size (typically front) face is oriented parallel to the projection plane.

Oblique sketches are relatively easy to draw. This is because complex features parallel to the projection plane project without distortion. For example, a circular edge will project as a circle, not an ellipse, as long as it is parallel to the projection plane.

Axis Orientation

In most oblique sketches, the object is oriented so that its front face is shown true size (parallel to the projection plane). Horizontal and vertical axes define the frontal plane. Width is measured along the horizontal axis, height along the vertical. The receding axis is then used to measure depth. The receding axis is typically drawn either up and to the left, or up and to the right. See Figure 5-33. The angle that the receding axis makes with the horizontal is usually drawn to be 30, 45, or 60 degrees.

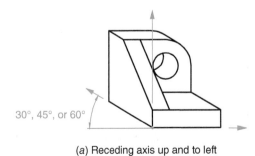

30°, 45°, or 60°

(a) Receding axis up and to left

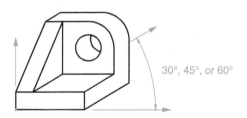

30°, 45°, or 60°

(b) Receding axis up and to right

Figure 5-33 Oblique sketches of the same object

Table 5-2 **Classes of oblique sketches**	
Oblique Type	Receding Axis Scale
Cavalier	1
Cabinet	$\frac{1}{2}$
General	Between $\frac{1}{2}$ and 1

Receding Axis Scale

Recall, from the section Classes of Oblique Projections, that there are three different classes of oblique pictorials: cavalier, cabinet, and general. The type of oblique is determined by the amount of scaling that occurs along the receding axis.[3] See Table 5-2.

In a cavalier oblique, the same scale factor is used along all three axes: horizontal, vertical, and receding. The resulting pictorial appears to be too long in the depth direction because of the lack of foreshortening (see Figure 5-34a). To correct

[3] Recall from the discussion of oblique projections earlier in the chapter that the amount of foreshortening along the receding axis is determined by the angle at which the parallel projectors pierce the projection plane. This angle is called the oblique projection angle (a). For a cavalier projection, a is 45 degrees; for a cabinet projection, a is ~64 degrees.

for this lack of foreshortening, a cabinet oblique employs a receding axis scale of $\frac{1}{2}$. Although it is convenient and also improves upon the appearance of a cavalier oblique, a cabinet oblique can appear to have too much foreshortening. To compensate for this in order to obtain a more pleasing pictorial, a general oblique that scales the receding axis between that of the cavalier and that of the cabinet is sometimes used.

Object Orientation Guidelines

Two rules apply when selecting the front true-size face to be used in an oblique pictorial. The first rule has already been discussed—that the projection plane should be chosen so that it is parallel to the principal face containing the most complex (circular, curved) or irregular shape. This rule was employed, for example, in Figure 5-33, where the curved features are parallel to the projection plane. The second rule is to use the longest face as the front true size view. Figure 5-35 shows two oblique pictorials of an angle beam, oriented according to the two rules.

In the event that the two rules are in conflict, the first rule takes precedence. Figure 5-36a is both easier to construct and less distorted than Figure 5-36b.

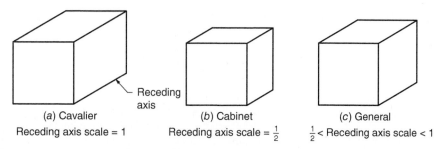

(a) Cavalier	(b) Cabinet	(c) General
Receding axis scale = 1	Receding axis scale = $\frac{1}{2}$	$\frac{1}{2}$ < Receding axis scale < 1

Receding axis

Figure 5-34 Cavalier, cabinet, and general obliques of a cube

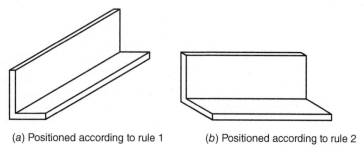

(a) Positioned according to rule 1   (b) Positioned according to rule 2

Figure 5-35 Object orientation for oblique pictorials; two rules

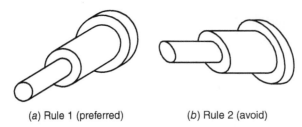

(a) Rule 1 (preferred)   (b) Rule 2 (avoid)

Figure 5-36 Object orientation for oblique pictorials; rule 1 preferred

Drawing an oblique pictorial of an *extrusion* is particularly easy.

1. Sketch the extruded profile as a front view; select and lightly sketch the direction and angle of receding axis.
2. Sketch construction lines extending from other front-face vertices, parallel to the receding axis.
3. Determine the depth, and then complete the back face of the object.
4. Go bold.

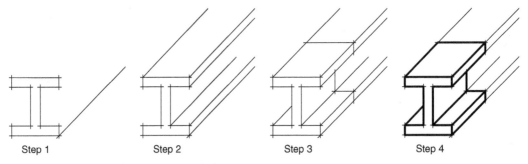

Step 1   Step 2   Step 3   Step 4

Figure 5-37 Oblique sketch of an extruded shape

1. PEB construction:
   a. Identify the most *complex* (irregular shape, circular features, etc.) object face as the front view.
   b. Lightly sketch a properly proportioned rectangle that just encloses object's front face.
   c. Lightly sketch a receding axis (direction, angle) from one vertex of the front rectangle.
   d. Mark off the appropriate depth along the receding axis (for cabinet, scale = $\frac{1}{2}$).
   e. Complete the bounding box using construction lines.
2. Using construction lines, block in feature details.
3. Go bold.

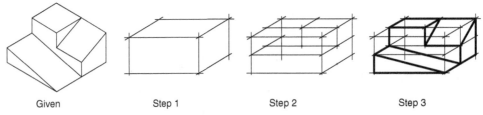

Given   Step 1   Step 2   Step 3

Figure 5-38 Multiple steps for cut block cabinet oblique sketch

1. PEB construction:
   a. Identify the most *complex* (irregular shape, circular features, etc.) object face as the front view.
   b. Lightly sketch a properly proportioned rectangle that just encloses object's front face.
   c. Lightly sketch a receding axis (direction, angle) from one vertex of the front rectangle.
   d. Mark off the appropriate depth along the receding axis (for cavalier, scale = 1).
   e. Still using construction lines, complete the bounding box.

2. Using construction lines, block in linear features.

3. Arc feature construction:
   a. Locate arc quadrants (front and back).
   b. Sketch the front face arc in bold.
   c. Use construction lines to sketch the back face arc.
   d. Sketch a line tangent to the two arcs in bold.
   e. Sketch the visible portion of the back arc in bold.

4. Hole feature construction (front face):
   a. Locate center and quadrant points.
   b. Sketch circle in bold.

5. Hole feature construction (back face):
   a. Locate center and quadrant points.
   b. Sketch circle using construction lines.
   c. Sketch visible portion of back edge in bold.

6. Go bold.

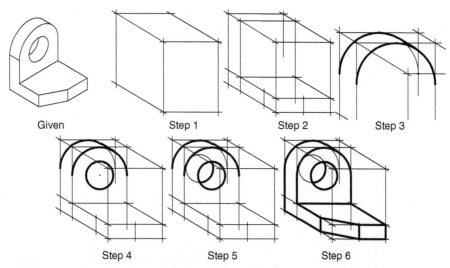

Given   Step 1   Step 2   Step 3

Step 4   Step 5   Step 6

Figure 5-39 Multiple steps for cavalier oblique sketch of object with circular features

ISOMETRIC SKETCHES

Introduction

Similar to obliques, isometric pictorials are parallel projections from which dimensional information can be obtained. In addition, isometric views can easily be created in CAD systems, something that is not true of oblique projections.

Axis Orientation

In sketching an isometric, the principal axes are aligned as shown in Figure 5-40. One axis is vertical, and the other two are inclined at 30 degrees to the horizontal. Note that this 30 degree angle is a direct outcome of the object's orientation with respect to the projection plane used to generate an isometric projection, which was described in the earlier section on isometric projections.

Isometric Scaling

Recall from the discussion earlier in the chapter that the most important property of an isometric projection is that all three principal axes are equally foreshortened. Therefore, for an isometric pictorial, all object edges parallel to a principal axis are scalable and can be directly measured.

For an isometric sketch, the foreshortening that would occur in a true isometric projection is generally ignored. Imagine, for example, an actual rectangular prism that measures 3 × 2 × 1. An isometric sketch of this prism is shown in Figure 5-41. When the sketch is created, the actual dimensions are laid out along the isometric axes, not the foreshortened lengths.

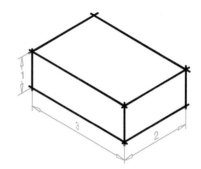

Figure 5-41 Isometric sketch of a rectangular prism

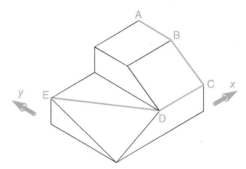

Figure 5-42 Isometric scaling

In an isometric sketch, object edges that are *not* parallel to a principal axis cannot be directly measured. In Figure 5-42, for example, the lengths of edges AB (parallel to the y axis) and CD (parallel to the x axis) can be directly measured. These lines parallel to a principal axis are referred to as ***isometric lines***. Edge lengths BC and DE cannot be scaled (i.e., are not measurable), because they are not parallel to any of the three principal axes. These are called ***nonisometric lines***. In order to sketch these edges, we must first locate their respective vertices.

Isometric Grid Paper

Isometric grid paper can be used when one is first learning to make isometric sketches. Figure 5-43 shows an example of isometric grid paper, along with superimposed isometric coordinate axes. Isometric grid paper consists of three sets of intersecting parallel lines: vertical, up and to the right at 30 degrees to horizontal, and up and to left at 30 degrees to horizontal. Figure 5-44 shows

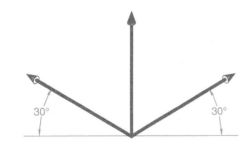

Figure 5-40 Isometric sketch axes

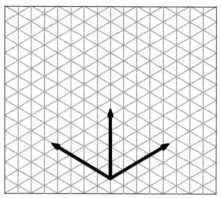

Figure 5-43 Isometric grid paper with superimposed isometric coordinate axes

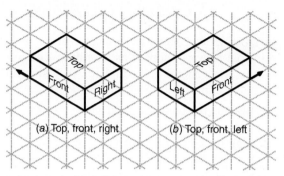

Figure 5-45 Two isometric views of a rectangular prism

an isometric sketch of a cut block using isometric grid paper.

Object Orientation Guidelines

Generally speaking, an object's longest principal dimension should appear as a horizontal dimension on the front face of the object. Assuming this to be the case, then an isometric view showing the top, front, and right side of an object is obtained by laying out the longest dimension on the leftmost *up and to the left* axis (Figure 5-45a). Alternatively, to see a top, front, and left isometric view, place the longest dimension along the rightmost *up and to the right* axis (Figure 5-45b).

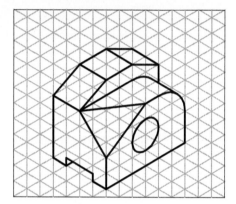

Figure 5-44 Isometric sketch of cut block using isometric grid paper

---

Step-by-step isometric sketch example for a cut block (see Figure 5-46)

1. PEB construction:
   a. Identify the front view of the object to be sketched.
   b. Using isometric axes (or isometric grid paper), lightly sketch a properly proportioned bounding box. To see the top, front, and right faces of the object, lay out the longest (horizontal) dimension along the *up and to the left* axis.
2. Still using construction lines, add feature details.
3. Go bold.

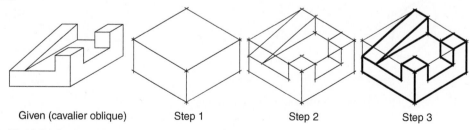

Given (cavalier oblique)   Step 1   Step 2   Step 3

Figure 5-46 Multiple steps for cut block isometric sketch

Circular Features in an Isometric View

In an isometric sketch, a circular feature on a principal face will appear as an ellipse. The following is a general procedure for sketching an isometric ellipse:

1. Identify the planar face on which the circular feature is to appear.
2. Lightly sketch the principal axes of the feature on the planar face.
3. Locate the quadrant points along these axes (equidistant from the intersection of the two axes).
4. If necessary, lightly sketch the bounding *rhombus* that just contains the circular edge.
5. Using bold lines, sketch in the elliptical shape. The ellipse should pass through the quadrant points and will be tangent to the rhombus sides.

Figure 5-47 demonstrates the use of this procedure in the construction of an isometric cylinder. In Figure 5-48, hole features are sketched on three faces of a box.

Step-by-step isometric sketch example for a cylinder (see Figure 5-47)

1. Sketch the axis, and locate the centers of the front and back faces of the cylinder. Note that in this example, the cylinder axis is horizontal.
2. Locate the quadrant points on the front face of the cylinder, and then sketch in the bounding rhombus. Note that the quadrant points are along the two axes in the plane of the face and equidistant from the center.
3. In bold, sketch the elliptical front face of the cylinder.
4. Locate quadrant points, sketch the bounding rhombus, and lightly sketch the elliptical back face of the cylinder.
5. Lightly sketch two lines to represent the *limiting elements* of the cylinder, parallel to the axis of the cylinder and tangent to the front and back elliptical faces.
6. Go bold, showing only the visible portion of the back edge.

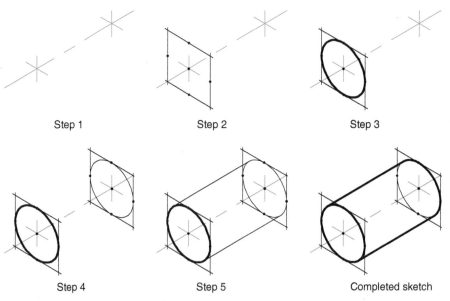

Step 1   Step 2   Step 3

Step 4   Step 5   Completed sketch

Figure 5-47 Construction of an isometric sketch of a cylinder

1. Locate the centers of the holes on each of the three visible faces of the box.
2. Locate the quadrant points of the holes, and then sketch in the bounding rhombus.
3. In bold, sketch the ellipses. A given ellipse should be tangent to its bounding rhombus.

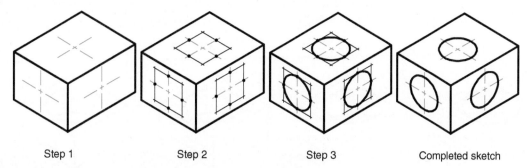

Step 1 Step 2 Step 3 Completed sketch

Figure 5-48 Construction of holes on three faces of an isometric box

1. PEB construction:
   a. Construct two boxes.
2. Locate hole centers:
   a. Upper-left through hole.
   b. Lower-right arc and through hole.
3. Locate top-face quadrants, rhombus construction:
   a. Locate quadrants for upper-left through hole.
   b. Use construction lines to create rhombus.
   c. Locate quadrants for lower-right arc.
4. Top-face ellipse construction:
   a. Sketch upper-left full ellipse in bold.
   b. Sketch lower-right partial ellipse in bold.
   c. Locate quadrants for lower-right through hole.
   d. Use construction lines to create rhombus.
5. Top-face ellipse construction continued:
   a. Sketch lower-right full ellipse in bold.
6. Bottom-face ellipse construction:
   a. Upper-left full ellipse using light construction lines.
   b. Lower-right full ellipse using light construction lines.
   c. Lower-right partial ellipse using light construction lines.
7. Go bold.

(continues)

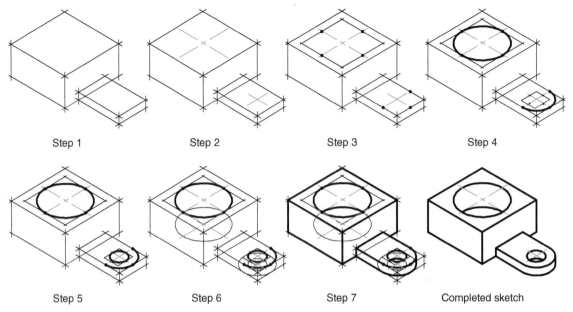

Step 1   Step 2   Step 3   Step 4

Step 5   Step 6   Step 7   Completed sketch

Figure 5-49 Multiple steps for isometric sketch of object with circular features

Chapter review: pictorial sketching scalability

A drawing is said to be *scalable* if dimensional information can be derived from it, even if the drawing itself is not to scale. If, for example, the actual measure of a certain distance depicted on the drawing is known, then other dimensions on the drawing can be approximated by forming proportional relationships between the actual and the measured distances.

$$\frac{x_{\text{actual}}}{y_{\text{actual}}} = \frac{x_{\text{measured}}}{y_{\text{measured}}}$$

Figure 5-50, for example, shows a front view of several well-known buildings, along with the actual building heights. Although the figure is not to scale, it can still be used to determine other (actual) building dimensions (e.g., building width). To do so, measure the building width and the building height on the drawing, divide the width by the height, and then multiply by the actual height to determine the actual width.

Multiview

- Parallel projection technique
- One PEB face is parallel to the projection plane

  Therefore, all edges that are parallel to the projection plane are scalable.

Oblique

- Parallel projection technique
- One PEB face is parallel to the projection plane

  Therefore, all edges that are parallel to the projection plane are scalable.

- For cavalier oblique, the receding axis is scaled the same as the other principal axes

  Therefore, all edges parallel to the receding axis are scalable on a cavalier oblique.

Isometric

- Parallel projection technique
- The PEB is oriented such that all three principal axes are equally foreshortened

  Therefore, all edges parallel to any principal axis are scalable.

Trimetric

- Parallel projection technique
- The PEB is oriented such that all three principal axes are foreshortened by different amounts

Therefore, all edges parallel to a single principal axis are scalable. In effect, there are three separate scales in a trimetric projection, one for each principal axis.

Note: For any planar projection technique, if an object (or feature) is parallel to the projection plane, the feature will be projected true shape. For a parallel projection, these features will also be projected true size.

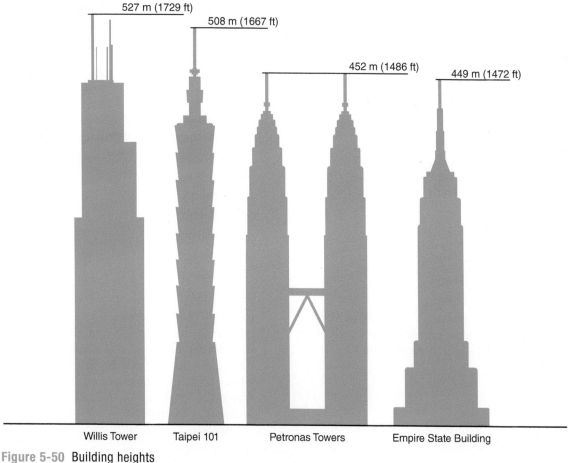

Figure 5-50 Building heights

TRUE OR FALSE

1. Parallel object edges always appear parallel in a parallel projection.
2. In a trimetric projection, none of the angles between the projected principal axes is equal.
3. Assuming that a projection plane is offset to a new parallel position, an axonometric projection of an object onto the projection plane will be identical (i.e., same size and shape) in either location.
4. In an isometric drawing, each principal axis is foreshortened by approximately 82% of its true length.
5. The receding axis angle of an oblique projection is governed by the out-of-plane angle α.

MULTIPLE CHOICE

SS 6. Which of the following is not considered to be a main element of a projection system?
   a. 3D object
   b. 2D cutting plane
   c. Projectors
   d. 2D projection plane
   e. 2D projected image

7. For the cavalier oblique drawing of the cut block shown in Figure P5-1, which edges are scalable (i.e., directly measurable)?
   a. BC and AC
   b. EF and DJ
   c. FH and HJ
   d. GJ and DJ
   e. All of the above
   f. None of the above

SS 8. For the isometric drawing of the cut block shown in Figure P5-2, which edges are scalable (i.e., directly measurable)?
   a. AB
   b. BF
   c. BD
   d. DF
   e. All of the above
   f. None of the above

9. Referring to Figure P5-3, if the sight lines are parallel to each other and also perpendicular

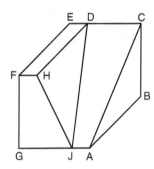

Figure P5-1 Cavalier oblique drawing of a cut block

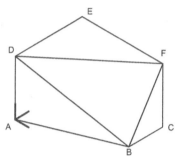

Figure P5-2 Isometric pictorial view of a cut block

to the infinite projection plane, β = 15° and φ = 30°, what is the type of projection of the block that will appear on the projection plane?
   a. One-point perspective
   b. Two-point perspective
   c. Trimetric
   d. Isometric
   e. Dimetric
   f. Cabinet oblique
   g. Cavalier oblique
   h. General oblique
   i. None of the above

10. Referring to Figure P5-3, if the sight lines SS are perpendicular to the infinite projection plane, β = 30° and φ = 20° initially, which of the following is true of the ratio between the projected length a' and the actual length a as φ is increased to 70°?
   a. Increases
   b. Decreases
   c. Does not change
   d. Cannot determine without additional information

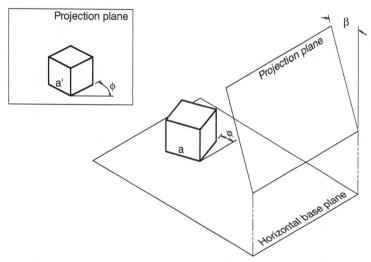

Figure P5-3 Pictorial projection where lengths a and a' represent the actual and projected lengths of the block, respectively (Figure adapted from the work of Michael H. Pleck)

SKETCHING

11. Given the isometric view of the cut block objects appearing in Figures P5-4 through P5-65:

   a. Indicate which visible face (left, right, or even top) to use as the front view of an oblique pictorial sketch. HINT: Review the Object Orientation Guidelines section on page 95.

   b. For Figures P5-4, 8, 12, 16, 20, 24, 28, 32, 36, 40, 44, 48, 52, 56, 60, and 64, use a blank sheet of paper to create a well-proportioned **cavalier oblique** sketch of the object. Use a receding axis oriented at an angle of 45 degrees from the horizontal.

   c. For Figures P5-5, 9, 13, 17, 21, 25, 29, 33, 37, 41, 45, 49, 53, 57, 61, and 65, use a blank sheet of paper to create a well-proportioned **cabinet oblique** sketch of the object. Use a receding axis oriented at an angle of 45 degrees from the horizontal.

   d. For Figures P5-6, 10, 14, 18, 22, 26, 30, 34, 38, 42, 46, 50, 54, 58, and 62, use a blank sheet of paper to create a well-proportioned **cavalier oblique** sketch of the object. Use a receding axis oriented at an angle of 135 degrees from the horizontal.

   e. For Figures P5-7, 11, 15, 19, 23, 27, 31, 35, 39, 43, 47, 51, 55, 59, and 63, use a blank sheet of paper to create a well-proportioned **cabinet oblique** sketch of the object. Use a receding axis oriented at an angle of 135 degrees from the horizontal.

12. Assume that the **even numbered** cut block objects appearing in Figures P5-66 through P5-95 are **cavalier oblique**. Use isometric grid or a blank sheet of paper to sketch an isometric view of the object.

13. Assume that the **odd numbered** cut block objects appearing in Figures P5-66 through P5-95 are **cabinet oblique**. Use isometric grid or a blank sheet of paper to sketch an isometric view of the object.

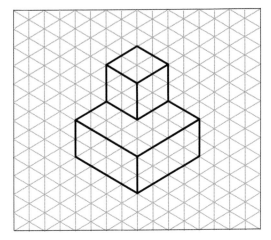

Figure P5-4

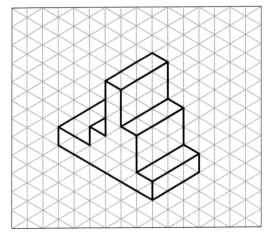

Figure P5-7

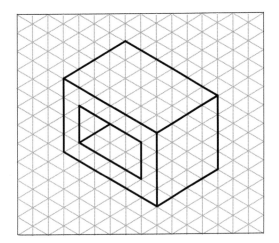

Figure P5-5

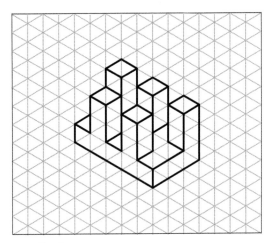

Figure P5-8

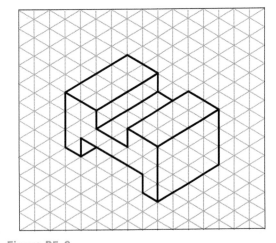

Figure P5-6

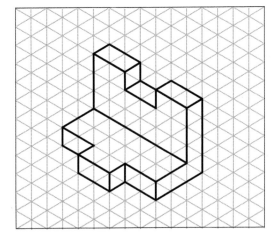

Figure P5-9

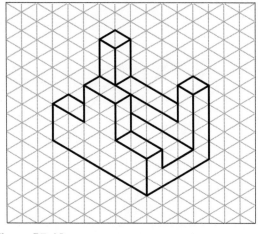

Figure P5-10

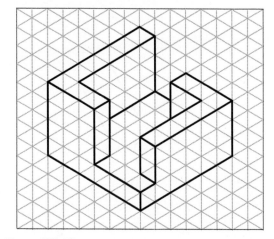

Figure P5-13

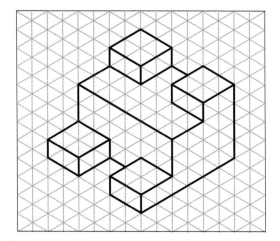

Figure P5-11

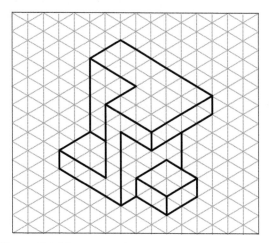

Figure P5-14

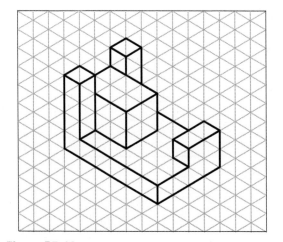

Figure P5-12

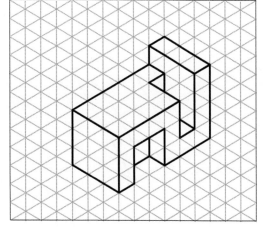

Figure P5-15

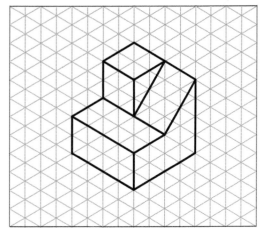

Figure P5-16

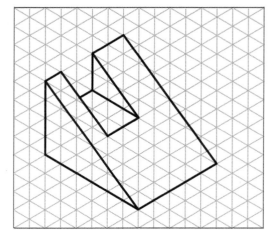

Figure P5-19

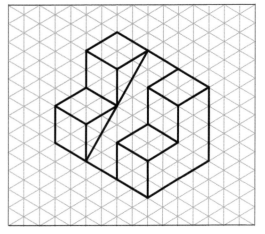

Figure P5-17

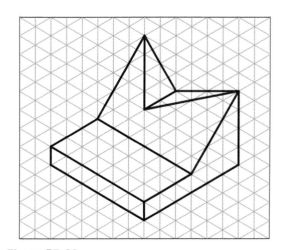

Figure P5-20

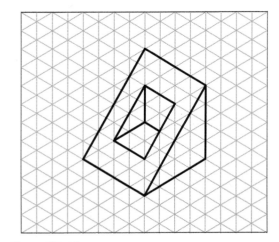

Figure P5-18

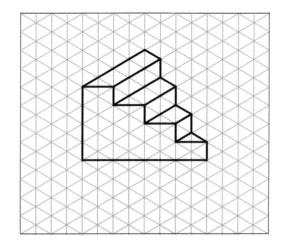

Figure P5-21

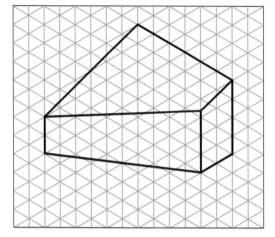

Figure P5-22

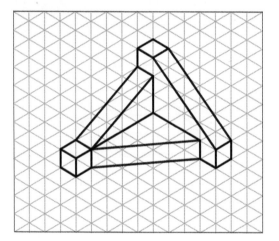

Figure P5-23

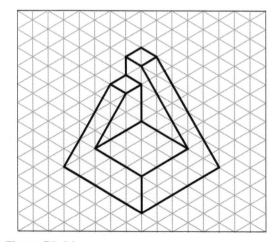

Figure P5-24

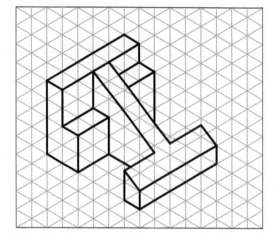

Figure P5-25

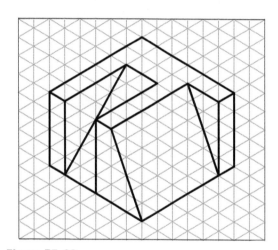

Figure P5-26

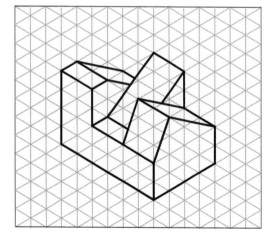

Figure P5-27

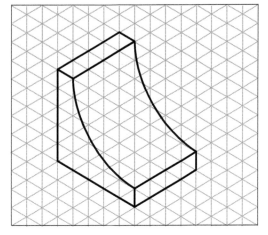

Figure P5-28

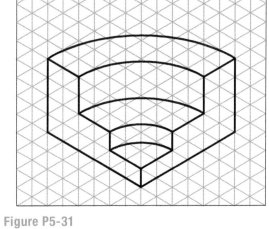

Figure P5-31

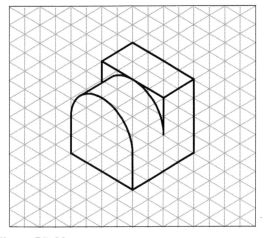

Figure P5-29

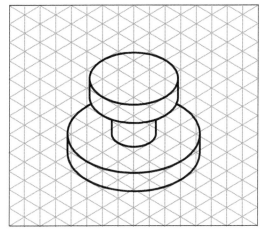

Figure P5-32

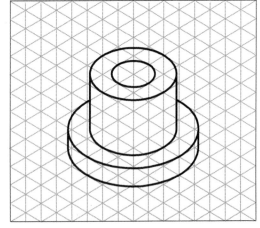

Figure P5-30

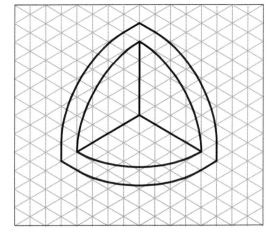

Figure P5-33

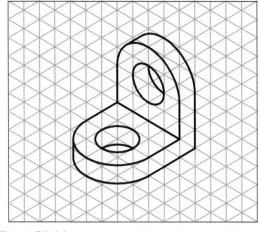

Figure P5-34

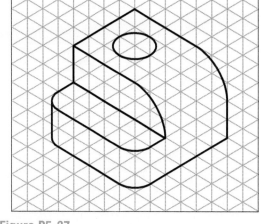

Figure P5-37

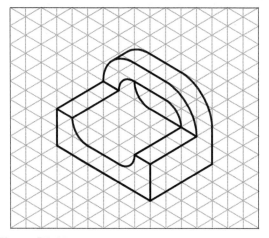

Figure P5-35

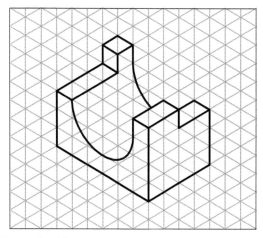

Figure P5-38

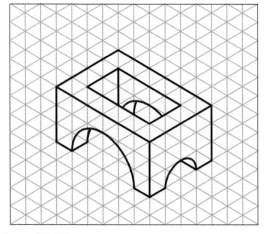

Figure P5-36

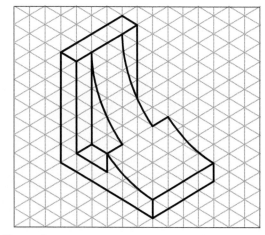

Figure P5-39

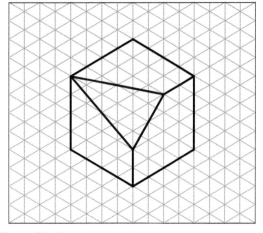

Figure P5-40

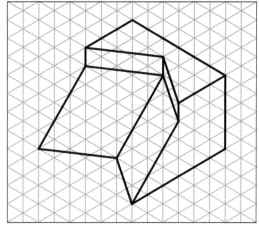

Figure P5-43

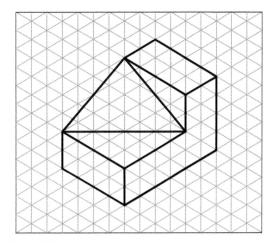

Figure P5-41

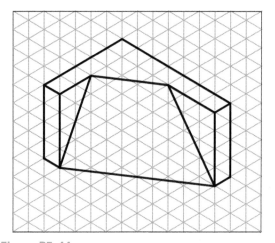

Figure P5-44

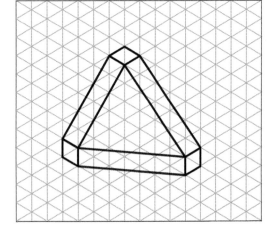

Figure P5-42

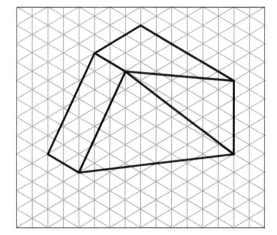

Figure P5-45

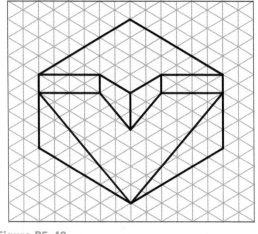

Figure P5-46

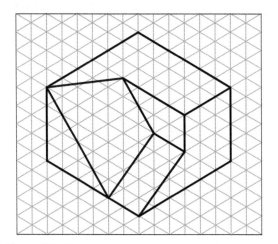

Figure P5-47

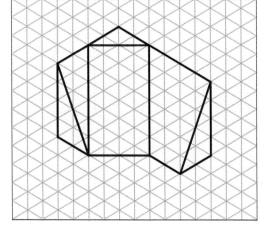

Figure P5-48

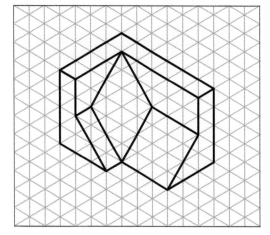

Figure P5-49

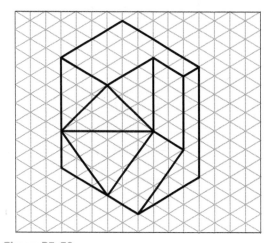

Figure P5-50

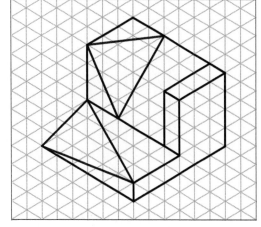

Figure P5-51

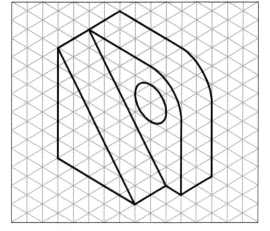

Figure P5-52

Figure P5-55

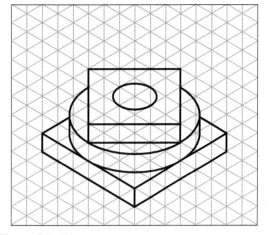

Figure P5-53

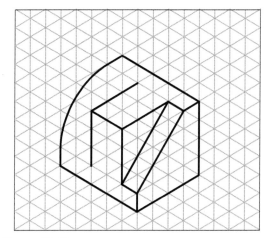

Figure P5-56

Figure P5-54

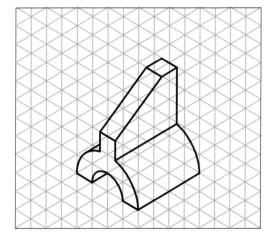

Figure P5-57

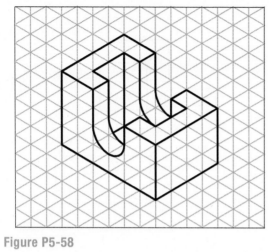

Figure P5-58

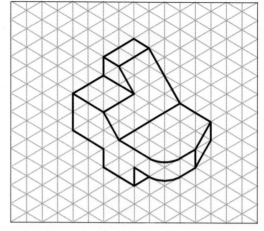

Figure P5-61

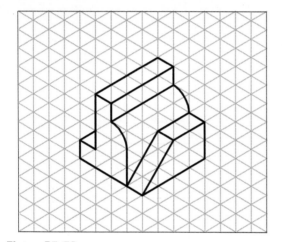

Figure P5-59

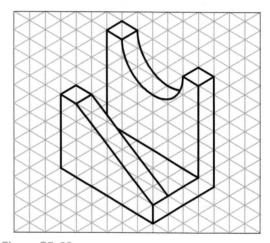

Figure P5-62

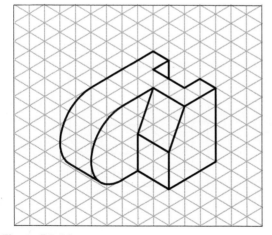

Figure P5-60

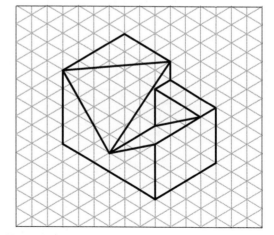

Figure P5-63

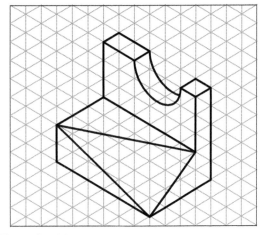

Figure P5-64

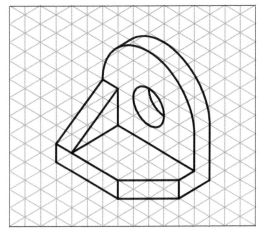

Figure P5-65

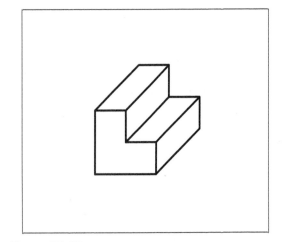

Figure P5-66

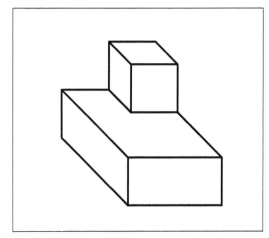

Figure P5-67

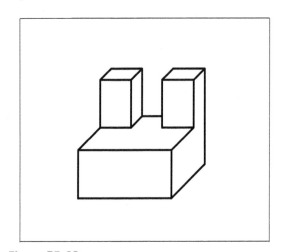

Figure P5-68

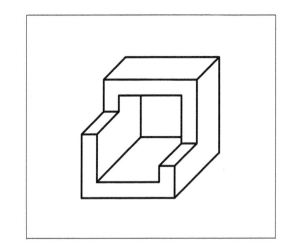

Figure P5-69

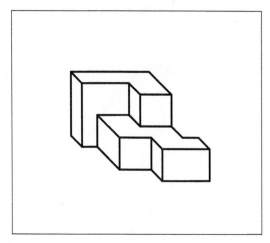

Figure P5-70

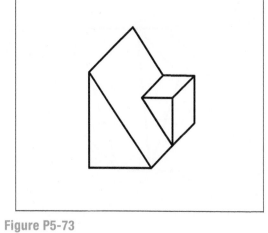

Figure P5-73

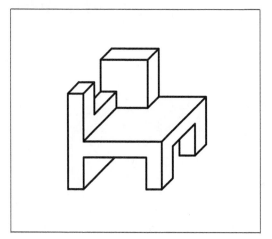

Figure P5-71

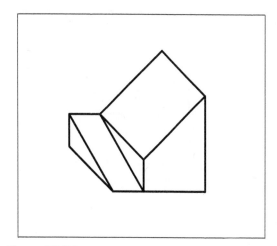

Figure P5-74

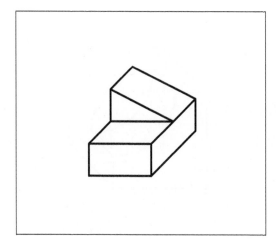

Figure P5-72

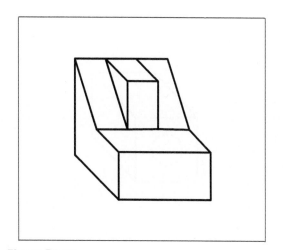

Figure P5-75

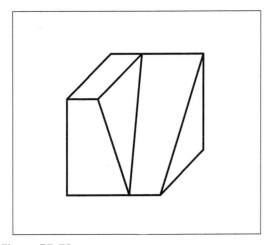

Figure P5-76

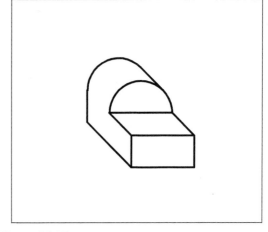

Figure P5-79

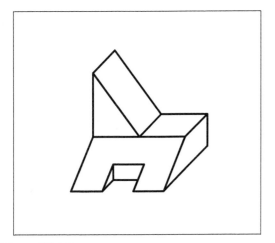

Figure P5-77

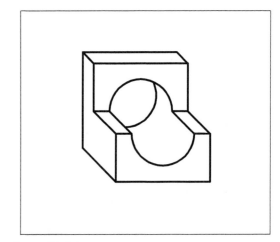

Figure P5-80

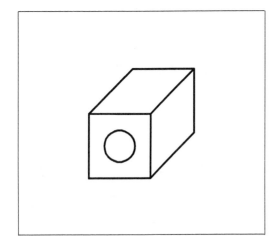

Figure P5-78

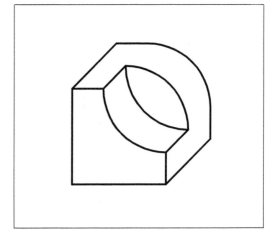

Figure P5-81

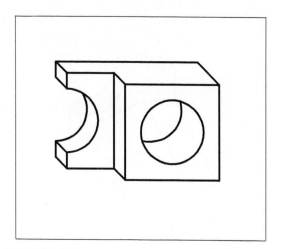

Figure P5-82

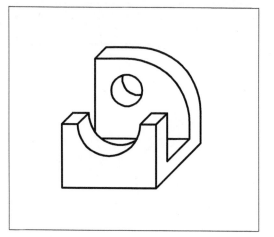

Figure P5-83

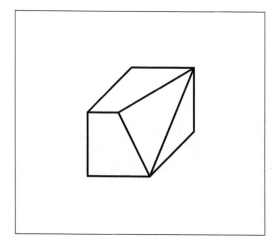

Figure P5-84

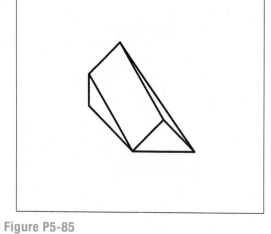

Figure P5-85

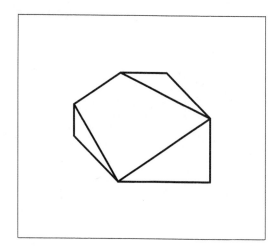

Figure P5-86

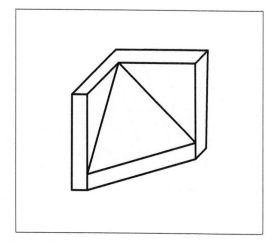

Figure P5-87

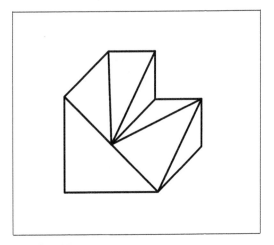

Figure P5-88

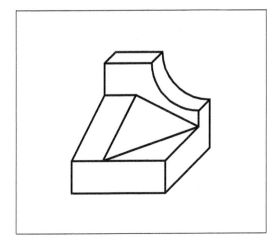

Figure P5-91

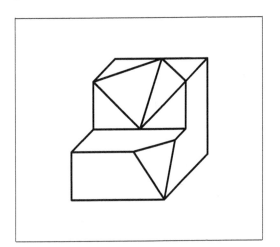

Figure P5-89

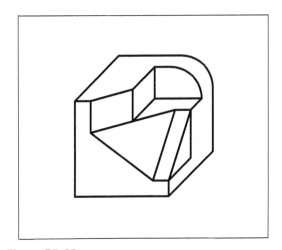

Figure P5-92

Figure P5-90

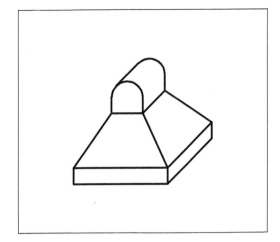

Figure P5-93

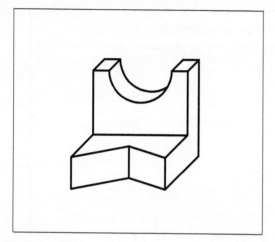

Figure P5-94

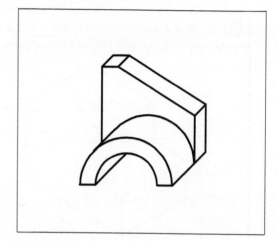

Figure P5-95

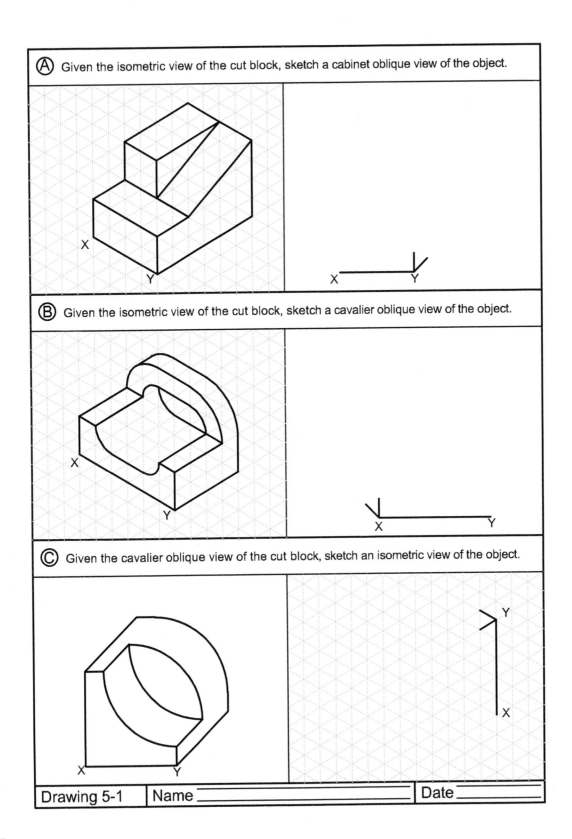

Ⓐ Given the isometric view of the cut block, sketch a cabinet oblique view of the object.

X

Y

X————Y

Ⓑ Given the isometric view of the cut block, sketch a cavalier oblique view of the object.

X

Y

X————Y

Ⓒ Given the cavalier oblique view of the cut block, sketch an isometric view of the object.

X   Y

Y

X

Drawing 5-1 | Name ————— | Date —————

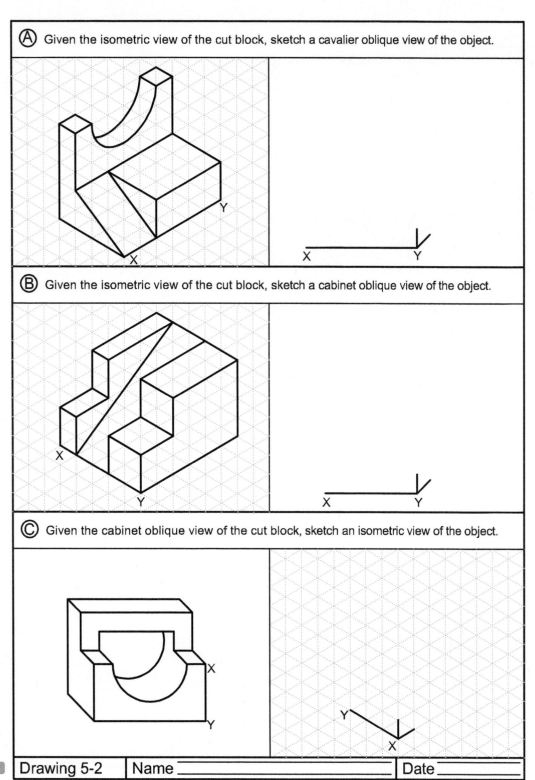

Ⓐ Given the isometric view of the cut block, sketch a cavalier oblique view of the object.

Ⓑ Given the isometric view of the cut block, sketch a cabinet oblique view of the object.

Ⓒ Given the cabinet oblique view of the cut block, sketch an isometric view of the object.

Drawing 5-2 | Name | Date

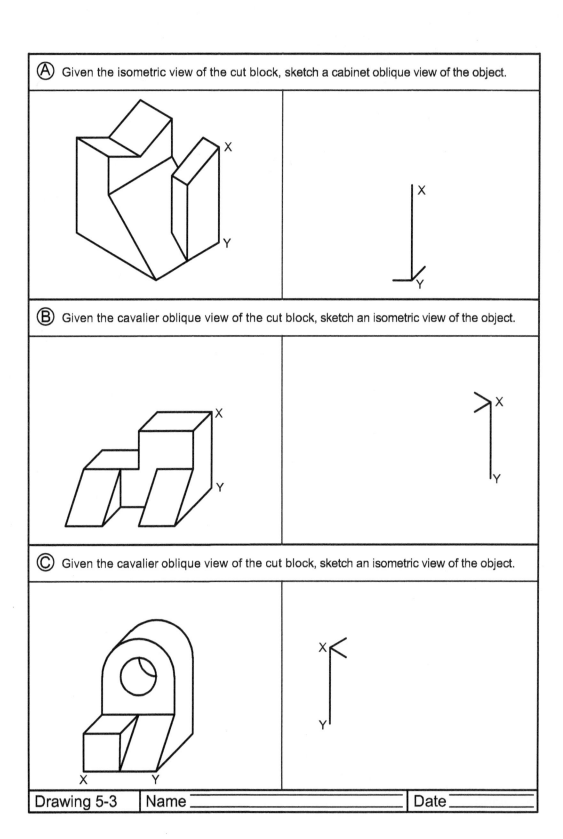

A Given the isometric view of the cut block, sketch a cabinet oblique view of the object.

X

X

Y

Y

B Given the cavalier oblique view of the cut block, sketch an isometric view of the object.

X

X

Y

Y

C Given the cavalier oblique view of the cut block, sketch an isometric view of the object.

X

X

Y

Y

| Drawing 5-3 | Name | Date |

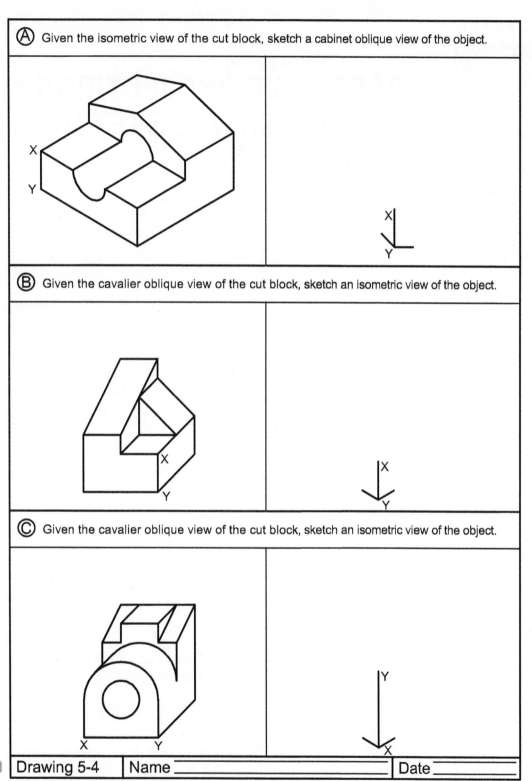

Ⓐ Given the isometric view of the cut block, sketch a cabinet oblique view of the object.

Ⓑ Given the cavalier oblique view of the cut block, sketch an isometric view of the object.

Ⓒ Given the cavalier oblique view of the cut block, sketch an isometric view of the object.

Drawing 5-4 | Name | Date

CHAPTER

6 PERSPECTIVE PROJECTIONS AND PERSPECTIVE SKETCHES

■ PERSPECTIVE PROJECTION

Historical Development

Perhaps the single most important development in Renaissance art is the use of perspective.[1] Just prior to that time, paintings like those of Duccio di Buoninsegna (1255–1319) tended to be rather flat and two-dimensional (see Figure 6-1). Artists had yet to achieve an adequate understanding of human anatomy, and they had not developed techniques like shading and perspective to create an illusion of depth. Giotto (1267–1377), actually a contemporary of Duccio's, is generally considered the first Renaissance painter. In Figure 6-2, Giotto employs converging lines to suggest spatial depth, although these lines do not systematically converge to a single vanishing point.

In the work of later Italian Renaissance artists such as Leonardo da Vinci[2] (1452–1519) and Raffaello Sanzio (1483–1520), we find paintings that employ one-point perspective to call the

[1]The development of perspective during the Renaissance was recently chosen by *Time* magazine as one of the *100 Ideas That Changed the World*, 2011.

[2]Two quotes regarding perspective are attributed to Leonardo, the first being "Perspective is the rein and rudder of painting." The second quote, "Perspective is nothing else than the seeing of an object through a sheet of glass, on the surface of which may be marked all the things that are behind the glass," goes a long way toward describing just how perspective works.

Figure 6-1 Duccio di Buoninsegna, *Maesta*, Siena, 1308–1311 (Courtesy of The Bridgeman Art Library International)

attention of the viewer to important details in the painting. See, for example, Figures 6-3 and 6-4. The mathematical rules of perspective were developed and documented by people like the German Albrecht Dürer (1471–1528) and several Italian artists, including Brunelleshi and Alberti. Filippo Brunelleshi (1377–1446), a Florentine, invented a systematic method for determining perspective projections in the early 1400s. Leon Battista Alberti (1404–1472) wrote the first treatise on perspective, *On Painting*.

Perspective Projection Characteristics

As we saw in Chapter 3, perspective differs from parallel projection in that in the former, the center of projection is a finite distance from the object. The projectors are therefore nonparallel

Figure 6-2 Giotto, *Franciscan Rule Approved*, Assisi, Upper Basilica, c.1295–1300 (Courtesy of The Bridgeman Art Library International)

Figure 6-3 Leonardo da Vinci, *Last Supper*, 1498 (Courtesy of The Bridgeman Art Library International)

rays that converge to the center of projection. As a consequence, when parallel object edges are not parallel to the projection plane, the edges converge to a *vanishing point* when projected. In addition, objects or features that are farther away from the projection plane are more *foreshortened* (i.e., smaller) than closer ones.

The principal advantage of perspective projection is that it produces a more realistic image. It closely approximates the view as seen by the human eye. A significant drawback of perspective projection, however, is that it does a

Figure 6-4 Raffaello Sanzio, *School of Athens*, Vatican, 1509 (Courtesy of The Bridgeman Art Library International)

poor job in preserving the scale of the object. Consequently, dimensional information often cannot be extracted. In addition, perspective projections are generally more difficult to execute than parallel projections.

Classes of Perspective Projection

Perspective views are categorized according to the orientation of the object with respect to the projection plane. This orientation determines the number of principal axes (refer to Figure 5-4 in Chapter 5) that are parallel to the projection plane. If an axis is not parallel to the projection plane, then object edges parallel to this axis will not be parallel when projected. Rather, they will converge to a single point, called a vanishing point. There are three possible cases:

1. One-point perspective (one principal vanishing point)
2. Two-point perspective (two principal vanishing points)
3. Three-point perspective (three principal vanishing points)

In a top view looking down, Figure 6-5 illustrates the orientation of three identical cubes (more generally, three cube-shaped bounding boxes), with respect to a vertical projection plane. Note that, because the scene is viewed from above, the vertical projection plane appears as a line, since it is viewed on edge. Note also that the principal axes of each cube are also represented.

Type of perspective projection

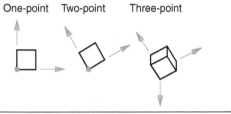

Projection plane on edge

Figure 6-5 Perspective classes

The cube on the left is oriented with one principal face parallel to the projection plane. If the principal axes of this principal enclosing box (PEB) are extended infinitely in both directions, only one axis intersects the projection plane; the other two axes (one vertical, one horizontal) are parallel to the projection plane. When the bounding box of an object is oriented in this way, a one-point perspective projection results.

The PEB in the middle has been rotated about a vertical axis so that only the vertical principal axis is parallel to the projection plane; the other two axes are inclined to the projection plane. In this orientation, a two-point perspective projection results.

Finally, imagine rotating the middle cube out of the plane of the paper about a horizontal axis. This is the position of the cube on the far right. Note that in this case, all three principal axes, when extended, intersect the projection plane. None of the three principal axes are parallel to the projection plane. In this orientation, a three-point perspective projection will result.

Vanishing Points

Before discussing these three cases in more detail, it is worth reiterating that, in a perspective projection, if parallel object edges are:

- Parallel to the projection plane, then the projected edges will also be parallel
- Inclined to the projection plane, then the projected edges will not be parallel; they will converge to a ***vanishing point***

Returning to Figure 6-5, it is apparent that a one-point perspective of a box has one vanishing point, a two-point perspective of a box has two vanishing points, and a three-point perspective of a box has three vanishing points. These vanishing points are called ***principal vanishing points***, since they are associated with the principal axes of the object. Table 6-1 provides a summary of the different perspective classes, along with the number of principal vanishing points of each.

Figure 6-6 illustrates the process of locating principal vanishing points (PVPs) for a two-point perspective projection. Figure 6-6a shows a pictorial view of the object, projection plane, and center of projection (CP). Figure 6-6b shows the same elements when viewed from above. We can see that this orientation of the object with respect to the projection plane will result in a two-point perspective, since two principal axes are inclined to the projection plane. Two dashed construction lines are drawn through the CP, each parallel to an inclined principal axis, until they intersect the projection plane. Each point of intersection locates a principal vanishing point. Each construction line is parallel to a set

Table 6-1 Classes of perspective projection

Perspective Type	Principal Vanishing Points (PVPs)	Principal Axis Orientation plane (PP)
One-point	1	• One principal axis perpendicular to projection • Two principal axes parallel to PP
Two-point	2	• Two principal axes inclined to projection plane (PP) • One principal axis parallel to PP
Three-point	3	• All three principal axes inclined to PP • No principal axes parallel to PP

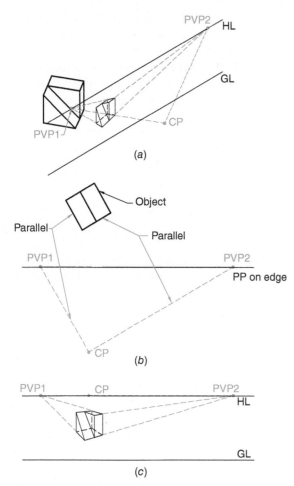

(a)

(b)

(c)

Figure 6-6 Locating principal vanishing points

of parallel edges on the object, with each edge set parallel to a principal axis. To summarize, taking a line parallel to an inclined object edge and passing it through the center of projection until the line pierces the projection plane locates the vanishing point for that object edge.

Figure 6-6*c* shows the projection plane and the resulting perspective projection. For clarity, the object is not shown. Note that the projected edges converge to the principal vanishing points. Figures 6-6*a* and 6-6*c* show the ground line (GL) and the horizon line (HL). We will use these lines later in this chapter when constructing perspective sketches. The **ground line** is a horizontal line formed by the intersection of the projection plane and the ground plane—that is, the plane on which the object rests. The **horizon**

line represents the eye level of the observer. The horizon line is formed by the intersection of the projection plane and the horizontal plane that passes through the center of projection.

As seen in Figure 6-6*c*, both principal vanishing points lie on the horizon line. Each of these vanishing points results from the projection of a set of parallel object edges. Both parallel edge sets are themselves parallel to the ground plane. Vanishing points for edges parallel to a plane always lie along a straight line in the projection plane, with the line parallel to the plane. Here the plane is the ground plane, and the straight line is the horizon line.

Any parallel group of inclined object edges will converge to a vanishing point in a perspective projection, not just principal axis edges. For example, the object depicted in the two-point perspective in Figure 6-6*c* actually has three vanishing points; see Figure 6-7. The edges of the inclined surface on the actual object are inclined (i.e., not parallel) to the projection plane. Consequently, when projected, the inclined surface edges also converge to a vanishing point.

One-Point Perspective Projection

In a one-point perspective projection, one object face is parallel to the projection plane. One principal axis is perpendicular to the projection plane, and the other two principal axes (horizontal, vertical) are parallel to the projection plane.

Figure 6-8 shows a one-point perspective drawing of a cube. Note that the vertical edges of the projected cube are parallel to one another, as

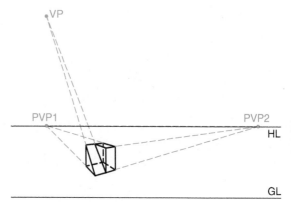

Figure 6-7 Another principal vanishing point

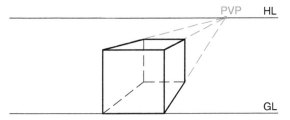

Figure 6-8 One-point perspective drawing of a cube

are the horizontal projected edges. Also note that the receding[3] edges of the cube are not parallel. Rather, they converge to a principal vanishing point.

Figure 6-9 shows the perspective arrangement used to obtain Figure 6-8. Once again, the top portion of the figure shows a view from above. The object, the projection plane, the center of projection, the projectors, and the construction lines used to locate the PVP are all depicted. Note that the front face of the object is coplanar with the projection plane. The projected image will appear within the encircled area on the projection plane.

---

[3]Object edges not parallel to the projection plane will appear to recede back into space when projected.

The bottom half of Figure 6-9 shows the resulting one-point projection (object not shown for clarity). The dotted vertical lines connecting the two portions of the figure are used to locate the projected image on the projection plane. Because the front face of the cube lies in the projection plane, it will be projected true size.

Figure 6-10 shows another example of a one-point perspective arrangement, this time with the object entirely behind the projection plane. Note that, because of this, the front face of the object is not projected true size.

Two-Point Perspective Projection

An example of a two-point perspective arrangement was shown in Figure 6-6. Let us review some of its characteristics:

- One set of principal edges (typically vertical) are parallel to the projection plane, causing the projected edges also to be vertical.

- The other two sets of principal edges, being inclined to the projection plane, will converge to vanishing points when projected. These principal vanishing points will lie on the horizon line.

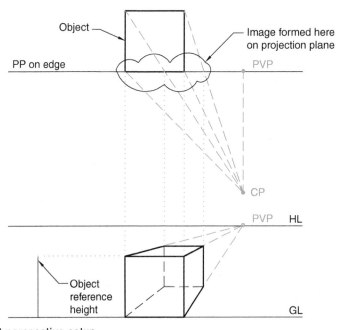

Figure 6-9 One-point perspective setup

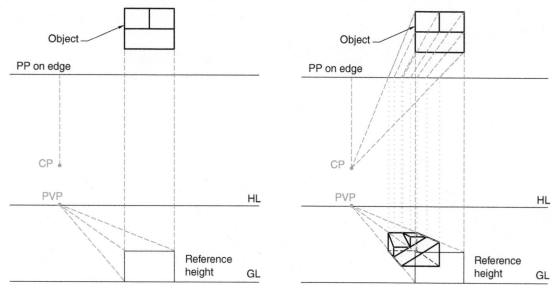

Figure 6-10 One-point perspective setup with object behind projection plane

- If the leading edge of the object lies behind the projection plane (as is the case in Figure 6-6), then none of the projected edges will appear true size.

  Figure 6-11 provides an example of a two-point perspective projection where the leading object

edge lies in the projection plane. In this case, the leading edge is projected true size.

Three-Point Perspective Projection

Because of the difficulty in their construction, three-point perspective drawings are rarely used. Figure 6-12 shows an example of a three-point

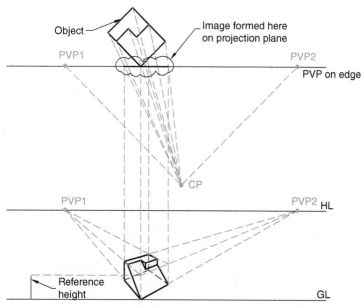

Figure 6-11 Two-point perspective projection setup; leading edge lies in projection plane

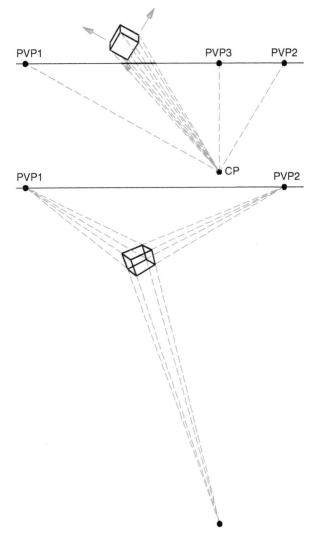

Figure 6-12 Three-point perspective setup

perspective arrangement. Note that all three principal axes are inclined to the projection plane and that, when projected, all three sets of edges converge to principal vanishing points. Two of the PVPs lie on the horizon line; one does not.

Perspective Projection Variables

There are several variables that influence the appearance of a perspective projection. Some of these variables are discussed next.

Perspective projection using a 3D CAD system

The procedure[*] described below may be used to create a perspective projection (or, for that matter, any of the planar projections described in Chapter 3) using a 3D CAD system like AutoCAD®. In this procedure, all of the

[*]This section is based on the work of Michael H. Pleck, who developed this technique at the University of Illinois at Urbana-Champaign.

common elements of a projection system (e.g., object, projection plane, center of projection, projectors), as well as the projection itself, are modeled.

1. Start by creating the object, the projection plane, and the center of projection. In Figure 6-13:
   a. The object is modeled as a solid.
   b. The projection plane is represented using line segments to draw a vertically oriented rectangle.
   c. The center of projection is modeled as a solid sphere or point.

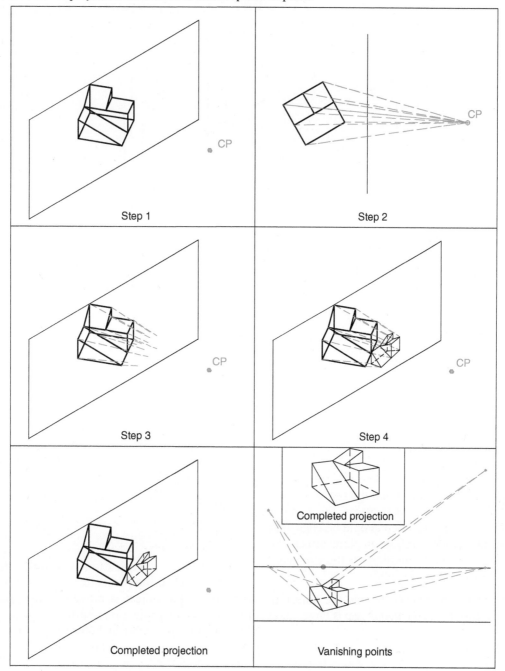

Figure 6-13 Perspective projection using a 3D CAD system

(*continues*)

2. Use the line command and object snap settings to draw the projectors from the vertices of the object to the CP.

3. Use the trim command to cut the projectors at the projection plane.

4. Use the line command to project each object edge onto the projection plane.

The completed projection is shown in the figure in the lower-left corner (projectors not shown). In the lower-right corner, projected edges are extended until they meet at vanishing points.

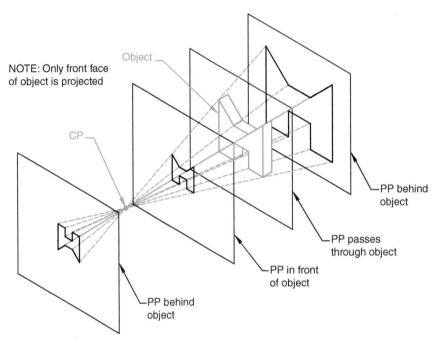

NOTE: Only front face of object is projected

Object

CP

PP behind object

PP passes through object

PP in front of object

PP behind object

Figure 6-14 Perspective projection plane projection

PROJECTION PLANE LOCATION

It has already been shown that in a perspective projection, the size of the projected image depends on the location of the projection plane with respect to the object and the center of projection. From Figure 6-14 (which is the same as Figure 5-13 in Chapter 5), it should be clear that the placement of the projection plane affects the size, and even the orientation, of the projected image. Here are the possibilities:

1. The projection place is behind the object, in which case the projected image is larger than the object.

2. The projection passes through the object, resulting in a projected image that is the same size as the object.

3. The projection plane is in front of the object, causing the projected image to be smaller than the object.

4. The projection plane is behind the center of projection. In this case, the projected image is inverted.

LATERAL MOVEMENT OF CP

If the center of projection is moved laterally with respect to the projection plane (or, equivalently, if the object is moved with respect to the center of projection), different projections will result, as shown in Figure 6-15. Generally speaking, it is recommended that the center of projection be placed in front of the object, slightly to one side.

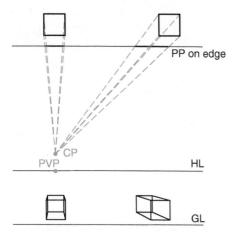

Figure 6-15 Lateral movement of an object with the same center of projection

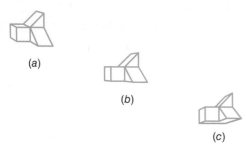

Figure 6-16 Vertical movement of the center of projection

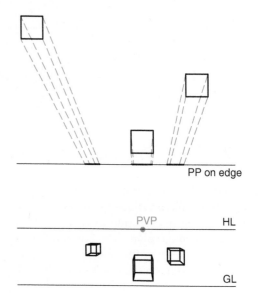

Figure 6-17 Varying object distance from projection plane

farther the object is from the projection plane, the smaller the projected image.

■ PERSPECTIVE SKETCHES

Introduction

Perspective sketches provide a more realistic representation of an object than parallel projection techniques, while sacrificing much of the latter's ability to preserve dimensional information. Perspective sketches are also more difficult to construct than either oblique or isometric sketches.

A perspective sketch represents an object as an observer would see it from a certain vantage point. Receding parallel object edges converge in a perspective pictorial, causing distant objects to appear smaller. In contrast, in a parallel projection, parallel edges remain parallel in the projected image. Using parallel projection, objects are projected as the same size, regardless of their distance from the projection plane.

Terminology

Key elements of a perspective sketch are shown in Figure 6-18. As we have already seen, these elements include the ground line, the horizon

VERTICAL MOVEMENT OF CP

Figure 6-16 shows how the projection of an object can change, depending on the vertical placement of the center of projection with respect to the ground plane. In Figure 6-16a, the center of projection is above the object. Figure 6-16b shows the same object, but with the center of projection at the same level as the object. Finally, in Figure 6-16c, the center of projection is below the object.

VARYING DISTANCE FROM CP

One of the strengths of perspective projection is that it results in a more realistic image than parallel projection. This is due to the fact that, much as in our own vision, the size of an image projected using perspective depends on the distance of the object from the projection plane. In Figure 6-17, cubes of the same size are projected using one-point perspective. Note that the

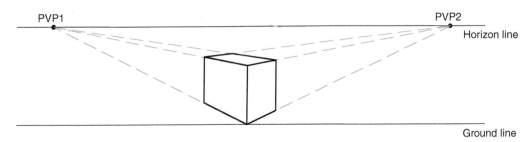

Figure 6-18 Perspective sketch terminology; two-point perspective

line, and vanishing points. The ground line represents the plane on which the object rests and is formed by the intersection of the ground plane with the projection plane. The horizon line represents the eye level of the observer.[4] A vanishing point is a position on the horizon to which depth projectors converge.[5]

One-Point Perspective Sketches

Recall that in a one-point perspective projection, one object face is parallel to the projection plane. This explains the similarity between a one-point perspective and an oblique sketch, which is also oriented with two principal axes parallel to the projection plane. See Figure 6-19 for a comparison between one-point perspective and oblique pictorials.

The main difference between the two pictorials is that the receding edges are parallel in oblique projection, whereas in the one-point perspective the receding edges converge to a vanishing point. In a one-point perspective projection, if the object's front face coincides with the projection plane, then the front face of the object is projected full scale (see Figure 6-14). Otherwise, if it lies behind (or in front of) the projection plane, the projected front face will be smaller (or larger) than the actual. In practical terms, though, when constructing a perspective

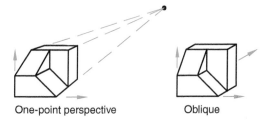

One-point perspective   Oblique

Figure 6-19 Comparison of one-point perspective and oblique sketches

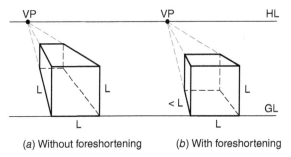

(a) Without foreshortening   (b) With foreshortening

Figure 6-20 One-point perspective of a cube, without and with foreshortening

sketch, we first choose a vertical edge length representing the height of the front face. The horizontal width dimension is then scaled proportional to this vertical dimension.

Scaling along the receding depth axis involves some visual approximation. Figure 6-20 shows two one-point perspective sketches of the same cube. On the left, the cube has been laid out using the same distance L along the horizontal, vertical, and converging axes. Clearly, the resulting pictorial appears to be too long in the receding axis direction. This distortion occurs because foreshortening along the depth axis has not been accounted for. On the right, the depth dimension has been reduced, resulting in an improved representation

[4] Recall from the discussion of perspective projection earlier in the chapter that the horizon line is at the same height as the center of projection.

[5] In a perspective projection, parallel object edges inclined to the projection plane converge when projected. If these edges are also parallel to the horizontal ground plane, they will be projected along the horizon line.

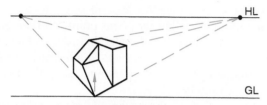

Figure 6-21 Two-point perspective sketch

of a cube. Although the amount of foreshortening depends on several variables, a good rule of thumb is to foreshorten the converging axis dimension on a one-point perspective sketch to approximately two-thirds of the actual.

Two-Point Perspective Sketches

In a two-point perspective sketch, the object is oriented so that only one principal axis, typically vertical, is parallel to the projection plane. The other two principal axes are inclined to the projection plane. As a consequence, vertical object edges remain parallel when projected, while the two other sets of principal edges converge to different vanishing points. Both of these principal vanishing points lie on the horizon line (see Figure 6-21).

If the projection plane passes through the leading vertical edge of the object, this edge will be projected full size (see Figure 6-14).

Otherwise, if the edge is behind (or in front of) the projection plane, the projected vertical will

be smaller (or larger) than the true length. When constructing a two-point perspective sketch, though, one simply chooses a vertical edge length representing the height of the leading edge of the bounding box, without regard for the location of this edge with respect to the projection plane. Convergence lines are then drawn from the leading edge end points to both principal vanishing points.

As was the case with one-point perspective, the amount of foreshortening along the receding axis must be estimated in order to create a well-proportioned sketch. In the case of a two-point perspective, however, there are two receding axes. Figure 6-22 shows a two-point perspective sketch of a cube, where the cube is placed at a 45-degree angle to the projection plane. If the horizontal distances from the leading edge to each principal vanishing point are equal, as is the case in Figure 6-22, then the amount of foreshortening will be the same along each receding axis. A good estimate of this foreshortening amount is provided in the figure.

Another scenario is provided in Figure 6-23. In this case, the cube is positioned at a 30-degree angle with respect to the projection plane. The cube is laterally positioned so that its leading edge is one-fourth the distance between the two principal vanishing points. Given this scenario, Figure 6-23 provides reasonable foreshortening estimates along both receding axes. Note that the

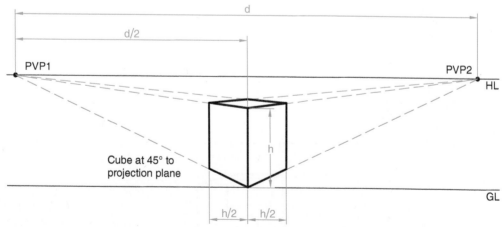

Figure 6-22 Two-point perspective pictorial of a cube at a 45-degree angle to the projection plane (Taken from Jerry Dobrovolny and David O'Bryant, *Graphics for Engineers, 2nd Edition*, John Wiley & Sons, 1984)

closer the PVP is to leading edge distance, the greater the amount of foreshortening.[6]

Proportioning Techniques

A useful proportioning technique when constructing perspective sketches is to sketch the

[6]Figures 6-22 and 6-23 are taken from *Graphics for Engineers*, 2nd Edition by Jerry Dobrovolny and David O'Bryant, John Wiley & Sons, 1984.

diagonals of a receding face in order to locate the midpoint of that face. This point can then be projected to an adjacent edge in order to locate other key vertices. Figure 6-24 illustrates this technique for both one-point and two-point perspective sketches.

In Figure 6-25 this technique is extended to allow for partitioning of a trapezoidal area into thirds and quarters. See the section on partitioning lines in Chapter 3 for additional information.

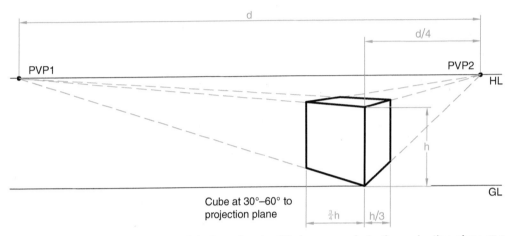

Figure 6-23 Two-point perspective pictorial of a cube at a 30-degree angle to the projection plane (Taken from Jerry Dobrovolny and David O'Bryant, *Graphics for Engineers, 2nd Edition*, John Wiley & Sons, 1984)

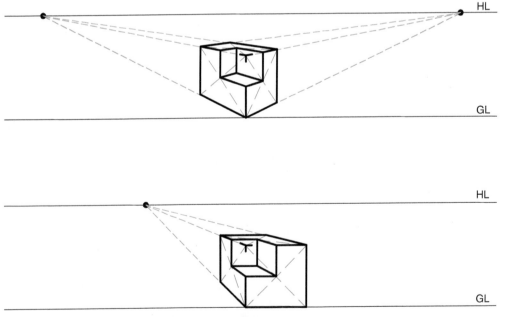

Figure 6-24 Use of diagonals to locate important vertices

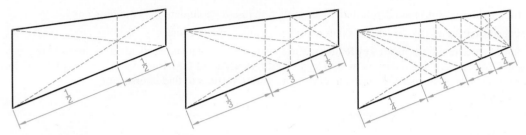

Figure 6-25 Partitioning trapezoidal areas

Given a cavalier oblique pictorial (PEB is a cube), a reference edge height and location, and a principal vanishing point, construct a one-point perspective sketch of the object.

1. Use construction lines to complete the front face of the bounding box.
2. Use construction lines to sketch convergence lines.

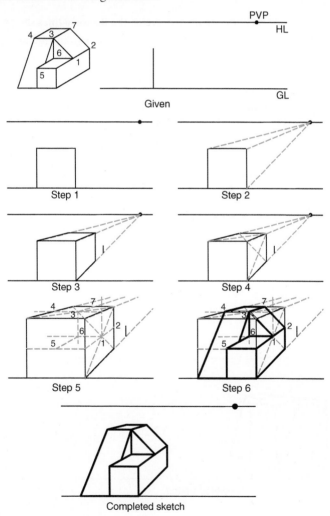

Figure 6-26 Multiple steps for one-point perspective sketch

(continues)

3. Estimate the foreshortened depth, and use construction lines to complete the bounding box.

4. Using construction lines, sketch diagonals on the receding (top, right) faces.

5. Using construction lines, locate key features. Midpoints of horizontal and vertical edges are used to locate 2, 5, and 7. Intersecting diagonals are used to locate mid-face vertices (1, 3). Remaining vertices (4, 6) are located by passing horizontal and/or vertical lines through existing vertices to find intersections.

6. Go bold.

Step-by-step two-point perspective sketch example (see Figure 6-27)

Given a cavalier oblique, the reference edge height and location, and the location of the principal vanishing points:

1. Use construction lines to sketch convergence lines to PVP1 and PVP2. Also lay out the unforeshortened dimensions of the object's PEB.

2. After estimating the foreshortened depths, complete the bounding box (use construction lines).

3. On the front face, sketch the diagonals of the face. Also pass a vertical line through the point of intersection of the diagonals.

4. In order to partition the front face into three segments, sketch the diagonal lines shown in Step 4.

5. Sketch two more vertical lines on the front face, each one passing through the intersection formed by the diagonal lines created in Steps 3 and 4.

6. Sketch the diagonals on the left and top faces.

7. Sketch a vertical line from the intersection of the left face diagonal to the upper-left edge of the bounding box, and then sketch a line from this intersection point to the intersection of the top face diagonals. Finally, extend this line until it intersects the right edge of the top face.

8. Go bold.

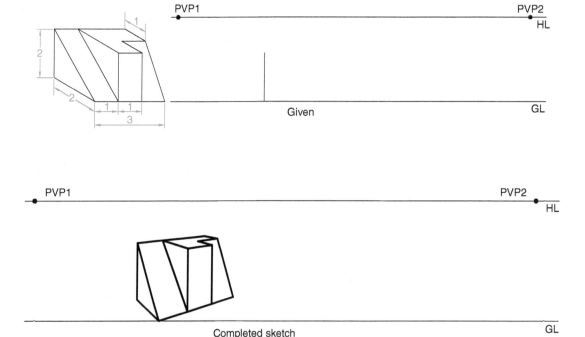

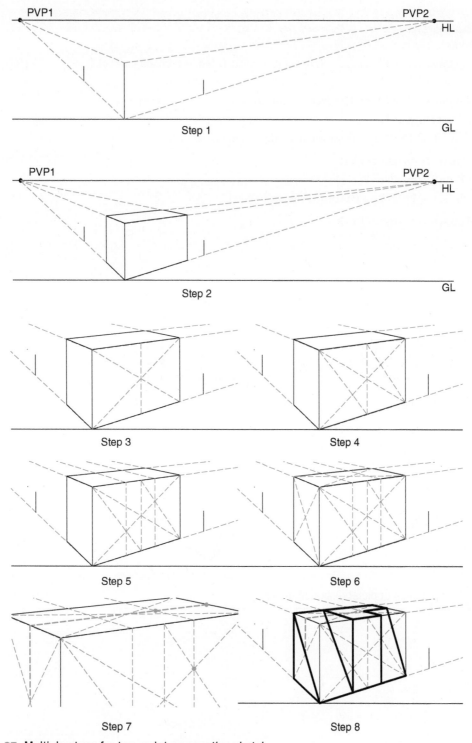

Figure 6-27 Multiple steps for two-point perspective sketch

(a) Oblique
- All three sets of PEB edges (horizontal, vertical, receding) remain parallel.

(b) Isometric
- All three sets of PEB edges (vertical, 30° to right, 30° to left) remain parallel.

(c) One-Point Perspective
- Two sets of PEB edges (horizontal, vertical) remain parallel.
- One set converges to PVP.

(d) Two-Point Perspective
- One set of PEB edges (vertical) remains parallel.
- Two sets converge to PVPs.

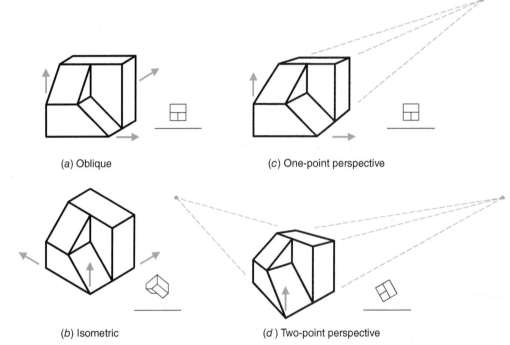

(a) Oblique

(c) One-point perspective

(b) Isometric

(d) Two-point perspective

Figure 6-28 Orientation of pictorial sketching axes

■ QUESTIONS

TRUE OR FALSE

1. A two-point perspective has two principal axes parallel to the projection plane.
2. Two-point perspective and oblique projections have the same number of principal axes inclined to the projection plane.
3. In a perspective projection, if the projection plane is located in front of the object, the projected image will be smaller than the object.

MULTIPLE CHOICE

SS 4. Figure P6-1 shows a perspective view of a vertical pole projected onto a projection plane. If the length of the pole is 30 feet, what is the approximate height of the observer (i.e., the distance from the ground to the observer's eye level)?
   a. 0 feet
   b. 3 feet
   c. 6 feet
   d. 12 feet
   e. 15 feet
   f. 30 feet
   g. Not determinable

SKETCHING

5. Given the isometric view of the cut block objects appearing in P5-4 through P5-65 in Chapter 5, use the one-point perspective set up in the back of the book (or download worksheet from the book website) to sketch a one-point perspective view of the object.
6. Given the isometric view of the cut block objects appearing in P5-4 through P5-65 in Chapter 5, use the two-point perspective set up in the back of the book (or download worksheet from the book website) to sketch a two-point perspective view of the object.

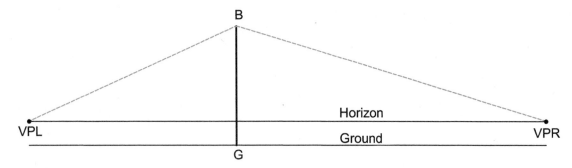

Figure P6-1 (Figure adapted from the work of Michael H. Pleck)

Ⓐ Given the isometric view of the object, sketch a one-point perspective view.

X

Y

———————————————— VP ————————————————

X

Y

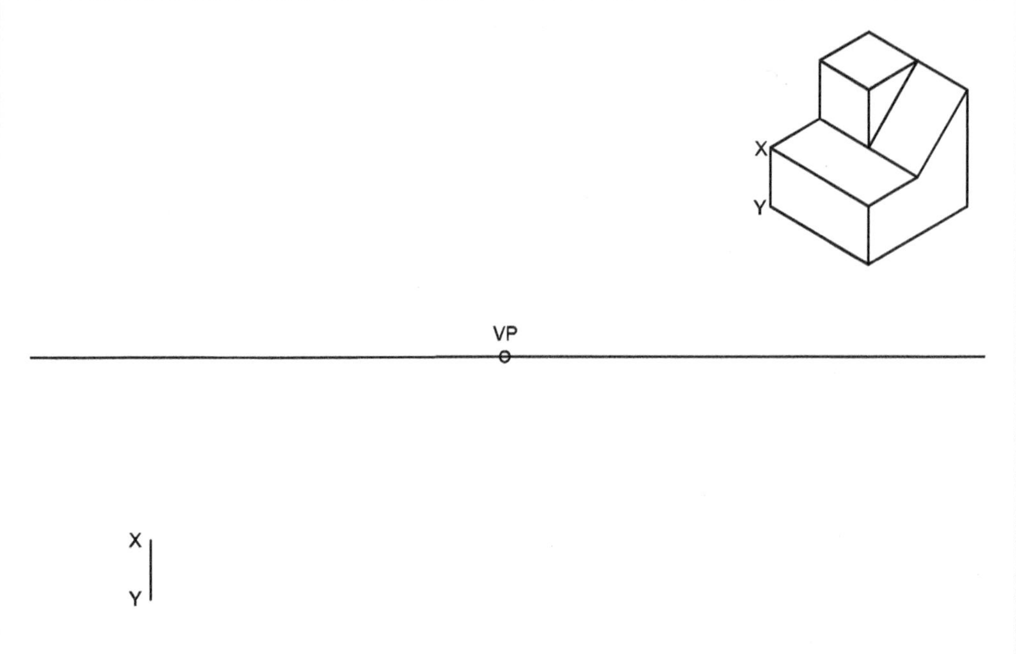

Ⓑ Given the isometric view of the object, sketch a two-point perspective view.

X

Y

VPL ———————————————————————————— VPR

X

Y

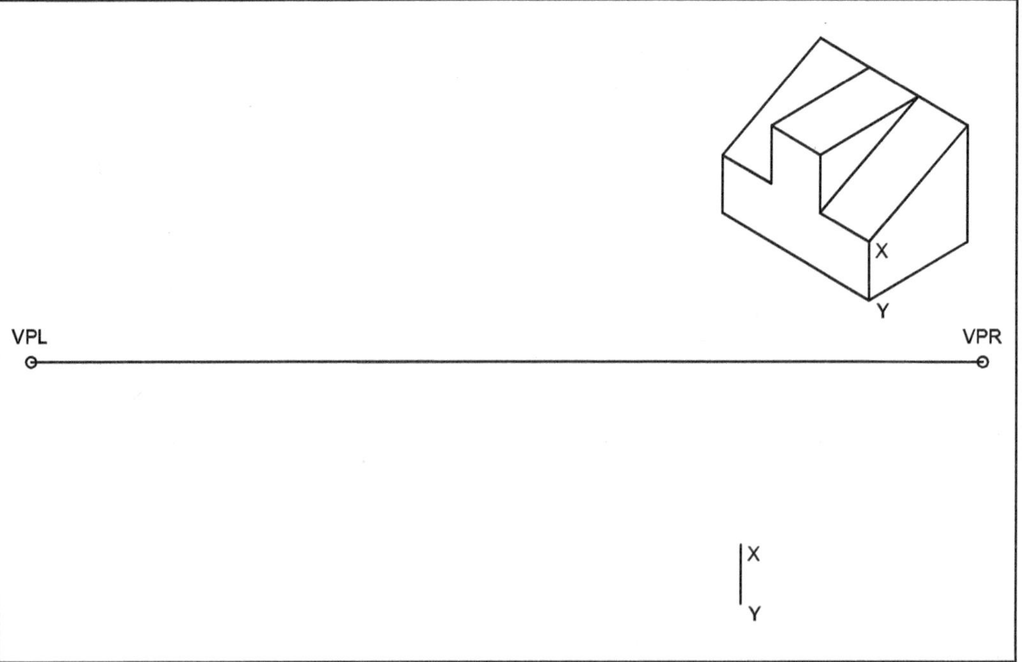

Drawing 6-1 | Name _____ | Date _____

Ⓐ Given the isometric view of the object, sketch a one-point perspective view.

VP

X
Y

X
Y

Ⓑ Given the isometric view of the object, sketch a two-point perspective view.

X
Y

VPL

VPR

X
Y

Drawing 6-2   Name _____   Date _____

(A) Given the isometric view of the object, sketch a one-point perspective view.

X
Y

VP

X
Y

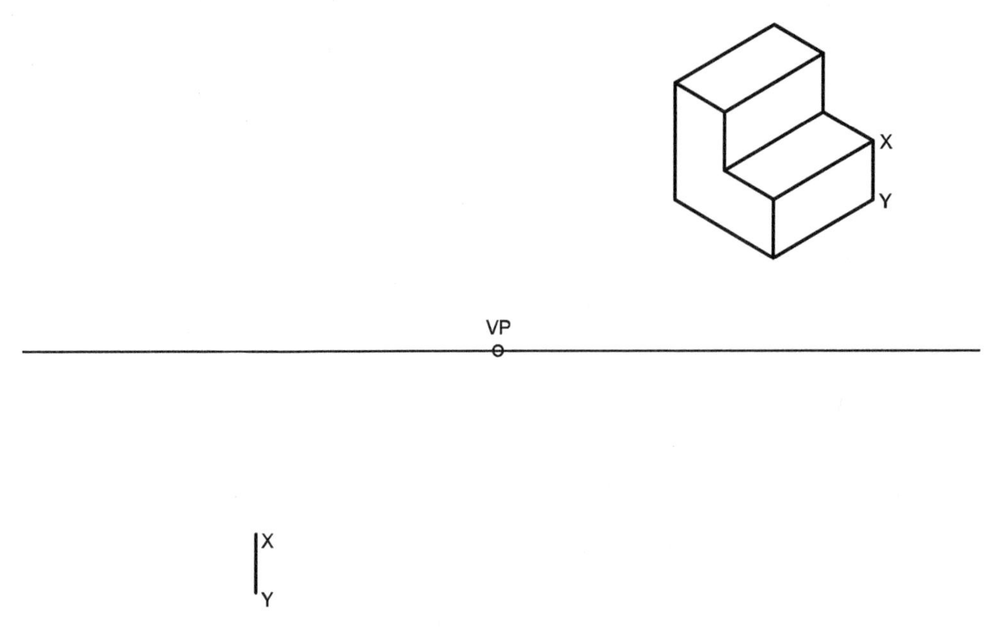

(B) Given the isometric view of the object, sketch a two-point perspective view.

X
Y

VPL

VPR

X
Y

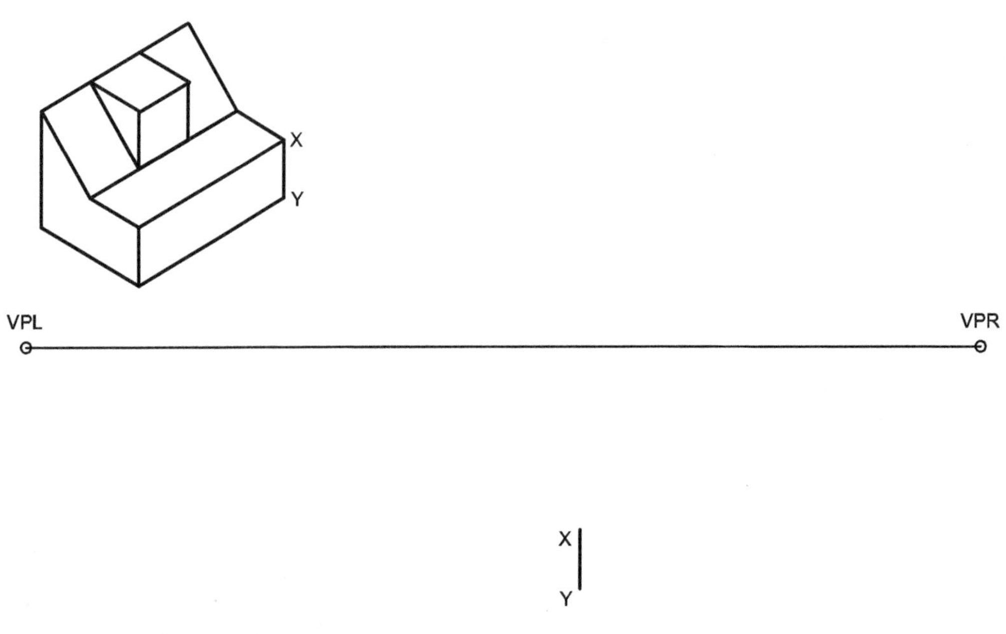

Drawing 6-3 | Name _____ | Date _____

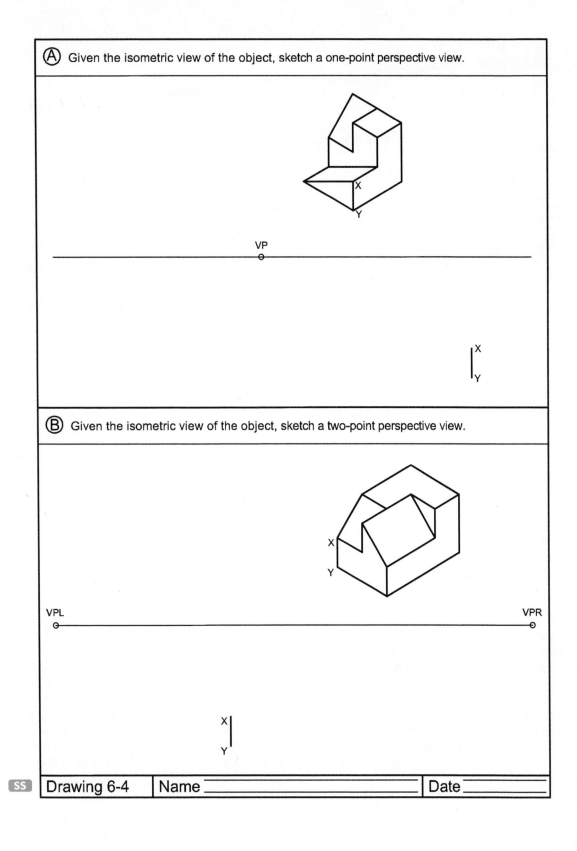

Ⓐ Given the isometric view of the object, sketch a one-point perspective view.

VP

X
Y

Ⓑ Given the isometric view of the object, sketch a two-point perspective view.

VPL

VPR

X
Y

Drawing 6-4 | Name _____ | Date _____

7

MULTIVIEWS

∎ MULTIVIEW SKETCHING

Introduction—Justification and Some Characteristics

Multiview drawings are at the core of what has traditionally been thought of as engineering graphics. The purpose of a multiview drawing is to fully represent the size and shape of an object using one or more views. Along with notes and dimensions, these views provide the information needed to fabricate the part.

Chapter 5 included a brief discussion of the characteristics of multiview projection. These characteristics, as seen in Figure 7-1, include (1) parallel projectors normal to the projection plane and (2) the object positioned so that one principal face is parallel to the projection plane.

As a consequence of this geometry, a multiview drawing can show only one object face. This means that in most cases, more than one view is needed to fully describe the object. It is for this reason that this orthographic projection technique is called multiview projection.

Although only two of the three sets of linear dimensions (i.e., width, depth, height) are projected in any one view, all of this projected information parallel to the projection plane is directly scalable.

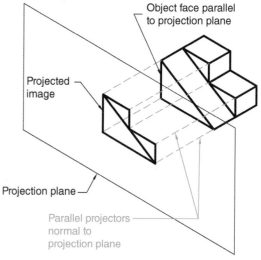

Figure 7-1 Pictorial view of the multiview projection process

Glass Box Theory

Because more than one view is typically needed to document an object using multiview projection, **glass box theory** is used to describe the arrangement of the different multiviews with respect to one another. Imagine that the object to be documented is placed inside a glass box, as shown in Figure 7-2. The object is positioned so that its sides are orthogonal to the sides of the

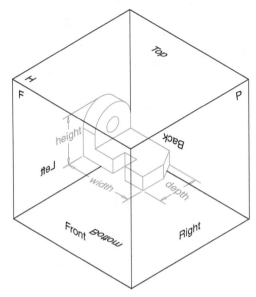

Figure 7-2 Object projected onto box sides

glass box. Note also that the principal dimensions of the object (width, depth, and height) are also indicated in this figure.

In Figure 7-3 the six sides of the glass box are used as projection planes, upon which the six principal views (top and bottom, front and back, right and left) of the object are projected.

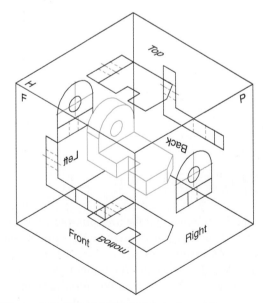

Figure 7-3 Object projected onto box sides

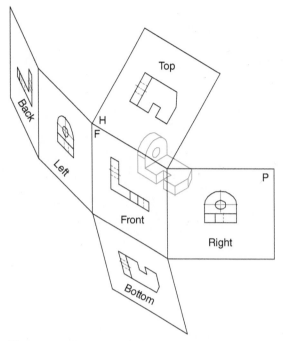

Figure 7-4 Glass box being opened

Imagine further that some of the glass box sides are hinged to one another. These hinges, or *fold lines*, when opened (as shown in Figure 7-4) and then laid flat, result in the six views being arranged as shown in Figure 7-5.

Note in Figure 7-5 that four of the views are hinged to the front view, which is traditionally treated as the primary view in multiview projection. Also note that the top, front, and bottom views are all aligned vertically, whereas the back, left, front, and right views are horizontally aligned.

All six principal views are not normally required to completely document an object. Note in Figure 7-5 the similarities between top and bottom, front and back, and left and right. Three principal views are sufficient to fully describe most objects. Most commonly these views are top, front, and right, as seen in Figure 7-6.

In line with this, we normally speak of three (not six) mutually perpendicular projection planes: Horizontal (H), Frontal (F), and Profile (P). The top and bottom views are projected onto H, front and back onto F, left and right onto P.

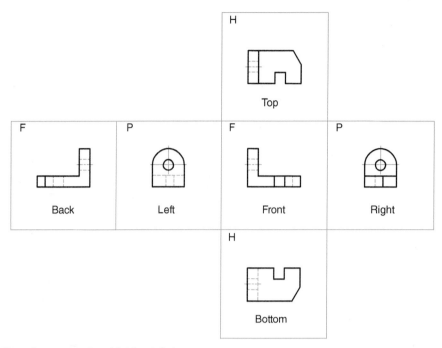

Figure 7-5 Glass box opened and laid out flat

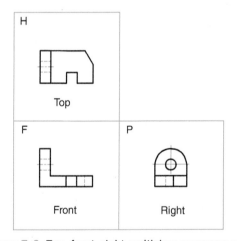

Figure 7-6 Top, front, right multiview arrangement

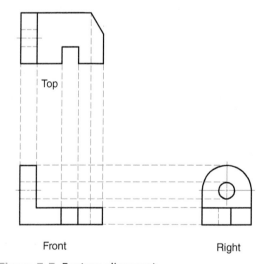

Figure 7-7 Feature alignment

Alignment of Views

Multiviews are always aligned according to the dictates of the glass box projection planes and their fold lines. As can be seen in Figure 7-7, not only the extents but also the internal features should be aligned.

Figure 7-8 shows that the top and front views are vertically aligned and share the width dimension, whereas the front and right views are horizontally aligned and share the height dimension. Aligned views that share a common dimension are said to be ***adjacent***. The top and right views,

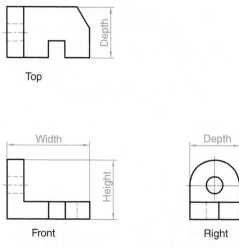

Figure 7-8 Three views with shared dimensions

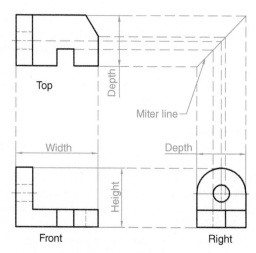

Figure 7-9 Transfer of depth using a miter line

though not aligned, share the depth dimension. Views that share a common dimension, but are not aligned, are said to be *related*.

Transfer of Depth

Every point or feature appearing in one view must be aligned along parallel projectors in their adjacent and related views. Between adjacent views, feature information can be transferred directly along parallel projectors (see Figure 7-7).

Either a trammel or a 45-degree miter line can be used to transfer depth information between related views (see Figure 7-9).

View Selection

The most descriptive view should be selected as the front view. In addition, the longest principal dimension should appear as a horizontal dimension in the front view. For example, in the multiview drawing of the boat in Figure 7-10, the

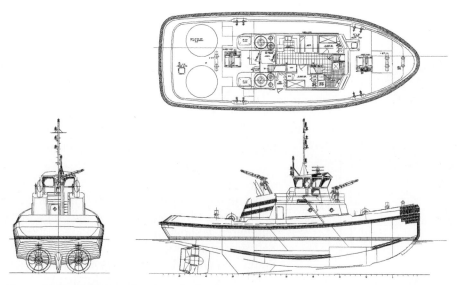

Figure 7-10 Multiview drawing of a boat (Courtesy of Jensen Maritime Consultants, Inc.)

side of the boat appears in the front view. This is because it is the most descriptive view and because the longest dimension appears in the front view as horizontal.

Another guideline related to view selection is to include the minimum number of views that allows for a complete, unambiguous representation of the object. For most objects, three views are required to fully document the part. In some cases, however, only two views are needed (see Figure 7-11). Simple extruded parts (e.g., washers, bushings) may require only a single view, along with a dimensional callout (see Figure 7-12).

In the event that two views provide the same information, choose the view that has the least number of hidden lines. In Figure 7-13 the right view is preferable to the left view because the right view has fewer hidden lines.

Third-Angle and First-Angle Projection

Both third-angle and first-angle projection are used to determine how the principal views are arranged with respect to one another. To explain the difference between the two, use a horizontal

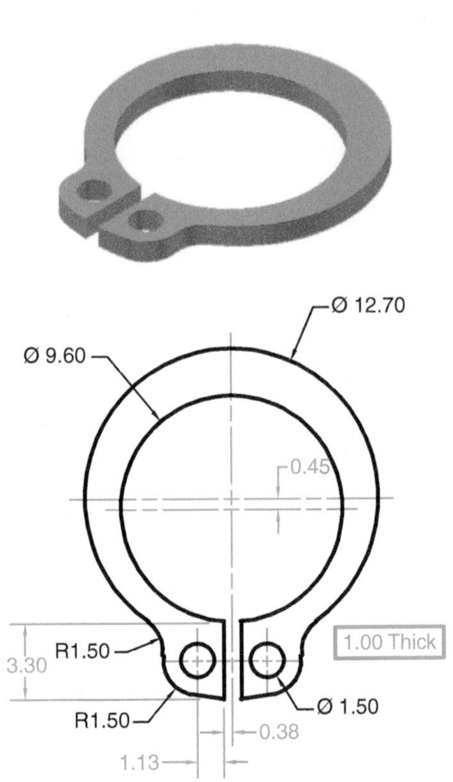

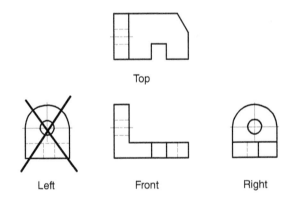

Figure 7-12 One-view drawing

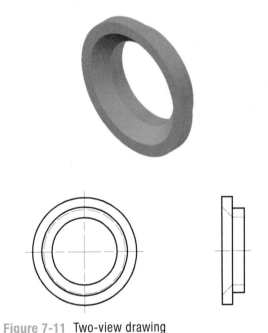

Figure 7-13 Select the view with fewer hidden lines

plane and a (vertically oriented) frontal plane to divide three-dimensional space into quadrants, numbered as shown in Figure 7-14.

Third-angle projection, used in the United States and Great Britain, assumes that an object to be projected resides in the third quadrant.

Figure 7-11 Two-view drawing

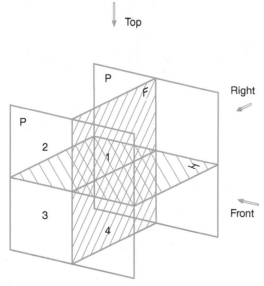

Figure 7-14 Horizontal and front plane, two profile planes

Using the viewing directions for top, front, and right shown in Figure 7-14, the object can be isolated as shown in Figure 7-15a. Note that in third-angle projection, the projection planes are between the viewer and the object. After projecting the views and opening the projection planes, Figure 7-15b results. Figure 7-15c shows the resulting view arrangement employed in third-angle projection.

In first-angle projection, used in the rest of Europe and Asia, the object to be projected is placed in the first quadrant (see Figure 7-14). Using the same viewing directions, the object and its projection planes have been isolated as shown in Figure 7-16a. Note that in first-angle projection, the projection planes are all placed behind or below the object. After projecting and then unfolding the projection planes, Figure 7-16b results.

To indicate whether third- or first-angle projection has been used, the international symbol of a truncated cone shown in Figure 7-17 is used.

Line Conventions

In a drawing view, dark thick *continuous lines* are used to represent the:

1. Edge view of a surface
2. Edge between two intersecting surfaces
3. Extent of a contoured surface

Examples of each of these are shown in Figure 7-18.

Dark thin dashed lines, called *hidden lines*, are used to represent features that are hidden in a particular view. Similar to visible continuous lines, hidden lines are used to represent:

1. A hidden edge of a surface
2. A hidden change of planes
3. The hidden extents or limiting elements of a hole

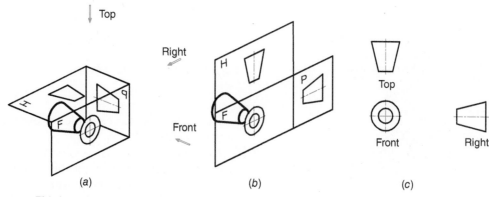

Figure 7-15 Third-angle projection

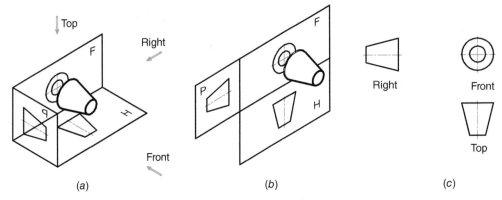

(a)

(b)

(c)

Figure 7-16 First-angle projection

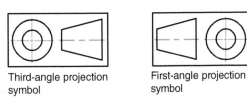

Third-angle projection symbol

First-angle projection symbol

Figure 7-17 Third- and first-angle projection symbols

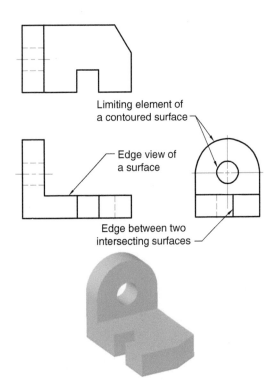

Limiting element of a contoured surface

Edge view of a surface

Edge between two intersecting surfaces

Figure 7-18 Uses of visible continuous line

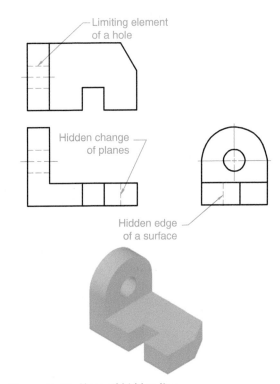

Limiting element of a hole

Hidden change of planes

Hidden edge of a surface

Figure 7-19 Uses of hidden line

See Figure 7-19 for examples of each of these occurrences.

Centerlines are used in a variety of situations. These thin dark lines typically extend about 5 millimeters beyond the feature being represented. Centerlines are commonly used to represent the axis of a cylinder or hole. In a circular

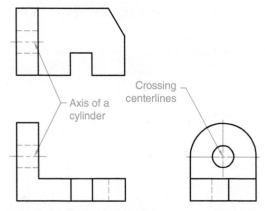

Figure 7-20 Uses of centerline

view, crossing centerlines are used. These centerlines should extend beyond the largest-diameter (or largest-radius) concentric circle (or arc) being represented. In the rectangular view, a single centerline represents the axis of the cylinder or hole. See Figure 7-20 for an example.

Centerlines are also used to indicate symmetry, to show a path of motion, or to represent bolt circles. See Figure 7-22.

Multiview drawing of a cylinder (see Figure 7-21)

A solid cylinder has two circular edges. In the rectangular view these edges project as straight lines. In order to complete the representation, the *limiting elements*, or extents, of the cylinder are also represented as continuous lines.

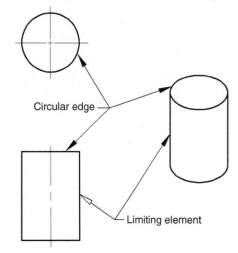

Figure 7-21 Multiview drawing of a cylinder

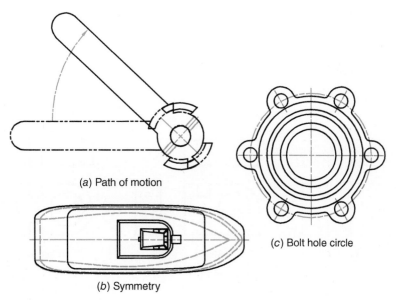

(a) Path of motion

(b) Symmetry

(c) Bolt hole circle

Figure 7-22 Other uses of centerline

Line Precedence

Different object features may sometimes coincide in a multiview sketch. When this occurs, the following order of line precedence is used to determine which lines are represented and which are not: (1) visible, (2) hidden, (3) center. Figure 7-23 illustrates three different collinear line combinations.

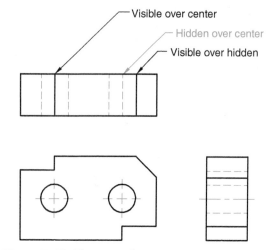

Figure 7-23 Line precedence

1. Using light construction lines, sketch three properly proportioned bounding box views.
2. Add front view feature details.
3. Project feature details from front view to adjacent views.
4. Starting with curved features, go bold.

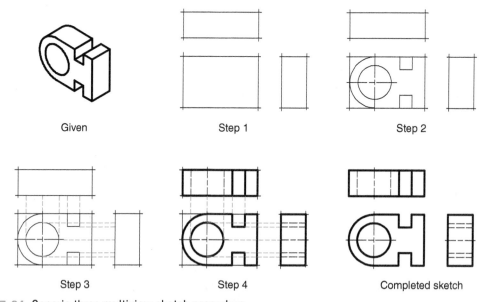

Figure 7-24 Generic three multiview sketch procedure

1. Using light construction lines, sketch three properly proportioned bounding box views.
2. Add visible edge details in top and front views.
3. Project feature details from front and top views to left view; use a miter line or trammel to transfer depth information from top to left view. Also project additional feature details between top and front views.
4. Using light construction lines, layout left view, as well as hidden lines in other views.
5. Starting with curved features, go bold.

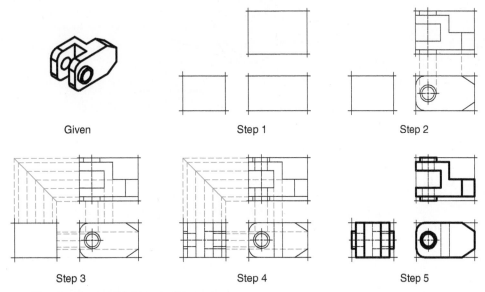

Figure 7-25 Step-by-step multiview sketch example

Intersections and Tangency

When a planar surface intersects a curved surface, a line is drawn to represent the intersecting surfaces. See Figure 7-26. In the event that the planar surface is tangent to a curved surface, no edge is shown to represent the line of tangency. See Figure 7-26 for examples of both situations.

Fillets and Rounds

In designing parts, sharp corners are avoided. Not only are they difficult to fabricate, but they can also lead to stress concentrations, resulting in weakened parts. A *fillet* is used to eliminate an internal corner, and a *round* removes an external corner. See Figure 7-27.

Cast parts are designed with fillets and rounds. Fillets prevent cracks, tears, and shrinkage at re-entry angles when casting. They also make it easier to extract a part from its mold. Once in service, fillets reduce stress concentrations in the cast part.

Because of the nature of this manufacturing process, cast parts have rough external surfaces. In order to mate a casting with another part, it is often necessary to machine the original surfaces in order to create a good mating surface. For this reason, a cast part with rounded corners generally indicates that the part is unfinished, whereas sharp corners indicate that the surface has been machined. See Figure 7-28.

Fillet and round features are displayed as small arcs in a multiview drawing, as shown in Figure 7-29. In their rectangular views, these fillet features are not shown.

A fillet or round connects two otherwise intersecting surfaces with a curved surface

MULTIVIEW SKETCHING

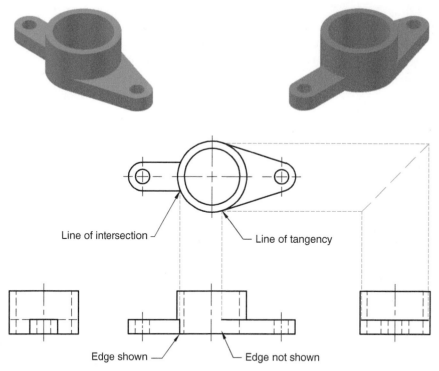

Figure 7-26 Intersection versus tangency

Line of intersection

Line of tangency

Edge shown

Edge not shown

tangent to the original surfaces. Because there is no real change in planes, the top view of the object shown in Figure 7-30 would normally be shown as a single surface (see Figure 7-30a). In order to provide a clearer representation, however, fillets and rounds are sometimes ignored in the rectangular view; edges are then drawn at the imaginary intersection of the two planes, as shown in Figure 7-30b.

Machined Holes

Machined holes are formed by various machining operations. These include drilling, boring, and reaming. The specific machining operation used to create the hole is not specified on the drawing,

Round

Fillet

Figure 7-27 Fillets and rounds

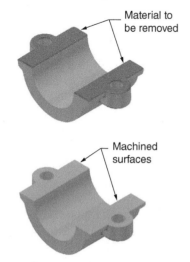

Material to be removed

Machined surfaces

Figure 7-28 Cast part surfaces before and after machining

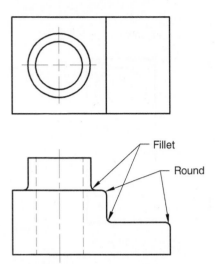

Figure 7-29 Fillet features display as arcs

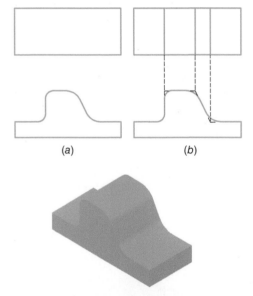

(a)   (b)

Figure 7-30 Fillet conventions

leaving this decision to the machinist. The diameter of a hole, not the radius, is specified, using a leader extending from the circular view. Figure 7-31 shows several different kinds of machined holes. A *through* hole, formed by drilling, goes all the way through the part. A *blind* hole, on the other hand, has a specific depth. Because a blind hole is also formed by drilling, the bottom of the hole comes to a conical point formed by the drill bit. Only the cylindrical portion of the hole should be dimensioned. The angle of the drill bit is 30 degrees.

A *counterdrilled* hole is formed by drilling a larger hole inside a smaller hole to enlarge the initial portion of the hole. As seen in Figure 7-31, a 120-degree shoulder is a byproduct of the counterdrill operation. The process of drilling and then conically enlarging a hole is called *countersinking*. A countersunk hole is used for flat head fasteners and may also serve as a chamfered guide for shafts and other cylindrical parts. In a countersunk hole, both the diameter and the angle of the countersink are specified. Although the angle of the countersink is typically 82 degrees, by convention it is often drawn as 90 degrees. *Spotfacing* is the process of machining the surface around a drilled hole, typically on a cast part, in order to provide a smooth mating surface for washers, bolt heads, nuts, and so on. The cylindrical diameter created by the spot-facing operation is specified; the required depth is left to the machinist. *Counterboring* is the process of cylindrically enlarging the initial portion of a drilled hole. The counterbore operation results in an enlarged hole with a flat bottom. A counterbore hole permits a bolt head to be flush with or recessed below the surface of the part. In a *threaded* hole, an internal thread is made by drilling a hole with a tap drill.

Conventional Representations: Rotated Features

A true orthographic projection of a part with radially distributed features like ribs, holes, and spokes can be confusing to visualize and difficult to construct. The front view of Figure 7-32a, for example, shows the true projection of a part with radially distributed ribs and holes. Note the lack of symmetry about the centerline. The rib(s) on the left side of Figure 7-32a will not be easy to draw, because it is not parallel to the frontal plane. Also, the holes are not symmetrical about the centerline.

To avoid this situation, by convention these views are simplified by rotating the radial features so that they are aligned in a single plane that is perpendicular to the line of sight. Looking at the top view of Figure 7-32b, imagine that a rib and a hole are rotated onto the horizontal axis, as indicated by the arrows. The front view of Figure 7-32b shows this conventional representation, with the revolved features now aligned.

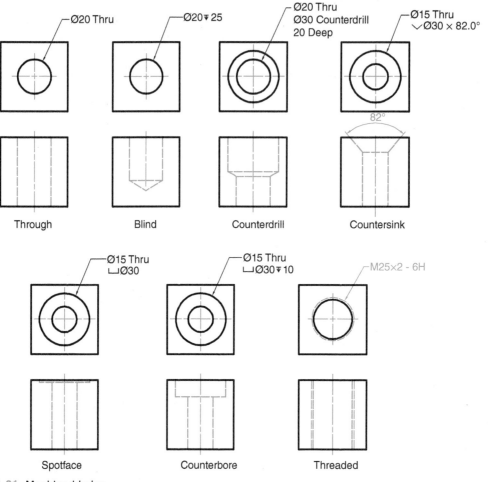

Figure 7-31 Machined holes

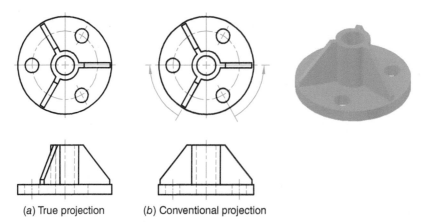

(a) True projection   (b) Conventional projection

Figure 7-32 Treatment of revolved features

1. Using light construction lines, sketch three properly proportioned bounding box views.
2. Add front-view feature details.
3. Project feature details from front view to adjacent views.
4. Starting with curved features, go bold.

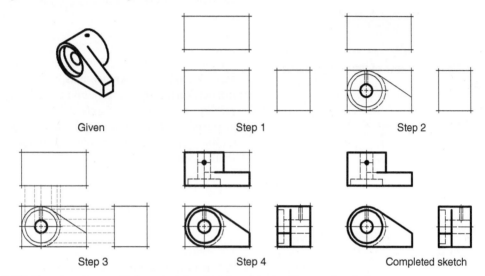

Given   Step 1   Step 2

Step 3   Step 4   Completed sketch

Figure 7-33 Step-by-step multiview sketch example: object with complex features

▌ VISUALIZATION TECHNIQUES FOR MULTIVIEW DRAWINGS

Introduction and Motivation

Visualization is a process by which shape information on a drawing is translated to give the viewer an understanding of the object represented. The part shown in the multiview drawing in Figure 7-34, for example, may not be easily recognizable, even to an experienced engineer. It is only after careful reading of the drawing that a mental image of the product begins to emerge (see Figure 7-35). In the remaining sections of this chapter, various spatial visualization techniques will be discussed.

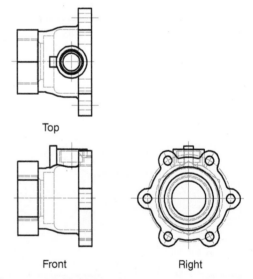

Top

Front   Right

Figure 7-34 Multiview drawing of a complex part

Treatment of Common Surfaces

NORMAL SURFACES

A rectangular prism like the one depicted in Figure 7-36 contains only normal surfaces. A ***normal surface*** is a planar surface that is orthogonal to the principal planes. If we look at the multiview projections of the prism, we see that

Figure 7-35 Rendered view of the complex part in Figure 7-34

surface A appears as an area in the top view, whereas in the other views, surface A appears as an edge. Also, note that because surface A is parallel to the horizontal projection plane, it is shown true size (TS) in the top view. In a three view drawing, a normal surface appears as a true size surface in one view and on edge in the other two views.

INCLINED SURFACES

Surface B in Figure 7-37 is called an inclined surface. An *inclined surface* can be described as a normal surface that has been rotated about a line parallel to a principal axis. An inclined surface is perpendicular to one principal plane and is inclined (i.e., neither parallel nor perpendicular) to the others. An inclined surface appears as a foreshortened (i.e., not true size) surface in two views, whereas in the third view the inclined surface appears on edge. This edge length is a true length.

OBLIQUE SURFACES

An *oblique surface* is a planar surface that has been rotated about two principal axes. An oblique surface is inclined to all three principal projection planes. Surface C in Figure 7-38 is an oblique surface. Note that surface C appears as an area in all three multiviews; in none of these views is the oblique surface true size.

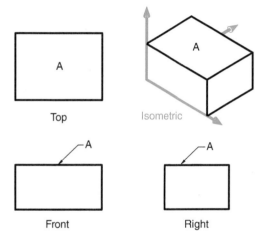

Figure 7-36 Normal surfaces

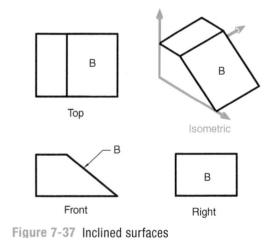

Figure 7-37 Inclined surfaces

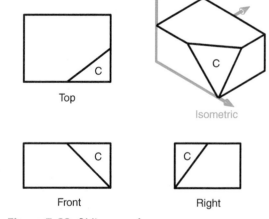

Figure 7-38 Oblique surfaces

Projection Studies

One way to improve visualization skills is to study four views (three multiviews and one pictorial view) of simple objects like those appearing in Figure 7-39. These projection studies improve one's ability to recognize common shapes and features in combination.

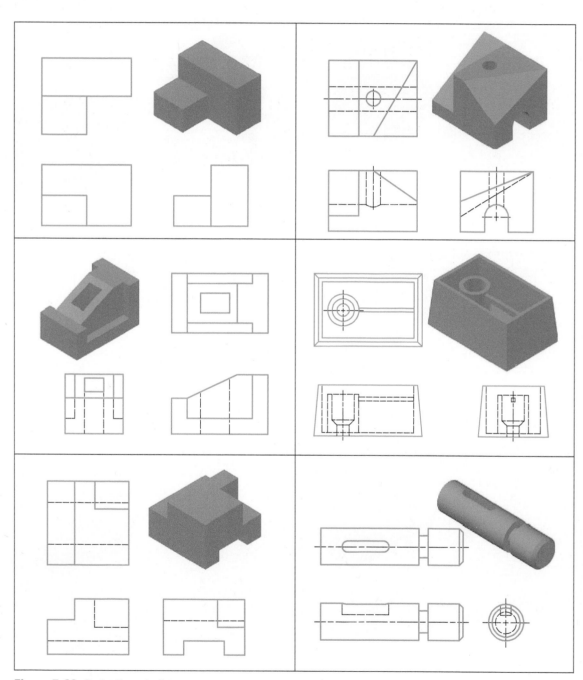

Figure 7-39 Projection studies

Adjacent Areas

The top view in Figure 7-40 shows three distinct areas. Because no two adjacent areas can lie in the same plane, these areas must represent different surfaces. Some possible objects matching this view are also shown in the figure.

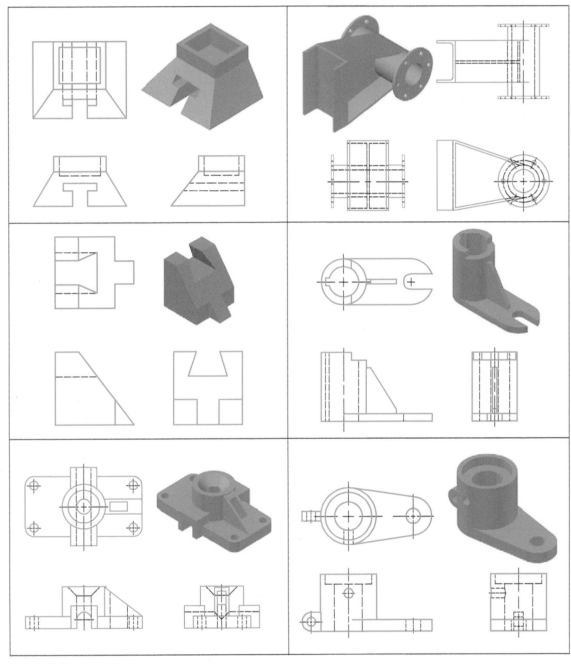

Figure 7-39 (*Continued*)

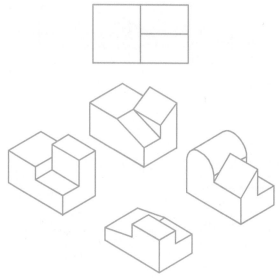

Figure 7-40 Adjacent areas

Surface Labeling

In order to help interpret a multiview drawing, it is sometimes useful to label these surfaces, as shown in Figure 7-41. Note that surface 1 is a normal surface. It appears as an area in the top view and as an edge in the other views. Surface 3 is an inclined surface, appearing on edge in the front view and as a foreshortened area in the other views. Surface 5 is an oblique surface and appears as a foreshortened area in all three views.

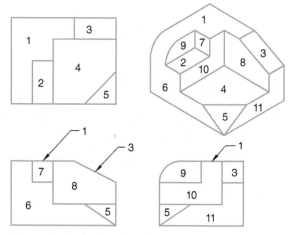

Figure 7-41 Surface labeling

Similar Shapes

Unless viewed on edge, a planar face will always be projected with the same number of vertices. In addition, these vertices will always connect in the same sequence, no matter what the view. These facts are useful when reading a multiview drawing containing either inclined or oblique surfaces (see Figure 7-42). Recall that an inclined surface appears as a foreshortened surface in two of three views and that an oblique surface appears as a foreshortened surface in all three views.

Vertex Labeling

In addition to labeling surfaces, it is also at times useful to label the vertices of a complicated surface in one view, and then to project these points into adjacent or related views. See Figure 7-43 for an example, where the vertices of an oblique surface are labeled in one view and then projected into the other views. Note the similar shape of this oblique surface in the different views.

Analysis by Feature

As we will see in greater detail in Chapter 10, parts are built up from features. These features include such three-dimensional shapes as extrusions, revolutions, holes, ribs, and chamfers. By combining features we arrive at a completed part. See, for example, Figure 7-44, where a part is built up from various features, including an extrusion, a boss (raised cylinder), a counterbore hole, a rib, and fillets and chamfers. Figure 7-45 shows a multiview drawing of the part shown in Figure 7-44. Note how these different manufacturing features appear in the multiview drawing. For example, a counterbore feature always appears as two concentric circles in the circular view, while in the rectangular view this feature appears as two rectangles stacked one on top of the other. Knowledge of how common manufacturing features appear in a multiview drawing can be very helpful when you are called upon to interpret more complicated drawings.

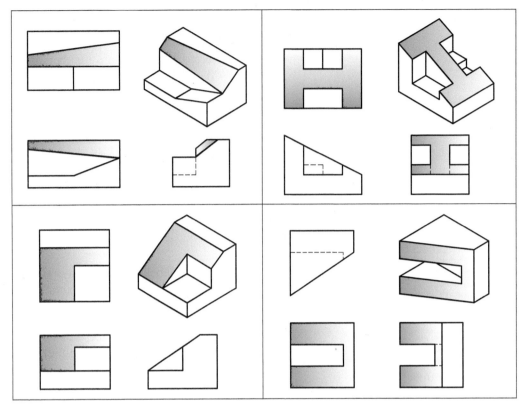

Figure 7-42 Similar shapes

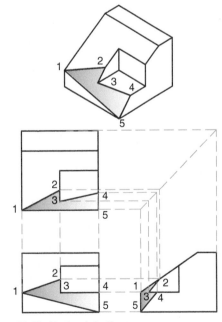

Figure 7-43 Vertex labeling

Now look at the multiview drawing shown in Figure 7-46. Without the benefit of a pictorial view, this object is difficult to visualize. However, if we break the part down into recognizable features, this task becomes more manageable. See Figure 7-47 for a breakdown of the features of the object depicted in Figure 7-46.

Missing-Line and Missing-View Problems

Two additional tools, or rather exercises, that are very useful in developing spatial reasoning are missing-line and missing-view problems. In a missing-line problem three views are given, but some lines are missing from the views. The objective is to identify the missing lines. This can be accomplished by identifying edges in one view that do not appear in an adjacent or related view. By projecting the location of these edges into the adjacent and related views, the location of the missing lines can be identified. Additional visual

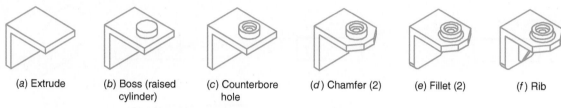

(a) Extrude (b) Boss (raised cylinder) (c) Counterbore hole (d) Chamfer (2) (e) Fillet (2) (f) Rib

Figure 7-44 Breakdown of a part by features

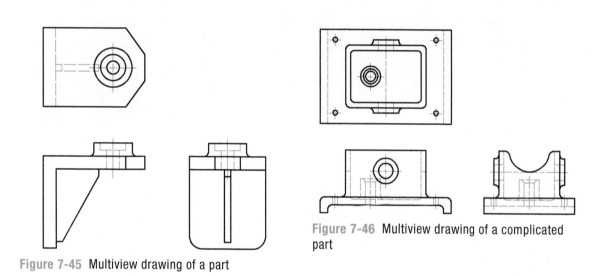

Figure 7-45 Multiview drawing of a part

Figure 7-46 Multiview drawing of a complicated part

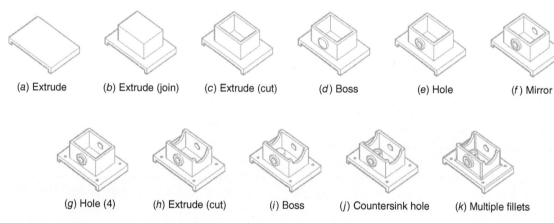

(a) Extrude (b) Extrude (join) (c) Extrude (cut) (d) Boss (e) Hole (f) Mirror

(g) Hole (4) (h) Extrude (cut) (i) Boss (j) Countersink hole (k) Multiple fillets

Figure 7-47 Breakdown of a complicated part by features

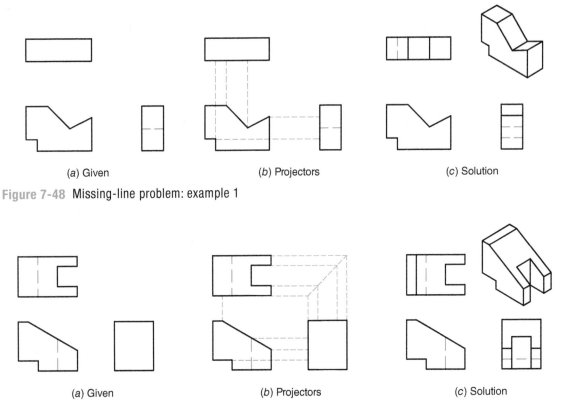

Figure 7-48 Missing-line problem: example 1

(a) Given   (b) Projectors   (c) Solution

Figure 7-49 Missing-line problem: example 2

(a) Given   (b) Projectors   (c) Solution

reasoning is required to identify the type (e.g., visible, continuous) and extent of the missing lines. Figures 7-48 and 7-49 provide two examples of missing-line problems.

More challenging still are missing-view problems. Here two of three views are given. The objective is to find the missing third view. As with missing-line problems, edge features in the given views can be projected to help identify the lines in the missing view. This technique is employed, for example, in Figures 7-50 and 7-51. In nearly all missing-view problems, however, it is even more helpful to sketch a well-proportioned pictorial view of the object. Start by sketching the object's bounding box. Next use the given views to identify prominent object features. The right view of the object in Figure 7-51, for example, suggests a backwards "C" shape extrusion. Add this feature to the pictorial sketch. The vertical hidden line in the right view still needs to be accounted for. By employing visual reasoning and some trial and error, one discovers that a wedge-shaped vertical cut can account for this hidden line, as well as for the internal vertical lines in the front view. The missing object is consequently composed of a "C" shape extrusion and the symmetrical wedge cut.

▊ QUESTIONS

TRUE OR FALSE

1. In a multiview drawing, it is always necessary to include at least three principal views to completely define the object.

2. In a multiview drawing, the right view should be used, even if it has more hidden lines than the left view.

3. In a three view drawing, an inclined planar surface will appear as a line in two of the principal views.

4. The angle at which the line-of-sight pierces the projection plane is the same for both multiview and axonometric projections.

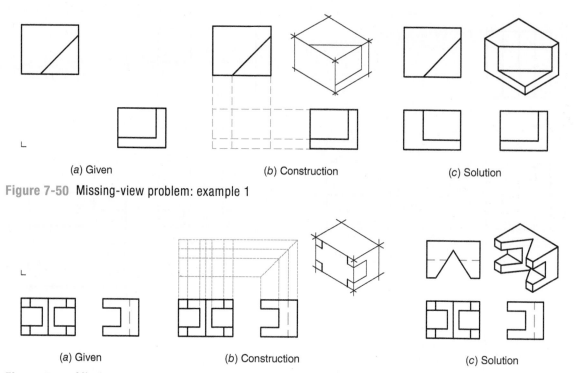

| (a) Given | (b) Construction | (c) Solution |

Figure 7-50 Missing-view problem: example 1

| (a) Given | (b) Construction | (c) Solution |

Figure 7-51 Missing-view problem: example 2

MULTIPLE CHOICE

5. Which planar surface appears as a fore-shortened surface in all of the standard multiviews?
   a. Normal
   b. Inclined
   c. Oblique
   d. Single curved

SS 6. The glass box theory is used to describe:
   a. First angle projection of a part
   b. How multiviews are arranged with respect to one another
   c. How orthographic projections are made
   d. How isometric views are created

SKETCHING

7. Given the isometric view of the cut block objects appearing in Figures P5-4 through P5-65 in Chapter 5, use rectangular grid paper in the back of the book (or download worksheet from the book website) to sketch a multiview set (three views) of the object.

8. Given the isometric view of the cut block objects appearing in Figures P5-4 through P5-65 in Chapter 5, use a blank sheet of paper to create a well-proportioned multiview sketch (three views) of the object.

9. Given the multiview set of the cut block objects appearing in Figures P7-1 through P7-71, use isometric grid paper in the back of the book (or download worksheet from the book website) to sketch an isometric view of the object.

10. Given the multiview set of the cut block objects appearing in Figures P7-1 through P7-71, use a blank sheet of paper to create a well-proportioned isometric sketch of the object.

11. Given the two views appearing in Figures P7-72 through P7-102, use rectangular and isometric grid paper in the back of the book (or download worksheets from the book website) to sketch the missing view and an isometric view of the object.

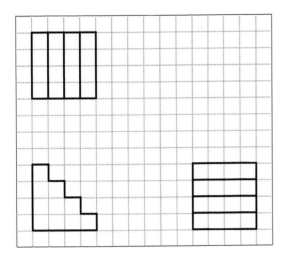

Figure P7-1

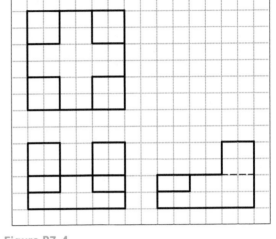

Figure P7-4

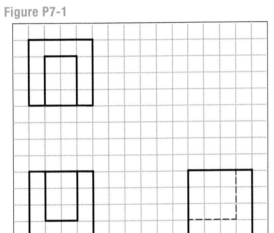

Figure P7-2

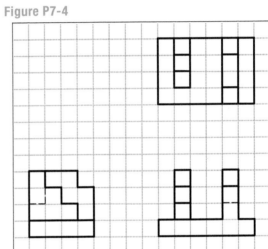

Figure P7-5

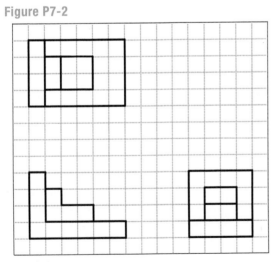

Figure P7-3

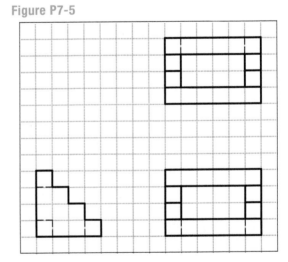

Figure P7-6

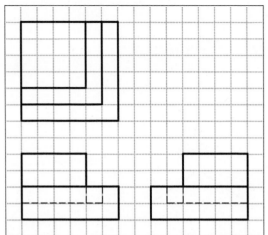

Figure P7-7

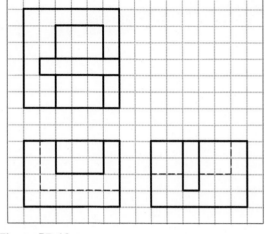

Figure P7-10

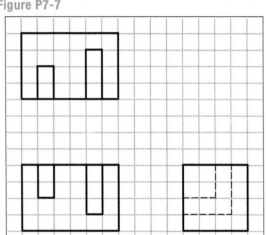

Figure P7-8

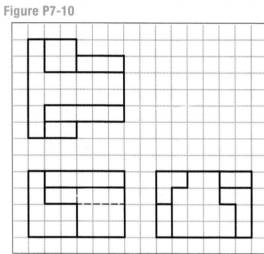

Figure P7-11

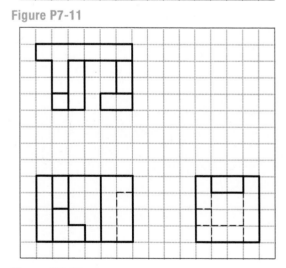

Figure P7-9

Figure P7-12

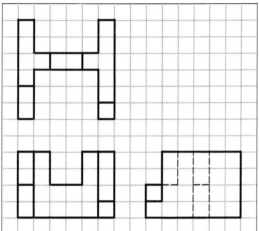

Figure P7-13

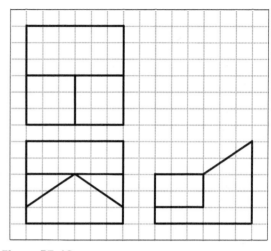

Figure P7-16

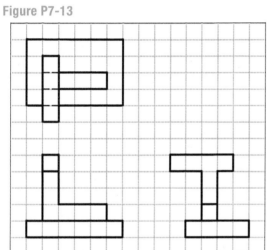

Figure P7-14

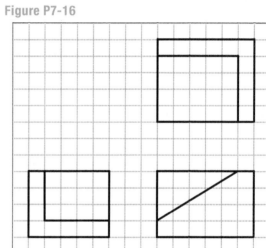

Figure P7-17

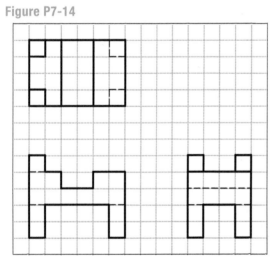

Figure P7-15

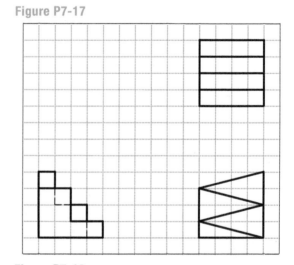

Figure P7-18

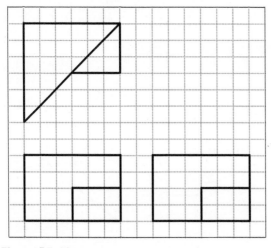

Figure P7-19

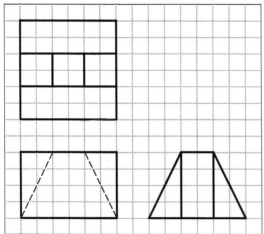

Figure P7-20

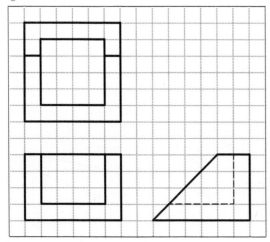

Figure P7-21

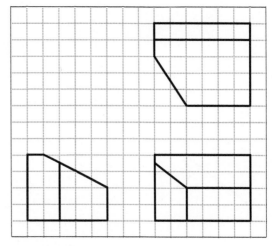

Figure P7-22

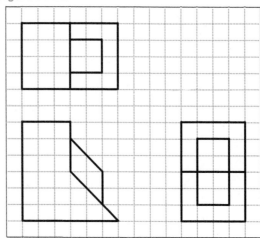

Figure P7-23

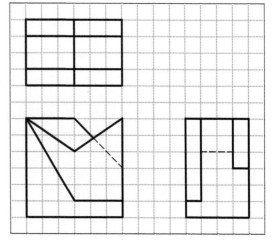

Figure P7-24

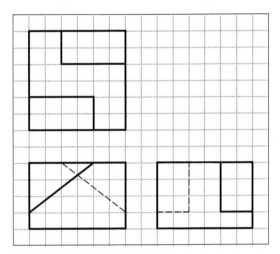

Figure P7-25

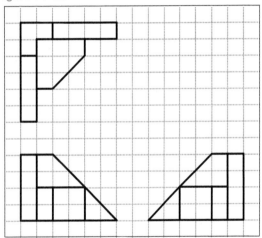

Figure P7-26

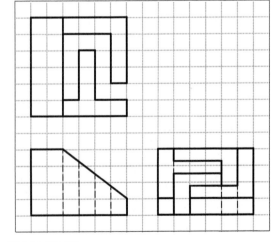

Figure P7-27

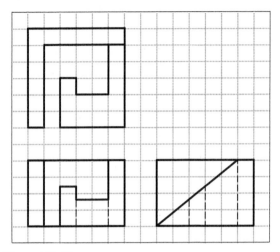

Figure P7-28

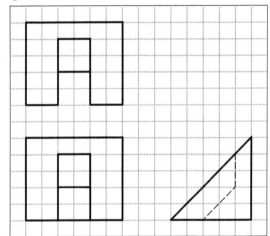

Figure P7-29

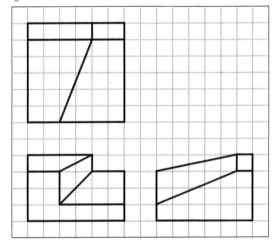

Figure P7-30

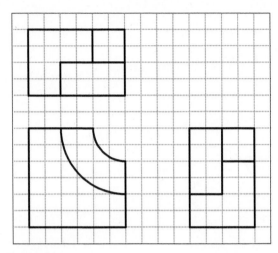

Figure P7-31

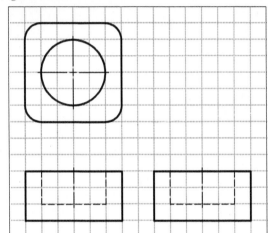

Figure P7-32

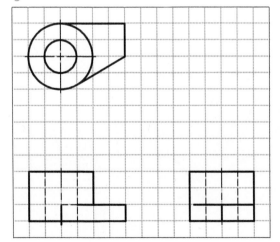

Figure P7-33

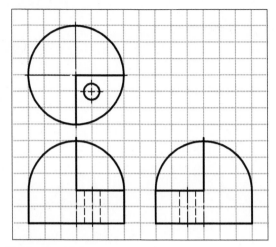

Figure P7-34

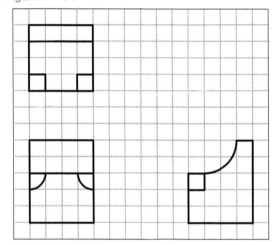

Figure P7-35

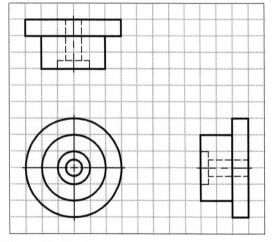

Figure P7-36

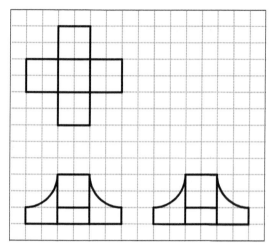

Figure P7-37

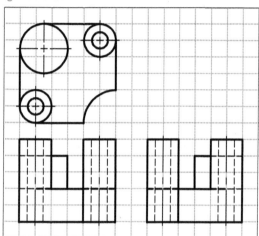

Figure P7-38

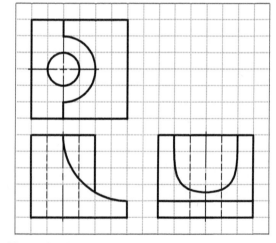

Figure P7-39

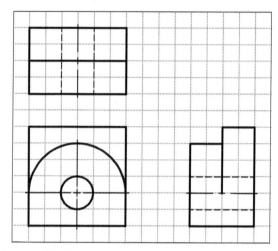

Figure P7-40

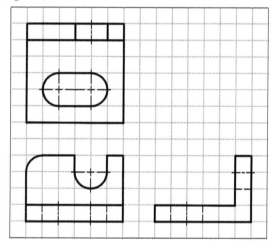

Figure P7-41

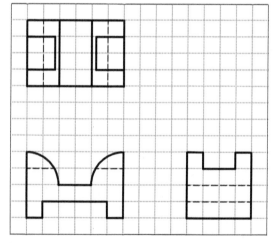

Figure P7-42

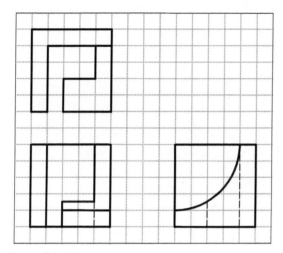

Figure P7-43

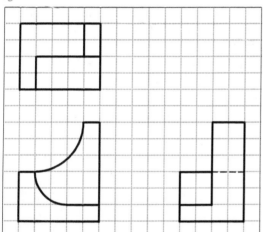

Figure P7-44

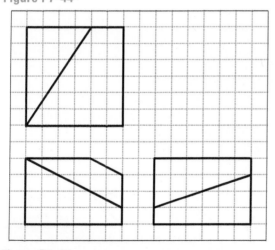

Figure P7-45

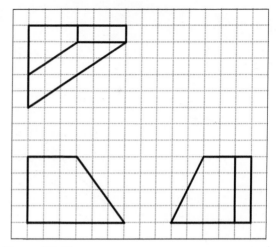

Figure P7-46

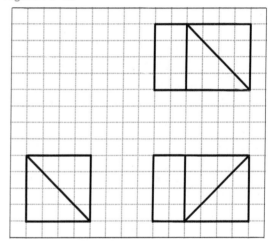

Figure P7-47

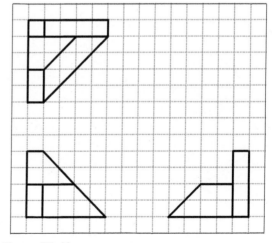

Figure P7-48

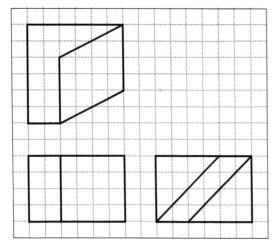

Figure P7-49

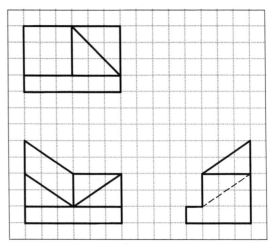

Figure P7-52

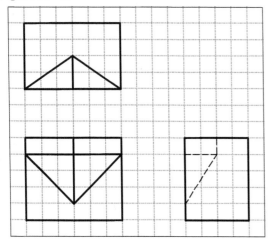

Figure P7-50

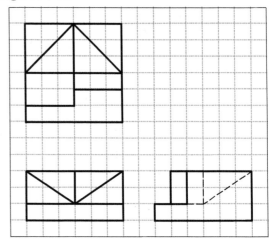

Figure P7-53

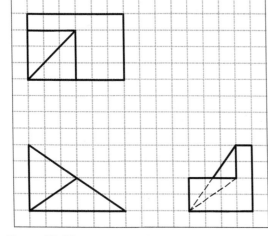

Figure P7-51

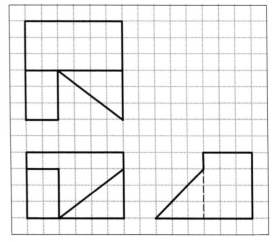

Figure P7-54

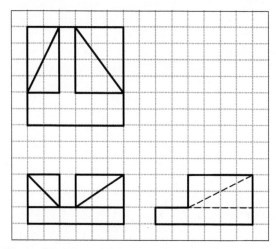

Figure P7-55

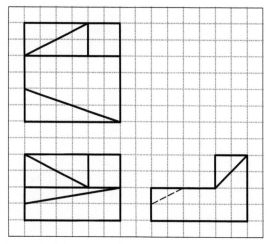

Figure P7-58

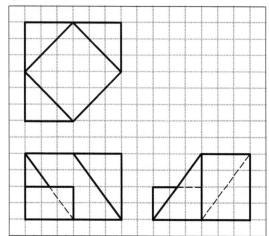

Figure P7-56

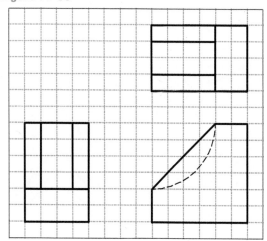

Figure P7-59

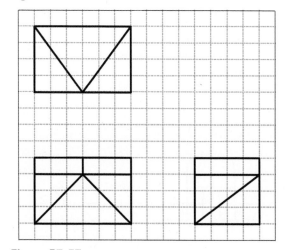

Figure P7-57

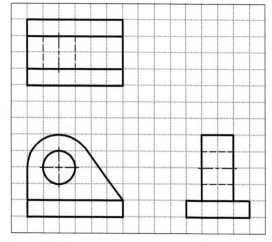

Figure P7-60

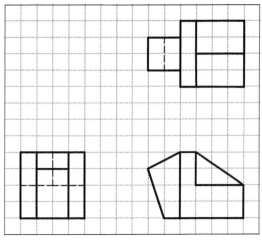

Figure P7-61

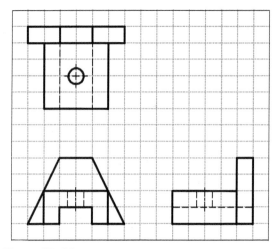

Figure P7-64

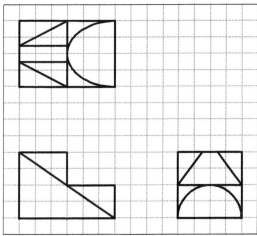

Figure P7-62

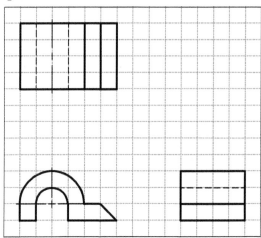

Figure P7-65

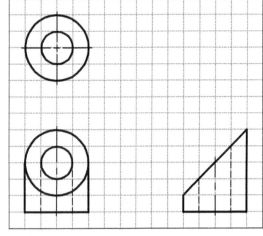

Figure P7-63

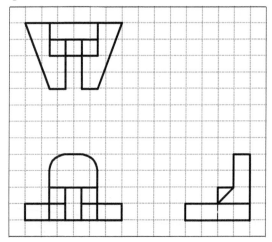

Figure P7-66

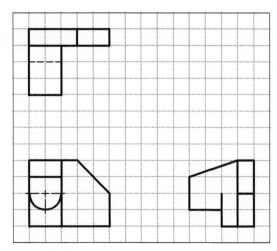

Figure P7-67

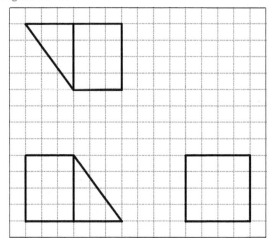

Figure P7-68

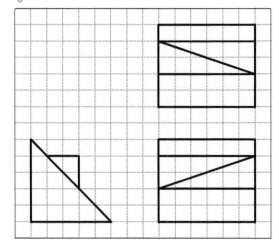

Figure P7-69

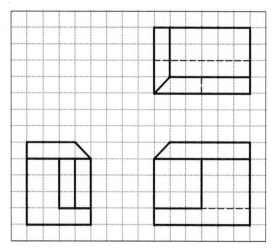

Figure P7-70

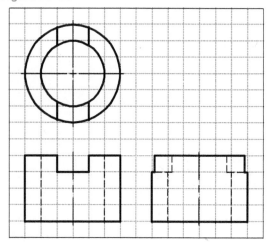

Figure P7-71

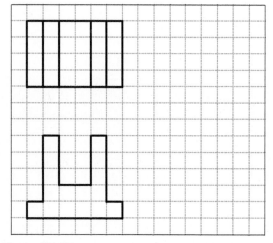

Figure P7-72

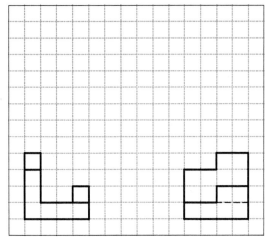

Figure P7-73

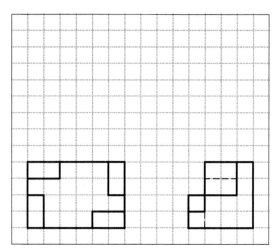

Figure P7-76

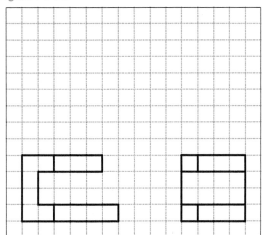

Figure P7-74

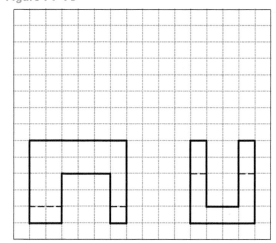

Figure P7-77

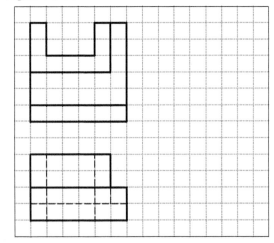

Figure P7-75

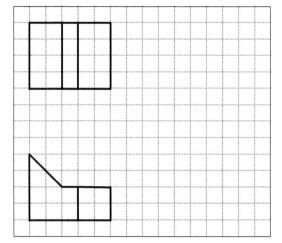

Figure P7-78

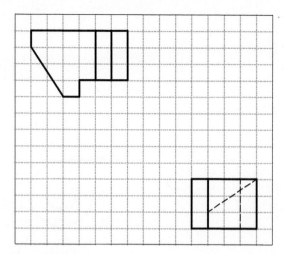

Figure P7-79

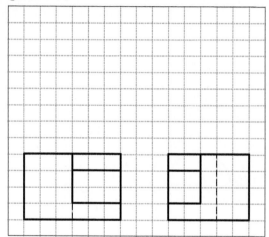

Figure P7-80

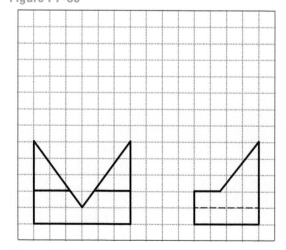

Figure P7-81

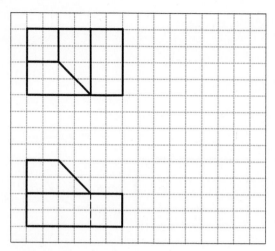

Figure P7-82

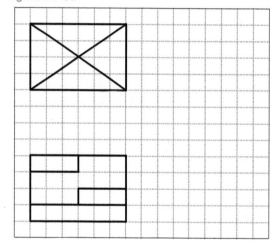

Figure P7-83

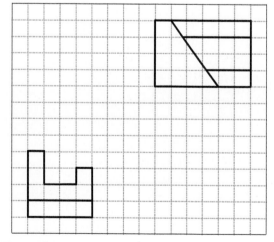

Figure P7-84

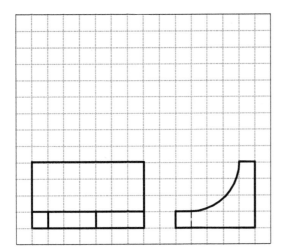

Figure P7-85

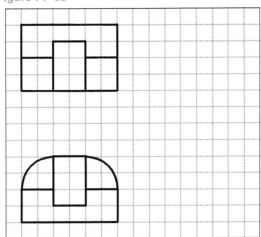

Figure P7-86

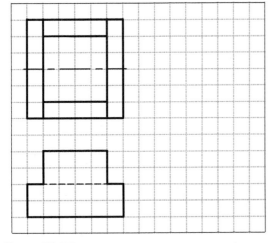

Figure P7-87

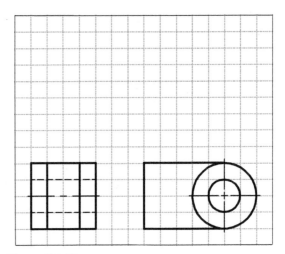

Figure P7-88

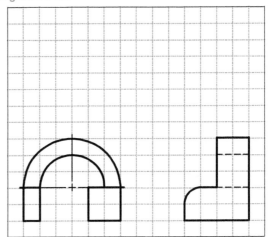

Figure P7-89

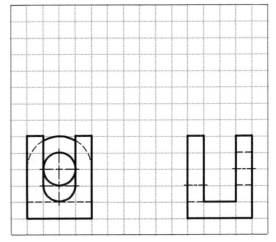

Figure P7-90

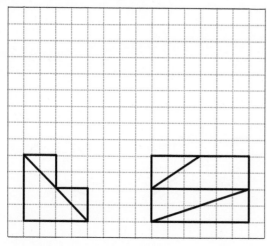

Figure P7-91

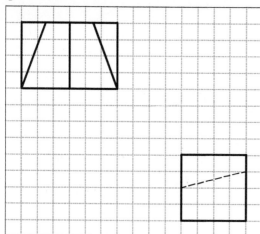

Figure P7-92

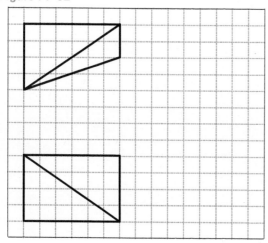

Figure P7-93.

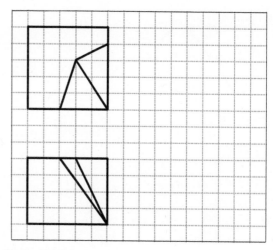

Figure P7-94

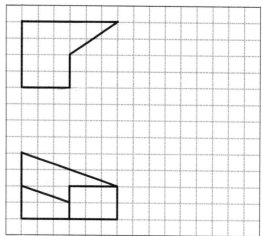

Figure P7-95

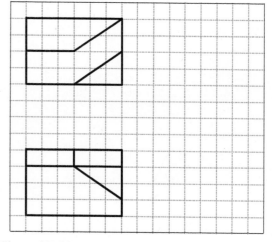

Figure P7-96

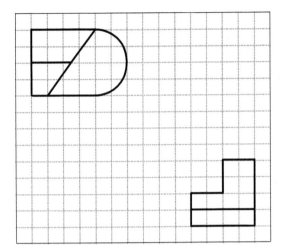

Figure P7-97

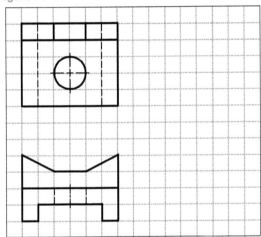

Figure P7-98

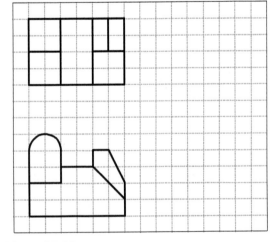

Figure P7-99

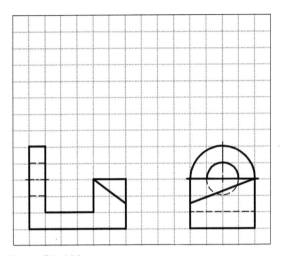

Figure P7-100

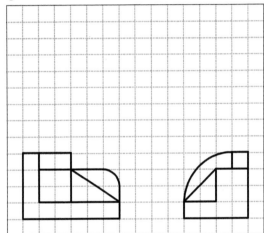

Figure P7-101

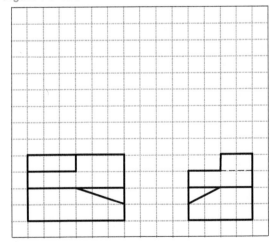

Figure P7-102

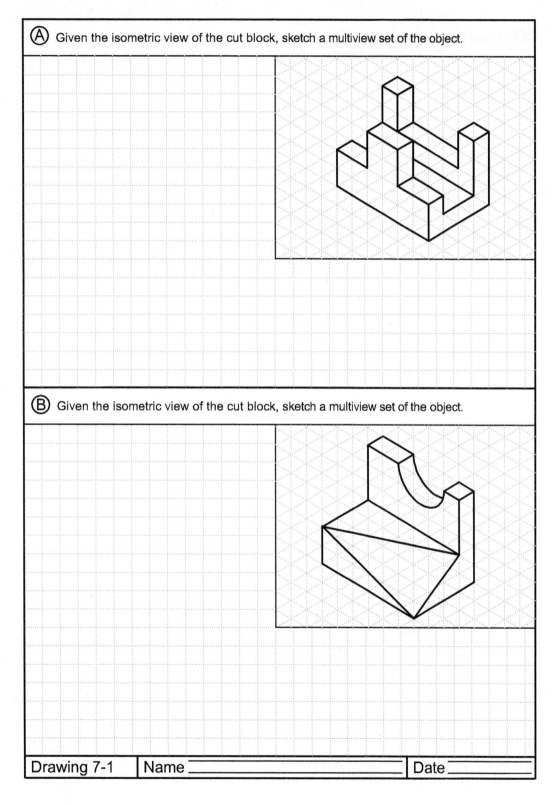

Ⓐ Given the isometric view of the cut block, sketch a multiview set of the object.

Ⓑ Given the isometric view of the cut block, sketch a multiview set of the object.

| Drawing 7-1 | Name | Date |

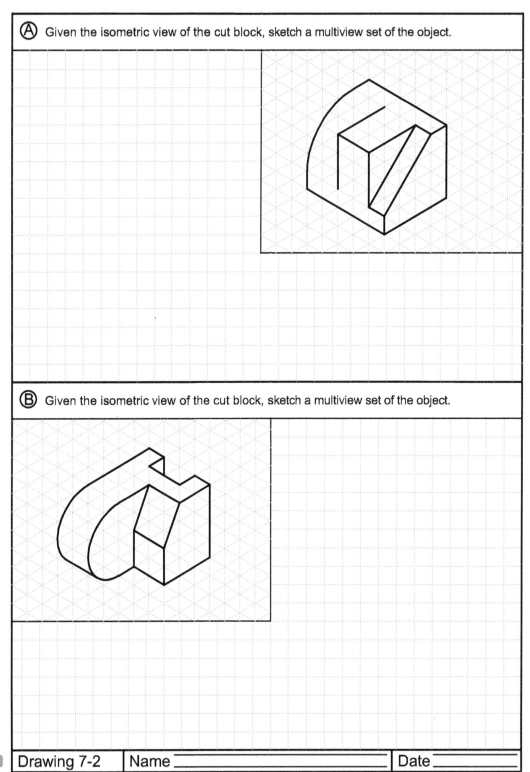

Ⓐ Given the isometric view of the cut block, sketch a multiview set of the object.

Ⓑ Given the isometric view of the cut block, sketch a multiview set of the object.

SS

| Drawing 7-2 | Name _____ | Date _____ |

Ⓐ Given the isometric view of the cut block, sketch a multiview set of the object.

Ⓑ Given the isometric view of the cut block, sketch a multiview set of the object.

| Drawing 7-3 | Name _____ | Date _____ |

Ⓐ Given the isometric view of the cut block, sketch a multiview set of the object.

Ⓑ Given the isometric view of the cut block, sketch a multiview set of the object.

| Drawing 7-4 | Name | Date |

(A) Given the isometric view of the cut block, sketch a multiview set of the object.

(B) Given the isometric view of the cut block, sketch a multiview set of the object.

| Drawing 7-5 | Name | Date |

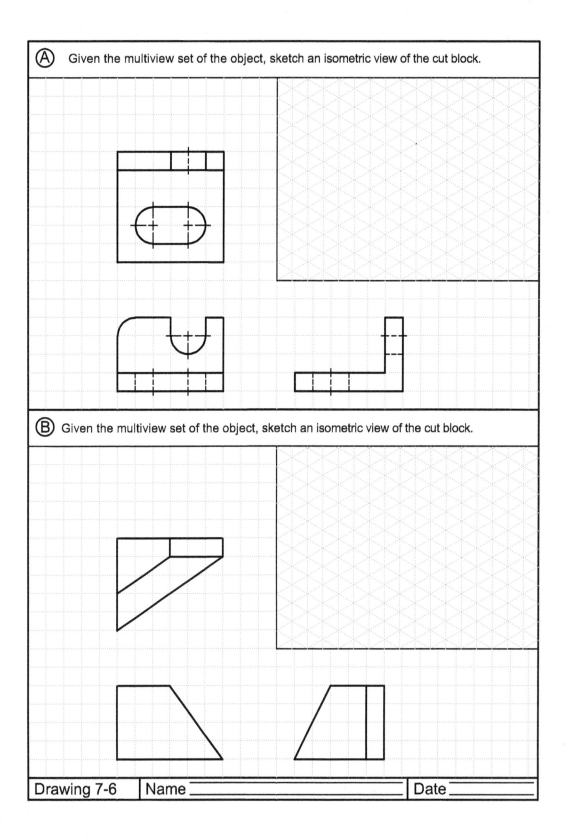

Ⓐ Given the multiview set of the object, sketch an isometric view of the cut block.

Ⓑ Given the multiview set of the object, sketch an isometric view of the cut block.

Drawing 7-6 | Name ⎯⎯⎯⎯⎯⎯⎯⎯⎯⎯⎯ | Date ⎯⎯⎯⎯⎯⎯

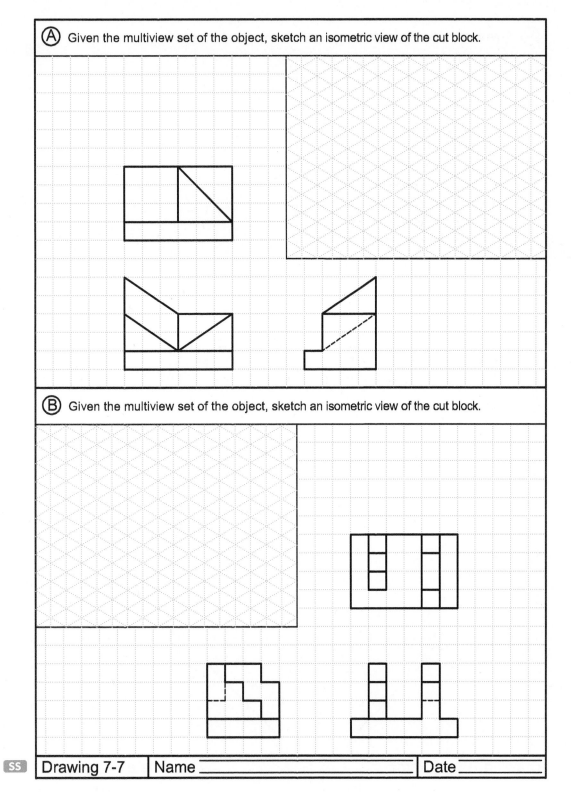

Ⓐ Given the multiview set of the object, sketch an isometric view of the cut block.

Ⓑ Given the multiview set of the object, sketch an isometric view of the cut block.

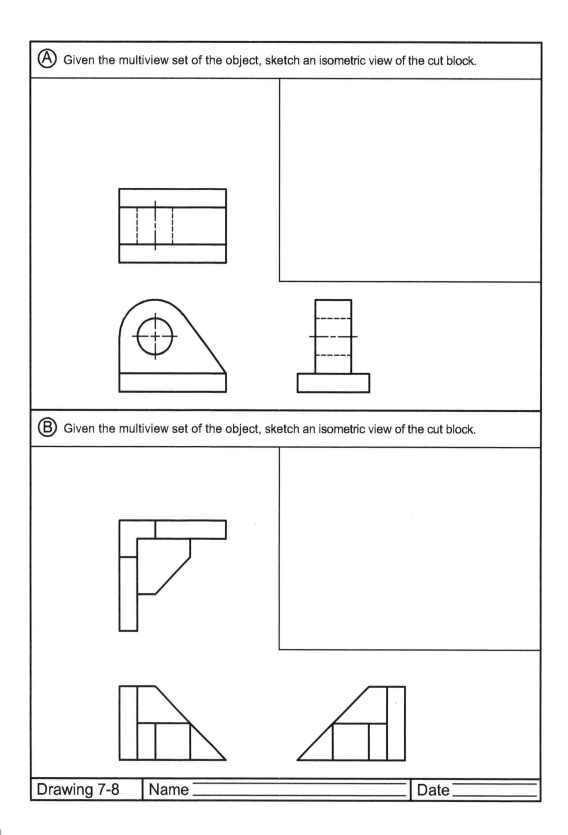

Ⓐ Given the multiview set of the object, sketch an isometric view of the cut block.

Ⓑ Given the multiview set of the object, sketch an isometric view of the cut block.

| Drawing 7-8 | Name | Date |

(A) Given the multiview set of the object, sketch an isometric view of the cut block.

(B) Given the multiview set of the object, sketch an isometric view of the cut block.

Ⓐ Given the multiview set of the object, sketch an isometric view of the cut block.

Ⓑ Given the multiview set of the object, sketch an isometric view of the cut block.

| Drawing 7-10 | Name | Date |

(A) Given the two views, sketch the missing view and an isometric view of the object.

(B) Given the two views, sketch the missing view and an isometric view of the object.

| Drawing 7-11 | Name | Date |

Ⓐ Given the two views, sketch the missing view and an isometric view of the object.

Ⓑ Given the two views, sketch the missing view and an isometric view of the object.

| Drawing 7-12 | Name | Date |

(A) Given the two views, sketch the missing view and an isometric view of the object.

(B) Given the two views, sketch the missing view and an isometric view of the object.

(A) Given the two views, sketch the missing view and an isometric view of the object.

(B) Given the two views, sketch the missing view and an isometric view of the object.

Drawing 7-14　　Name _____　　Date _____

A Given the two views, sketch the missing view and an isometric view of the object.

B Given the two views, sketch the missing view and an isometric view of the object.

| Drawing 7-15 | Name | Date |

8 SECTION AND AUXILIARY VIEWS

▌ SECTION VIEWS

Introduction

Parts like the ones shown in Figure 8-1 contain several internal features. These interior construction details show up as hidden lines when in a standard multiview projection. Because hidden lines can be difficult to interpret and visualize, *section views* are frequently used to expose the internal features of a part.

Section View Process

In a section view, an imaginary *cutting plane* is passed through a part, often along a part's plane of symmetry. The portion of the part between the viewer and the plane is removed, and the part is then viewed normal to the cutting plane (see Figure 8-2). A *section lining*, or hatch pattern, is applied to the surfaces that make contact with the cutting plane.

The cutting plane on edge appears in a view adjacent to the section view, and shows the location of the section. The cutting-plane edge is typically represented as a thick dashed line. Note that a cutting-plane line has precedence over a centerline, should the two coincide. The viewing direction is indicated by arrows drawn perpendicular to the cutting plane.

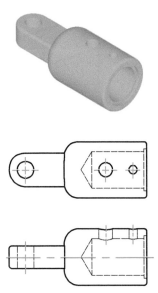

Figure 8-1 Views of a part with multiple internal features

Lines that would be visible after making a cut are also shown in a section view; see Figure 8-3. By convention, hidden lines are normally not shown in a section view. Exceptions are occasionally made, however, if it is felt that the clarity of the drawing is improved.

Section views are typically labeled using indexed capital letters—for example, Section

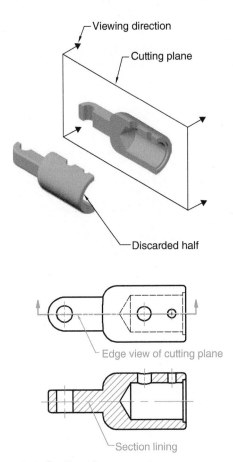

Figure 8-2 Section view process

A-A, Section B-B, and so on. The capital letters are also used to label the cutting-plane line and to clearly associate the two views, as shown in Figure 8-4.

Section Lining (Hatch Patterns)

As we have already seen, section lining is applied to solid areas on the part that have been exposed by the cutting plane. As shown in Figure 8-5, types of section lining are often associated with different materials. The most commonly employed section lining consists of uniformly spaced continuous lines, set at a 45-degree angle. This section-lining angle, however, should be adjusted to avoid the section lines being too close

to parallel with or perpendicular to the visible lines that bound them, as seen in Figure 8-6. In CAD software programs, section lines are typically referred to as hatch patterns.

Full Sections

In a *full section*, the cutting plane passes all the way through the object. Figure 8-7 shows another example of a full-section view that appears in the front view.

Another possibility—one that is typically more difficult to visualize—uses a horizontally oriented cutting plane. In this situation (see Figure 8-8), the cutting plane appears on edge in the front view, while the section appears in the top view. Note the direction of the cutting-plane arrows in the front view. The arrow direction represents the viewing direction; the observer is looking down at the bottom portion of the object, the top portion having been removed.

Yet another possible orientation of a full-section view appears in Figure 8-9. Here the section appears in the (left) side view, and the cutting-plane edge appears in the top view. Once again, note the direction of the sight arrows. Also note that the cutting edge could just as well have appeared in the front view.

Half Sections

With symmetrical or very nearly symmetrical objects, it is not always necessary to pass the cutting plane all the way through the part. In a *half section*, the cutting plane passes only halfway through the part, as shown in Figure 8-10. In a half-section view, one-quarter of the part is removed.

Half sections possess the advantage of showing both the interior and the exterior of the part in the same view. External features are included on the unsectioned half. A centerline is used to separate the two halves. Hidden lines are normally omitted in both halves, but they may be shown in the unsectioned half.

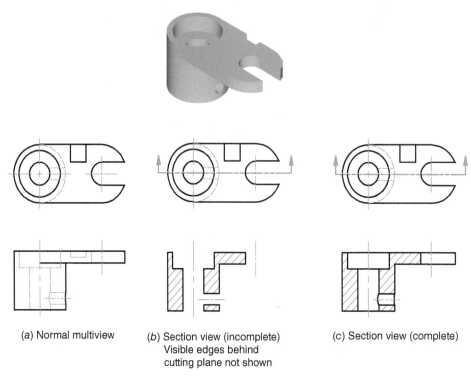

(a) Normal multiview

(b) Section view (incomplete)
Visible edges behind
cutting plane not shown

(c) Section view (complete)

Figure 8-3 Treatment of visible edges behind the cutting plane in a section view

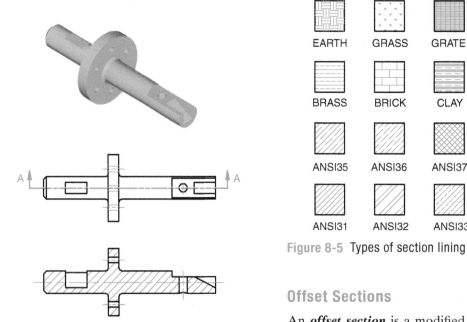

Section A-A

Figure 8-4 Section view labeling

EARTH	GRASS	GRATE	GRAVEL
BRASS	BRICK	CLAY	CORK
ANSI35	ANSI36	ANSI37	ANSI38
ANSI31	ANSI32	ANSI33	ANSI34

Figure 8-5 Types of section lining

Offset Sections

An *offset section* is a modified full section that is used when important internal features do not lie in the same plane. In an offset section the

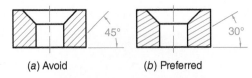

(a) Avoid   (b) Preferred

Figure 8-6 Adjusting the section lining angle

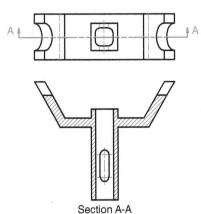

Section A-A

Figure 8-7 Full section appearing in the front view

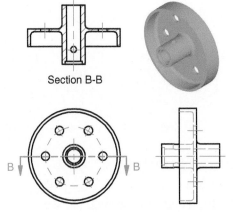

Section B-B

Figure 8-8 Full section appearing in the top view

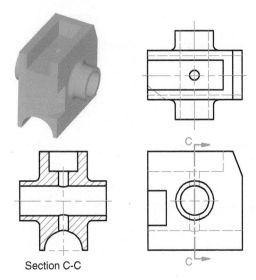

Section C-C

Figure 8-9 Full section appearing in the side view

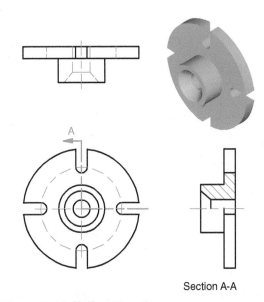

Section A-A

Figure 8-10 Half-section view

cutting plane is stepped, or offset, in order to pass through these features. Figure 8-11 shows an example of an offset section. Note that any steps (90-degree bends) in the cutting-plane line are not shown in the section view. The section is drawn as if these offsets all lie in the same plane. Also note that the offsets should be located in regions where there are no features.

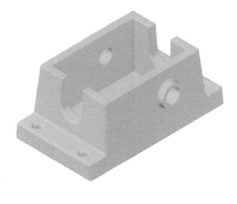

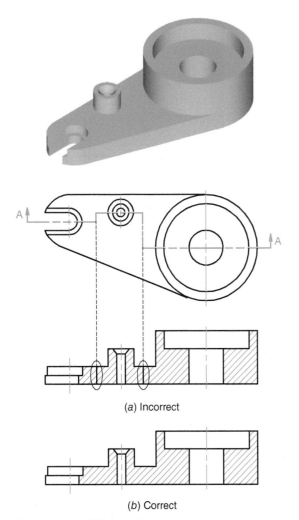

(a) Incorrect

(b) Correct

Figure 8-11 Offset section view

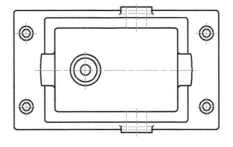

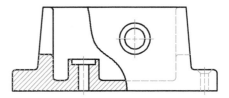

Figure 8-12 Broken-out section view

Broken-Out Sections

A **broken-out section** is used when only a portion of the part needs to be sectioned. See Figure 8-12 for an example of a broken-out section. A jagged, freehand break line is used to separate the sectioned from the unsectioned portion of the drawing. Like half sections, broken-out sections have the advantage of showing internal and external features in the same view. In addition to being used on multiview drawings, broken-out sections are also used on pictorial views, particularly when they are executed in CAD (see Figure 8-13).

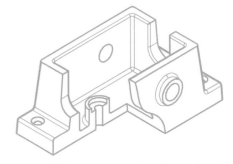

Figure 8-13 Broken-out section view—pictorial

Revolved Sections

In all of the sections discussed thus far (full, half, offset, and broken-out), the section view is projected from the adjacent view in which the cutting-plane line appears. A ***revolved section***, on the other hand, is created by passing a cutting plane perpendicular to the longitudinal axis of an elongated symmetrical feature, and then revolving the resulting cross section 90 degrees into the plane of the drawing. This results in the cross section being superimposed on the original view. Figure 8-14 shows an example of a revolved section. The original section may be shown with (Figure 8-14*a*) or without conventional break lines (Figure 8-14*b*). A centerline is used to represent the axis of the revolved section.

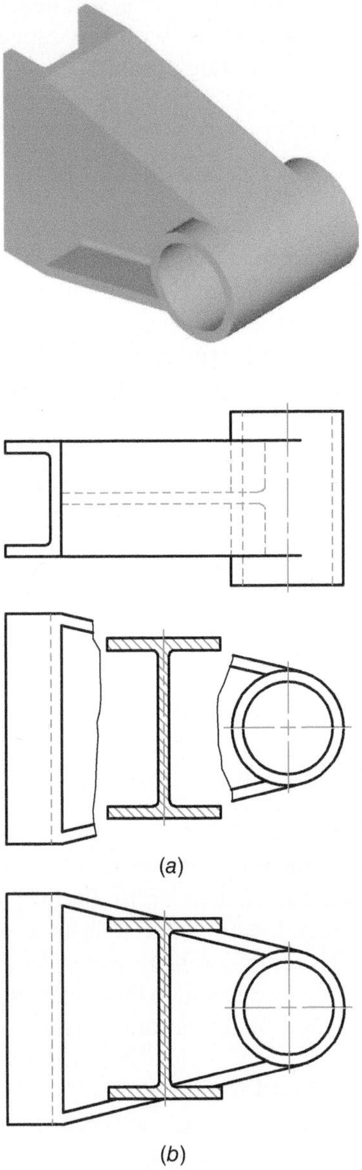

Removed Sections

A ***removed section*** is similar to a revolved section, except that the cross section is not superimposed on the view. Rather, the removed section is placed at some convenient location. Standard section view labeling practices are used to relate the cutting-plane location to the resulting section. See Figure 8-15 for an example of a removed section. Removed sections are used when there is insufficient room for a revolved section, and

(a)

(b)

Figure 8-14 Revolved section

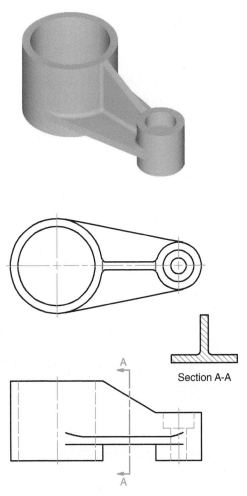

Section A-A

Figure 8-15 Removed section

when several cross sections are needed to show the transition of an elongated feature from one shape to another (see Figure 8-16).

Conventional Representations: Section Views

In order to simplify the construction and improve the clarity of section views, conventional representations are sometimes employed in place of true orthographic projections. These simplified representations are widely recognized and accepted as being a part of standard drawing practice. Conventional representations associated with section views include the treatment of thin features, radially distributed and off-angle features, and section lining in assembly sections. Note that when using CAD, it may actually be easier to obtain a true projection, rather than a simplified representation. Although this has to some extent reduced the usage of some conventional representations, it is still necessary for engineers to be familiar with this aspect of the language of engineering graphics.

Conventional Representations: Thin Features

In an effort to make some section views more readable, section lining is not applied to the outline of thin features like ribs, webs, and lugs when the cutting plane passes along the length of the feature. Figure 8-17 provides an example of this convention applied to a rib feature.

Note that without this convention, the section shown in Figure 8-18 could be incorrectly interpreted as depicting a part with uniform thickness (Figure 8-18*a*), rather than as a ribbed part (Figure 8-18*b*).

Figure 8-19 provides an example of this thin-feature convention applied to both a lug and a web feature.

Section View Construction Process—Example 1

Figure 8-20 illustrates the process of constructing a full-section view of a part. In this particular problem, only the top and right-side principal views are given, and we are asked to draw the cutting plane on edge and to find the front-section view. In a less difficult variant of this problem, a pictorial view of the object is given, along with the cutting-plane location.

In Step 1, the cutting plane is drawn in the top view along the object's axis of symmetry. The arrows should point as shown, to indicate that the section will appear in the front view. Parallel

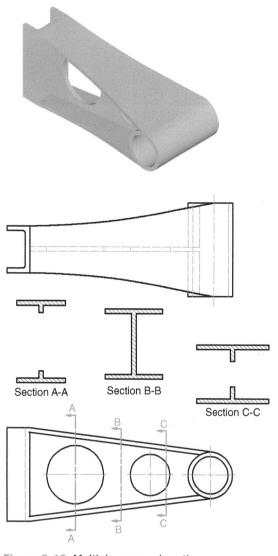

Section A-A   Section B-B

Section C-C

Figure 8-16 Multiple removed sections

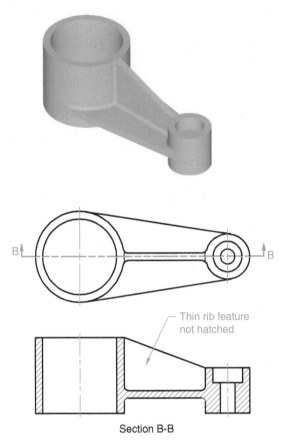

Figure 8-17 Conventional treatment of a thin rib feature in a section view

Thin rib feature
not hatched

Section B-B

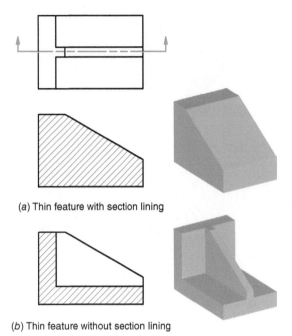

(a) Thin feature with section lining

(b) Thin feature without section lining

Figure 8-18 Justification of thin-feature convention

construction lines, or projectors, are drawn from the given views to help locate the extents of the object features in the front view.

Without the benefit of a pictorial view, one must now employ visual reasoning, reading between the given top and side views, to help piece together a mental image of the object. In the top view on the left there are three concentric circles. A raised cylinder with either a counterdrilled, counterbore, or countersunk hole feature (see Figure 7-31 in Chapter 7, for example) will appear like this in the circular view. Looking now at the right view, it appears that this is a counterdrilled hole feature. Reading between the two given views, it also appears that a horizontally oriented hole is drilled through the raised cylinder, intersecting with the vertically oriented counterdrill feature. Note that these two holes account for all of the hidden lines appearing in the two given views.

Moving now to the right side of the object, as shown in the top view on the right, we see a feature tangent to the raised cylinder on the left, with a fillet and a semicircular cut on the right. But what are the vertical extents of this feature? Looking at the right view, we see three horizontally aligned rectangles a little way up from the bottom. Together these views suggest a plate, rounded at one end with a semicircular cutout, attached to the raised cylinder. At this point (Step 2), these features can be laid out. The only remaining feature, represented by the thin rectangles centered on the centerline in the given views, top and right, suggests a support rib. Note that, based on the information provided in the given views, the rib profile could have been either straight-sided (as shown in Figure 8-20c) or contoured (curved).

In Step 3, section lining is applied to the solid areas through which the imaginary cutting plane passes, except for the rib feature, which is left without hatching by convention. Note that the section-lining angle has been adjusted from the default 45 to 30 degrees, as shown in Figure 8-6, because of the inclined lines describing the counterdrill feature. Centerlines have also been added in Step 3.

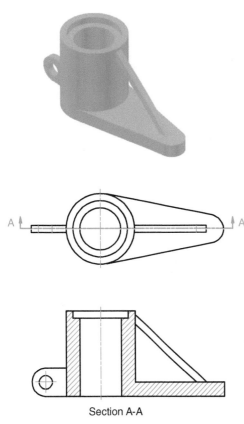

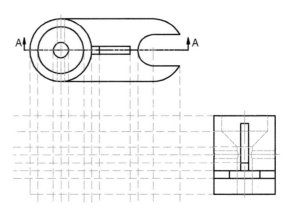

Figure 8-20*b* Step 1

A ↕   ↕ A

Section A-A

Figure 8-19 Thin-feature convention applied to a part with a lug and web

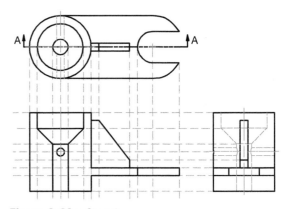

Figure 8-20*c* Step 2

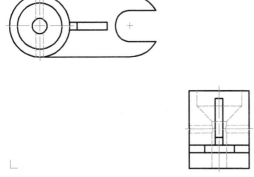

Figure 8-20*a* Full-section view construction process: given

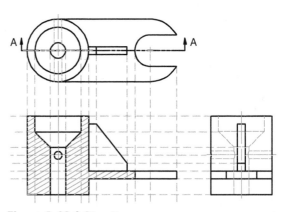

Figure 8-20*d* Step 3

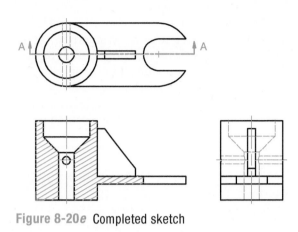

Figure 8-20e Completed sketch

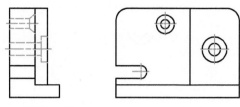

Figure 8-21a Given

Figure 8-20f Shaded view

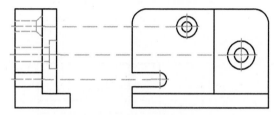

Figure 8-21b Projection of cut features

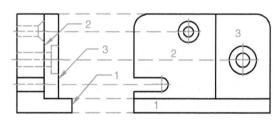

Figure 8-21c Surface labeling

Section View Construction Process—Example 2

In this example, two views are given, as shown in Figure 8-21a. The problem calls for finding an offset section in the top view.

Start with the visualization of the object. Cut features include, from top to bottom, a countersink hole, a counterbore hole, and an open slot, as seen in Figure 8-21b.

In Figure 8-21c, three (normal) surfaces are labeled in the front view. The corresponding edge views of these surfaces are also labeled in the left view.

From this, the object can now be visualized, as seen in Figure 8-21d.

Next, draw an offset cutting plane in the front view, as shown in Figure 8-21e. Note that in moving from left to right, the offsets (90 degree bends) are chosen so that they occur after one

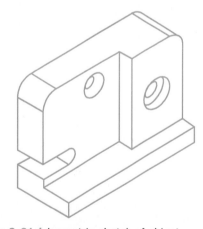

Figure 8-21d Isometric sketch of object

internal feature ends and before the next feature begins. Note also the arrow direction. The arrows point down because a top view is seen when looking down.

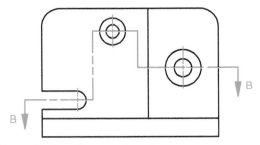

Figure 8-21 *e* Offset cutting plane on edge

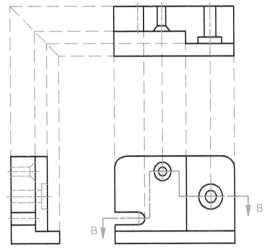

Figure 8-21 *f* Top view construction using projection from adjacent and related views

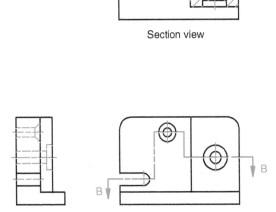

Section view

Figure 8-21 *g* Solution

The top view is next constructed by projecting from the front and left views, as shown in Figure 8-21*f*. A miter line is used to project from the left to the top view. All features in the top view are visible because it is an offset section view.

Figure 8-21*g* shows the solution. Section lining is applied to those areas where the cutting plane passes through solid material.

Conventional Representations: Aligned Sections

This convention is used to simplify the construction of section views containing radially distributed and off-angle features like holes, ribs, and lugs. In the object shown in Figure 8-22*a*, for example, a true section view based on orthographic projection results in a difficult-to-interpret, foreshortened view of the rib on the left. To eliminate these problems, by convention this feature is rotated into the cutting plane, or, alternatively, the cutting plane is bent to pass

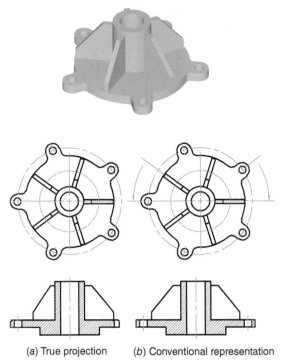

(a) True projection   (b) Conventional representation

Figure 8-22 Conventional representation of radially distributed features

through the feature. Similarly, the lug and hole features shown on the right side in Figure 8-22*a* are rotated into the cutting plane. The result is a clearer representation of the geometry, as shown in Figure 8-22*b*.

Another example of features being rotated or aligned to simplify the representation is shown in Figure 8-23.

Assembly Section Views

In a section view of an assembly, different hatch patterns are applied to different parts. See, for example, Figure 8-24, where several different section linings are used to represent different parts. Note that thin-walled parts (such as shafts and nuts) are not sectioned. In addition, parts like

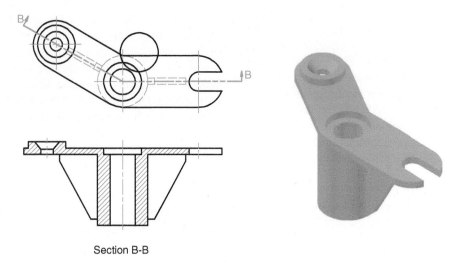

Section B-B

Figure 8-23 Conventional representation of an off-angle feature

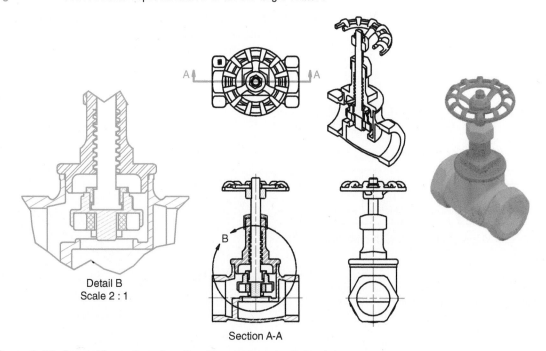

Detail B
Scale 2 : 1

Section A-A

Figure 8-24 Assembly section view (Courtesy of Alexander H. Hays)

washers, bushings, gaskets, bolts, screws, keys, rivets, pins, bearings, spokes, and gear teeth are not sectioned where the cutting plane lies along the longitudinal axis of the part.

■ AUXILIARY VIEWS

Introduction

Recall that in a multiview drawing of an object with an inclined surface, in one multiview the inclined surface is seen on edge, whereas in the other two multiviews the inclined surface appears as a foreshortened (i.e., not true-size) surface. In Figure 8-25, for example, the inclined surface labeled A is seen on edge in the top view, and as a foreshortened surface in the front and right views. In some circumstances though, a view showing the true size of an inclined surface is useful.

From *descriptive geometry*[1] it is known that the true size and shape of a planar face (or the true length of a line) can be represented in an orthographic projection only if the line of sight is normal to the planar face, or, equivalently, if the projection plane is parallel to the face. This knowledge will be put to use to find the true size of an inclined surface.

Definitions

In earlier chapters we saw that multiview projection is an orthographic projection technique wherein a three-dimensional object is projected onto one of three mutually perpendicular planes.

These are the principal planes: horizontal, frontal, and profile. An *auxiliary view* is an orthographic view that is projected onto any plane other than one of the principal planes. A *primary auxiliary view* is an auxiliary view that is projected onto a plane perpendicular to one of the principal planes and inclined to the other two. A primary auxiliary view can be used to find the true size and shape of an inclined surface. A *secondary auxiliary view* is projected from a primary auxiliary view onto a plane that is inclined to all three principal projection planes. A secondary auxiliary view can be used to find the true size and shape of an oblique surface.

Auxiliary View Projection Theory

Figure 8-26 shows an object with an inclined surface that has been placed inside a glass box. Note that the glass box has been modified by adding

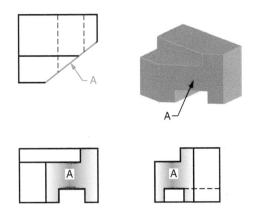

Figure 8-25 Views of cut block with an inclined surface

[1]The term *descriptive geometry* refers to the body of knowledge, developed over the centuries, consisting of mathematical-graphical procedures used to accurately describe 3D geometry within 2D media. The French mathematician Gaspard Monge (1746–1818) is considered the father of descriptive geometry.

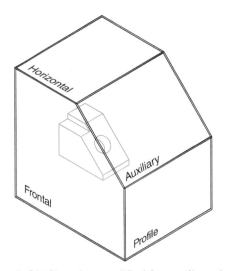

Figure 8-26 Glass box modified for auxiliary view

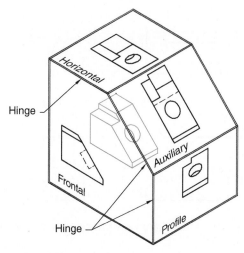

Figure 8-27 Views projected onto sides of glass box

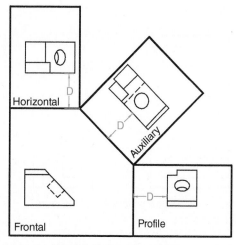

Figure 8-28 Modified glass box sides open and laid flat

an additional plane that is parallel to the inclined surface. For the situation shown in Figure 8-26, the inclined plane (i.e., the auxiliary plane) is perpendicular to the frontal plane and inclined to the horizontal and profile planes.

In Figure 8-27 orthographic projection is used to project the object onto the projection planes (i.e., the sides of the glass box), including the auxiliary plane. Because the auxiliary plane is parallel to the inclined surface, the resulting projection shows the true size and shape of the inclined surface, as well as foreshortened projections of any other visible surfaces. Also notice that the edge view of the inclined surface appears on the frontal projection plane.

Now imagine that the frontal plane is hinged to the horizontal, profile, and auxiliary planes. If the hinged views are then unfolded so that they lie in the same plane as the frontal view, Figure 8-28 results. Note that the distance D from the hinge line to the near side of the inclined surface is the same for all three projected views (horizontal, profile, and auxiliary). This fact will be used later in the construction of an auxiliary sketch of an inclined surface.

As shown in Figure 8-29, an auxiliary view is in alignment with the principal view that shows the inclined surface on edge. Because these inclined edges are projected true length, the perpendicular distances between the dashed lines shown

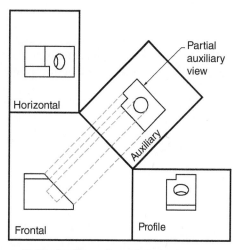

Figure 8-29 Auxiliary view alignment

in Figure 8-29 represent the actual inclined edge lengths on the inclined surface.

Note that the auxiliary view in Figure 8-29 shows only the inclined surface. This is called a ***partial auxiliary view***. Because they are both easier to execute and easier to visualize, partial auxiliary views are frequently employed. In creating an auxiliary projection with a CAD system, however, a full auxiliary view is obtained automatically. The resulting view can always be modified by either hiding or deleting lines in order to obtain a view that shows only the inclined surface of interest.

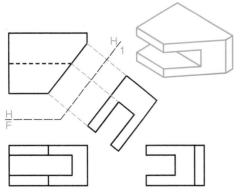

Figure 8-30 Auxiliary view projected from horizontal plane

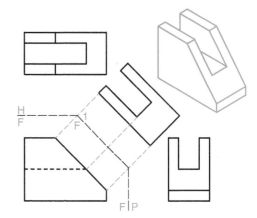

Figure 8-31 Auxiliary view projected from frontal plane

Auxiliary Views: Three Cases

The primary auxiliary view is projected from the principal plane containing the edge view of the inclined surface. If an object with an inclined surface is oriented as shown in Figure 8-30, then the edge view of the inclined surface appears on the horizontal projection plane. In this case, the auxiliary view is projected from the horizontal plane. Note how the hinge lines between the horizontal and frontal (H-F) and the horizontal and auxiliary (H-1) planes are represented.

With the same object oriented as shown in Figure 8-31, the inclined surface appears on edge in the frontal plane. In this case the primary auxiliary view is projected from the frontal plane.

In the third case, shown in Figure 8-32, the edge view of the inclined surface appears in the profile plane. The primary auxiliary view is consequently projected from the profile plane.

General Sketching Procedure for Finding a Primary Auxiliary View

In this section the sketching procedure for obtaining a primary auxiliary view is outlined. The problem can be stated as follows: Given a multiview drawing of an object with an inclined surface (see Figure 8-33a), find the primary auxiliary view showing the true size and shape of this

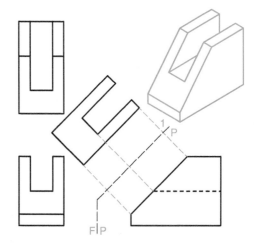

Figure 8-32 Auxiliary view projected from profile plane

inclined surface (see Figure 8-33b). The process takes advantage of the fact that the inclined surface is shown as a true-length edge in one of the multiviews. Perpendicular projectors are erected from the edge view to obtain these distances. All that is required to complete a partial auxiliary view are the edge lengths perpendicular to the inclined edge. These distances are available in the views adjacent to the view containing the inclined surface on edge. A trammel can be used to capture and then transfer these edge lengths

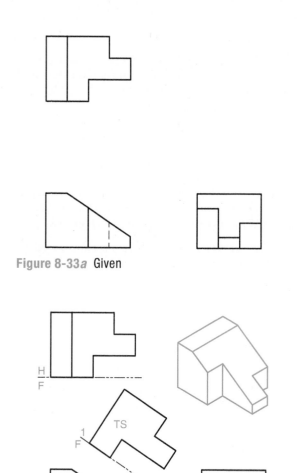

Figure 8-33a Given

Figure 8-33b Solution

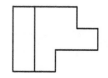

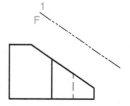

Figure 8-33c Step 1

STEP 1

Sketch a dashed reference line parallel to the edge view of the inclined surface to be projected (see Figure 8-33c).

- The perpendicular distance between the reference line and the inclined edge should be chosen such that the resulting auxiliary view does not interfere with other views.

- This reference line is labeled either H-1, F-1, or P-1, depending on whether the edge view of the inclined surface appears in the horizontal (H), frontal (F), or profile (P) plane. The "1" indicates that a primary auxiliary view is being constructed.

- The reference line represents the hinge that the auxiliary view is rotated 90 degrees about, thus causing it to lie in the same plane as the principal plane it is projected from.

STEP 2

Sketch perpendicular projectors from the inclined surface (see Figure 8-33d).

- The edge view of the inclined surface shows the true lengths of these edges, because these edges are parallel to the projection plane.

- Steps 1 and 2 may be performed in either order.

STEP 3

Draw a second dashed reference line at a convenient location between the view being projected

to the auxiliary view. Two reference edges, one for the adjacent/related view and the other for the auxiliary view, assist in the transferring of the distances. These reference edges are comparable to the glass box hinge lines shown in Figures 8-27 and 8-28, but they are more flexible. In the solution shown in Figure 8-33b, for example, the reference edge labeled H-F is conveniently located so that it is collinear with the near-side edge of the inclined surface, as shown in the top view. This is equivalent to setting D = 0 in Figure 8-28, reducing the number of required trammel dimensions by one.

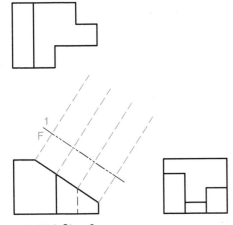

Figure 8-33*d* Step 2

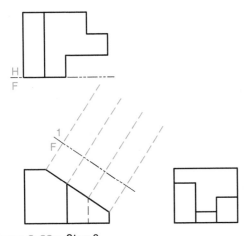

Figure 8-33*e* Step 3

STEP 4 (OPTIONAL)

In this optional step, label the vertices of the inclined surface in the adjacent/related view, and then transfer the vertex labels to the edge view (see Figure 8-33*f*).

- This step may be helpful in correctly orienting the auxiliary view.

STEP 5

Using a trammel or dividers, transfer the depth dimensions from the F-H reference line in the top view to the F-1 reference line (see Figure 8-33*g*).

- The vertex labels can also be transferred to the auxiliary view.

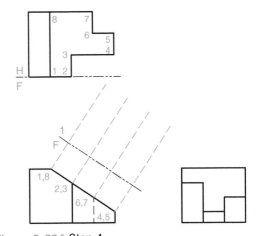

Figure 8-33*f* Step 4

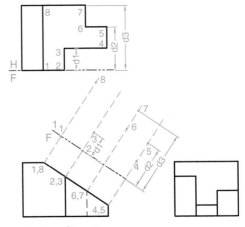

Figure 8-33*g* Step 5

from and an adjacent or related principal view (see Figure 8-33*e*).

- Label the second reference line such that the first letter indicates the view being projected from (either H, F, or P), and the second letter (H, F, or P) indicates the adjacent or related view from which the missing dimensional information is to be obtained.

- In Figure 8-33*e*, reference edge F-H will be used to capture depth information from the top view. Alternatively, a reference edge F-P could have been used to capture this same depth information from the profile view.

- This second reference line represents a hinge, where the adjacent/related view is rotated into the same plane as the edge view.

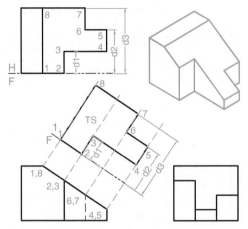

Figure 8-33h Step 6

STEP 6

Now that the placement of the edges is known, the inclined surface can be sketched. It is customary to label the projected surface with "TS," indicating that it is true size (see Figure 8-33h).

Finding a Primary Auxiliary View of a Contoured Surface

In the example shown in Figure 8-34, the procedure for finding an auxiliary view is repeated. However, this time the inclined surface has a curved profile.

Note that the partial symmetry of this surface is exploited by choosing the H-1 reference edge to lie along the axis of symmetry.

Finding a Partial Auxiliary View, an Isometric Pictorial, and a Missing View, Given Two Views

For the two-view drawing shown in Figure 8-35, find (1) an auxiliary view of the inclined surface, (2) an isometric pictorial of the object depicted, and (3) the missing right view.

The inclined surface appears on edge as the diagonal line in the top view in Figure 8-35a. Recall that an inclined surface will appear as an edge in one view and as a foreshortened surface in the other two views. Projecting from the inclined edge to the front view as seen in Figure 8-35b, it is clear that the inclined surface has the shape of a backwards "Z." This means that we can expect to see this same Z shape, now true size, in the auxiliary view, as well as a foreshortened Z shape surface in the missing right view.

To find the partial auxiliary view, we follow the steps described in the earlier section "General Sketching Procedure for Finding a Primary Auxiliary View." The results are shown in Figure 8-35c.

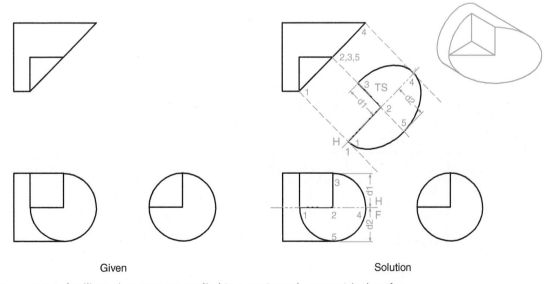

Given   Solution

Figure 8-34 Auxiliary view process applied to a contoured, symmetrical surface

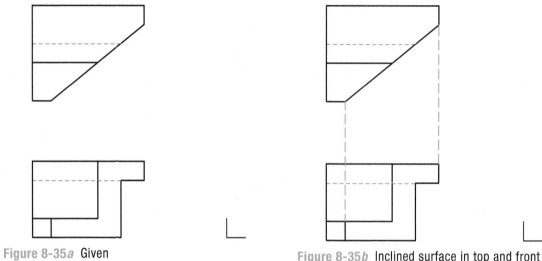

Figure 8-35a Given

Figure 8-35b Inclined surface in top and front views

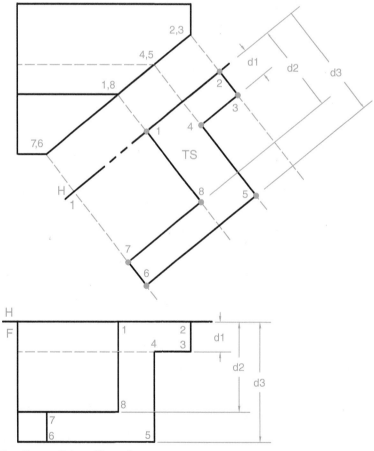

Figure 8-35c Finding the partial auxiliary view

These steps are as follows:

1. Sketch an H-1 reference line parallel to the edge view of the inclined surface in the top view. This reference line is used as a hinge about which the inclined surface is rotated 90 degrees into the plane of the paper.

2. To obtain one set of true length dimensions that define the inclined surface, sketch four perpendicular projectors extending from the edge view of the inclined surface in the top view. Since the edge view of an inclined surface is shown true length in an orthographic projection, the distances between these projectors are the true horizontal edge lengths of the inclined surface. Use the adjacent (in this case, front) view to find the second set of dimensions defining the inclined surface. In this case, they are height dimensions (i.e., d1, d2, d3, shown in Figure 8-35c).

3. Sketch a second reference line, H-F, between the two views. This reference line is used when trammeling the missing height dimensions. It is convenient to locate this horizontal reference line so that it passes through the upper edge of the inclined surface. In this way, the vertices defining the upper edge of the inclined surface will not need to be trammeled, because they will lie on the hinge line.

4. Label the vertices of the inclined surface in the front view, and then project them to the top view.

5. Use a trammel to transfer the height dimensions from H-F in the front view to H-1 in the auxiliary view. Vertex labels can also be transferred to the auxiliary view.

6. Sketch the true size inclined surface by connecting the vertices.

Having completed the partial auxiliary view, the next step is to sketch an isometric pictorial of the object. Because the inclined surface has the shape of a backwards "Z," it makes sense to start with an isometric sketch of this profile, and then extrude it, as seen in Figure 8-35d. Figure 8-35e shows a multiview drawing of this extrusion. Note the similarity between the top view in Figure 8-35e

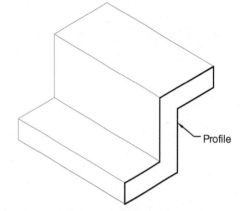

Figure 8-35d Z shape profile and extrusion

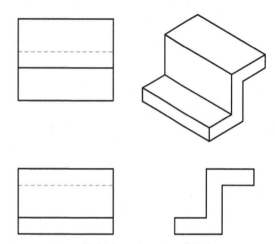

Figure 8-35e Multiview drawing of Z shape extrusion

and the top view in Figure 8-35a. This suggests that the given object in Figure 8-35a results from passing an inclined cutting plane through the extrusion (see Figure 8-35f), and then trimming away the portion of the object closest to the viewer.

To sketch this, draw lines on the isometric sketch where the cutting plane passes through the extrusion, as seen in Figure 8-35g. Whenever there is a change in planes, the cut line will change direction. Once the closed profile formed by the intersection of the plane with the solid is drawn, erase everything in front of the profile to reveal the object, as seen in Figure 8-35h.

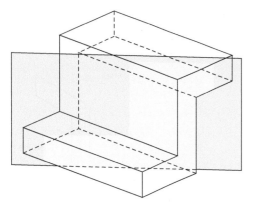

Figure 8-35*f* Cutting plane passed through Z shape extrusion

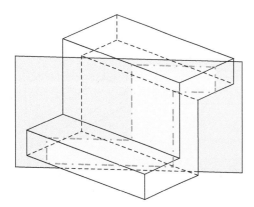

Figure 8-35*g* Profile of cut revealed

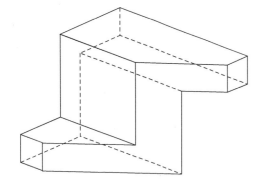

Figure 8-35*h* Pictorial view of object

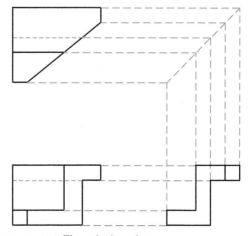

Figure 8-35*i* The missing view

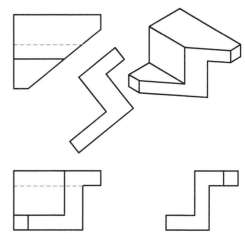

Figure 8-35*j* Complete solution

Having sketched a pictorial of the given object, constructing the missing right view is straightforward. From the isometric sketch, it is clear that there will be two visible surfaces in the right view, one being the inclined "Z" shaped surface, and the other, a square surface in the upper right corner of the view. The dimensions of these surfaces in the right view can be found by projecting from the adjacent front and related top views, as seen in Figure 8-35*i*. Figure 8-35*j* shows the completed solution.

▮ QUESTIONS

TRUE OR FALSE

1. Primary auxiliary views are used to find the true size and shape of oblique surfaces.
2. Only the inclined face of an object projected onto an auxiliary plane is shown in a partial auxiliary view.

3. When a cutting plane is passed through the length of a thin feature such as a rib, section lining is not applied.

MULTIPLE CHOICE

SS 4. The projection plane for a primary auxiliary view is:
   a. Parallel to one of the principal projection planes and perpendicular to the other two
   b. Perpendicular to one of the principal projection planes and inclined to the other two
   c. Perpendicular to two of the principal projection planes and inclined to the other one
   d. Inclined to all three of the principal projection planes

5. Which of the following is not a legitimate type of section view?
   a. Full
   b. Half
   c. Quarter
   d. Removed
   e. Revolved
   f. Aligned
   g. Offset
   h. Broken-out

SKETCHING AND MODELING

6. Given the two views appearing in Figures P8-1 through P8-6, use rectangular grid paper in the back of the book (or download worksheet from the book website) to sketch the section view of the object.

7. Given the two views appearing in Figures P8-7 through P8-19, use rectangular grid paper in the back of the book (or download worksheet from the book website) to sketch the section view of the object with the cutting plane shown in the appropriate view.
   a. Full section (Figures P8-7 through P8-11)
   b. Half section (Figures P8-12 through P8-15)
   c. Offset section (Figures P8-16 through P8-19)

8. Given the two views appearing in Figures P8-20 through P8-34, use rectangular grid paper in the back of the book (or download worksheet from the book website) to sketch the section view of the object with the cutting plane shown in the appropriate view.
   a. Full section (Figures P8-20 through P8-22)
   b. Half section (Figures P8-23 through P8-24)
   c. Offset section (Figures P8-25 through P8-26)
   d. Broken-out section (Figures P8-27 through P8-28)
   e. Removed section (Figures P8-29 through P8-31)
   f. Revolved section (Figures P8-29 through P8-31)
   g. Aligned section (Figures P8-32 through P8-34)

9. Given the dimensioned isometric view (Figures P10-1, P10-3, P10-7, P10-10, P10-13, P10-20, P10-21, P10-22, P10-24 in Chapter 10), create a solid model of the object and generate a multiview drawing that includes a section view.

10. Given the two views appearing in Figures P8-35 through P8-60, use rectangular grid paper in the back of the book (or download worksheet from the book website) to sketch the partial auxiliary view of the object.

11. Given the two views appearing in Figures P8-35 through P8-60, use rectangular and isometric grid paper in the back of the book (or download worksheets from the book website) to sketch the partial auxiliary, missing, and isometric views of the object.

12. Given a dimensioned isometric view (Figures P10-7, P10-10, P10-12, P10-15, P10-23 in Chapter 10), create a solid model of the object and generate a multiview drawing with the corresponding auxiliary view.

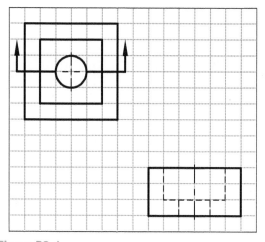

Figure P8-1

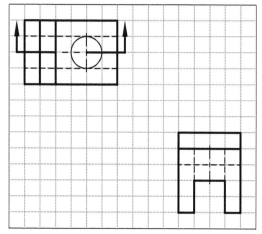

Figure P8-2

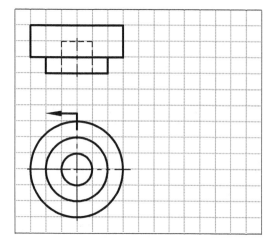

Figure P8-3

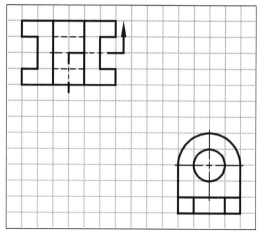

Figure P8-4

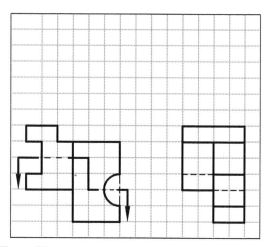

Figure P8-5

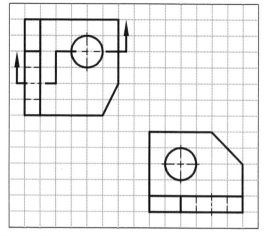

Figure P8-6

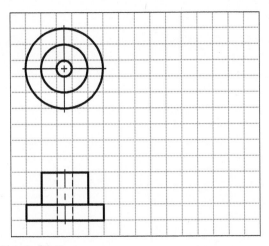

Figure P8-7

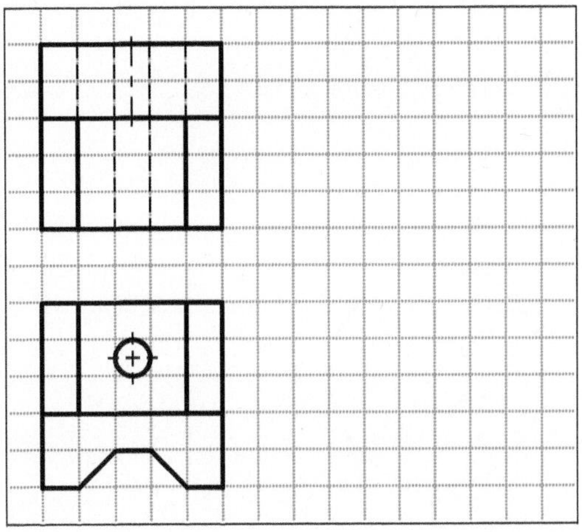

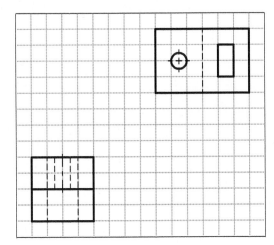

Figure P8-8

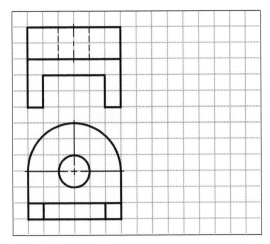

Figure P8-9

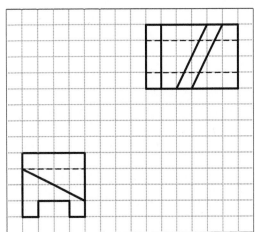

Figure P8-10

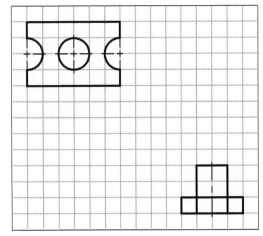

Figure P8-11

Figure P8-12

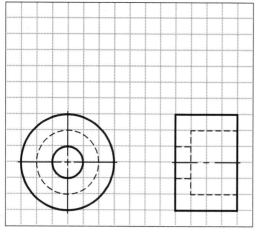

Figure P8-13

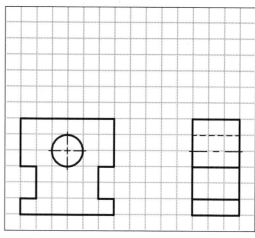

Figure P8-16

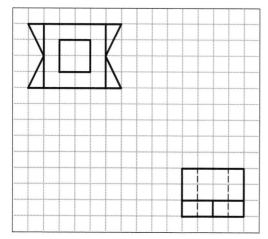

Figure P8-14

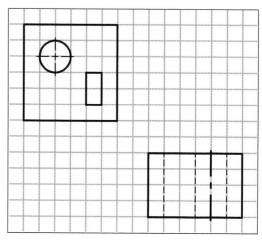

Figure P8-17

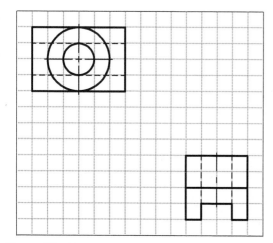

Figure P8-15

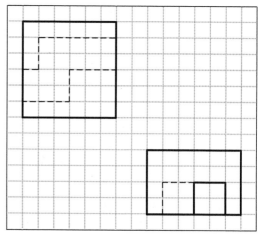

Figure P8-18

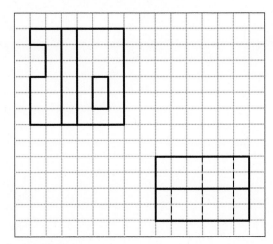

Figure P8-19

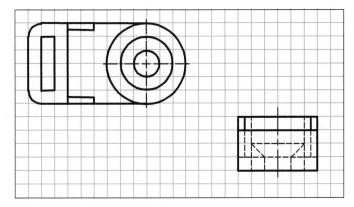

Figure P8-20

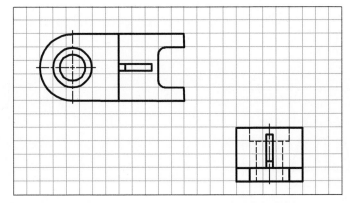

Figure P8-21

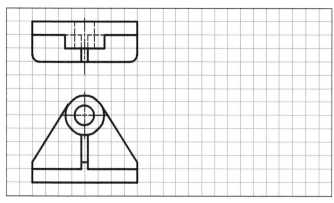

Figure P8-22

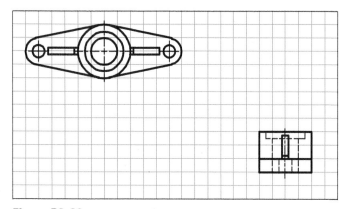

Figure P8-23

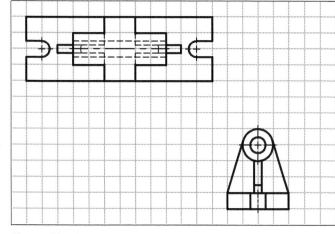

Figure P8-24

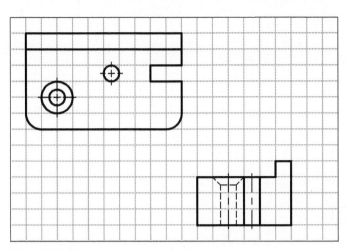

Figure P8-25

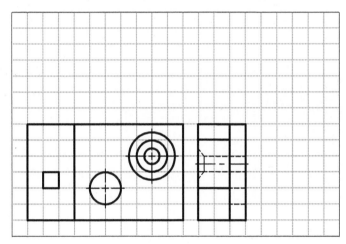

Figure P8-26

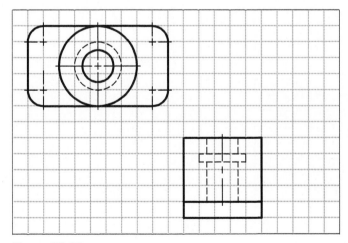

Figure P8-27

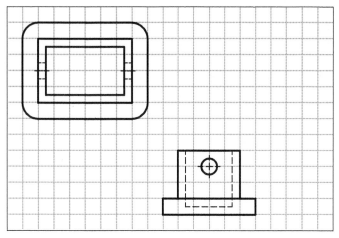

Figure P8-28

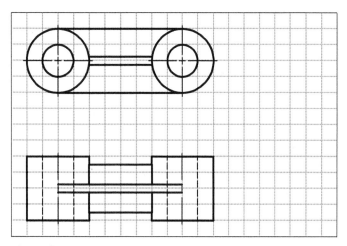

Figure P8-29

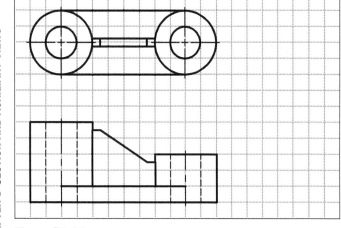

Figure P8-30

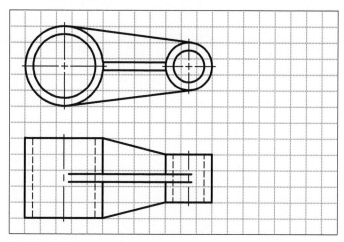

Figure P8-31

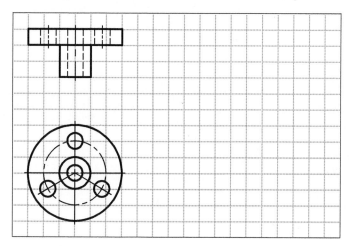

Figure P8-32

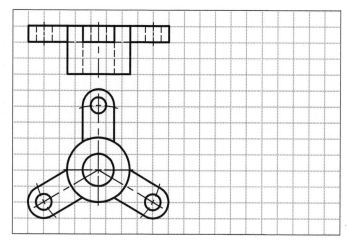

Figure P8-33

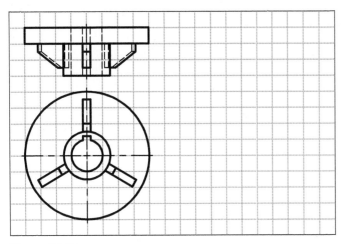

Figure P8-34

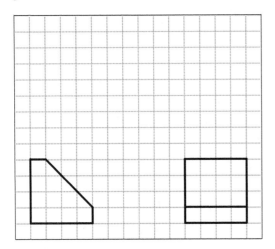

Figure P8-35

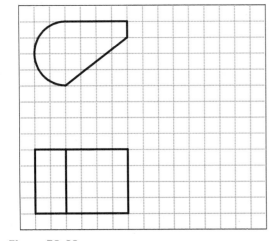

Figure P8-36

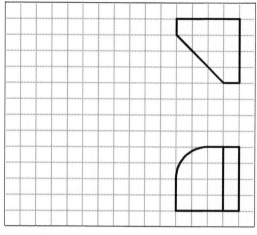

Figure P8-37

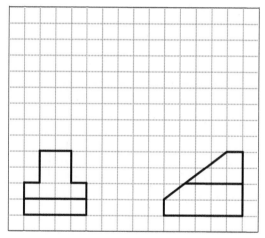

Figure P8-38

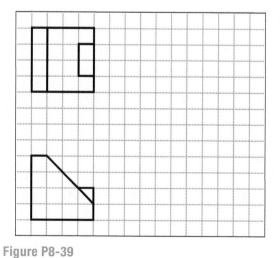

Figure P8-39

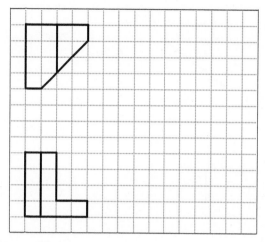

Figure P8-40

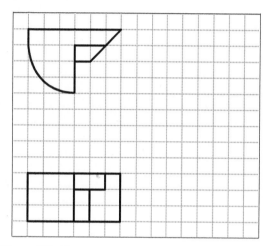

Figure P8-41

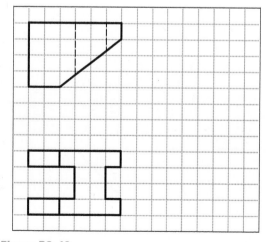

Figure P8-42

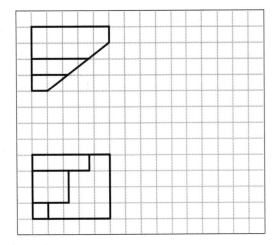

Figure P8-43

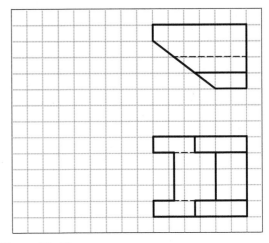

Figure P8-46

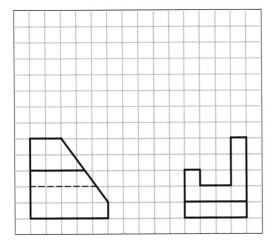

Figure P8-44

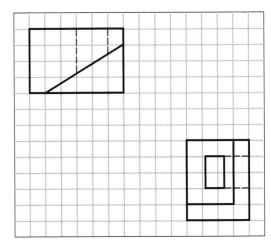

Figure P8-47

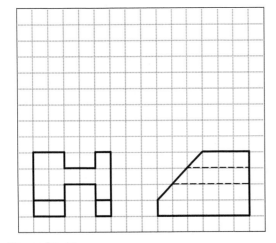

Figure P8-45

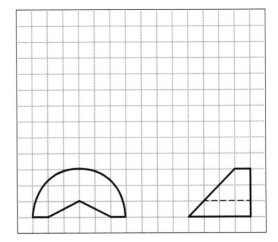

Figure P8-48

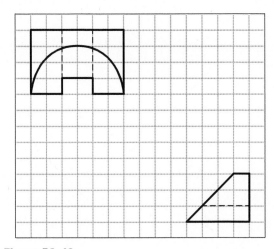

Figure P8-49

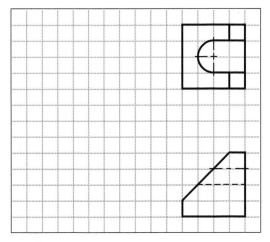

Figure P8-50

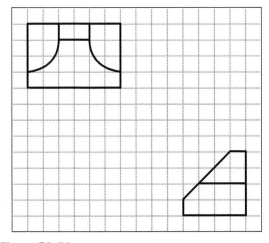

Figure P8-51

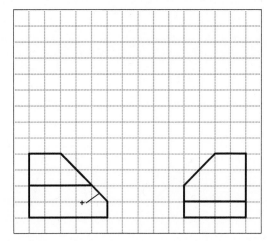

Figure P8-52

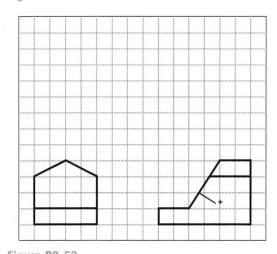

Figure P8-53

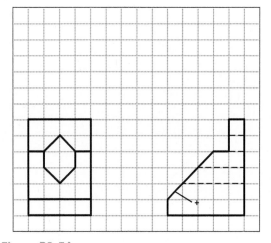

Figure P8-54

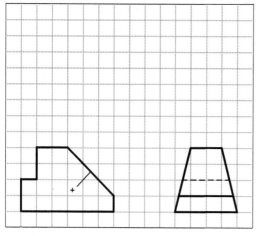

Figure P8-55

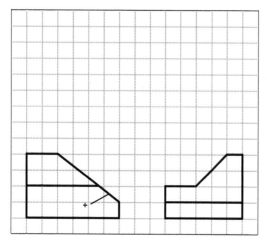

Figure P8-56

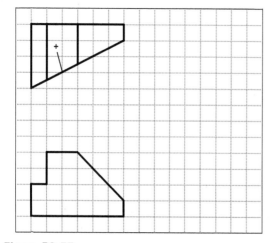

Figure P8-57

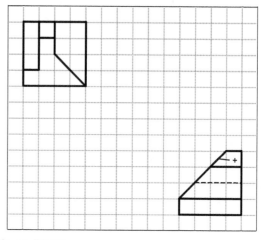

Figure P8-58

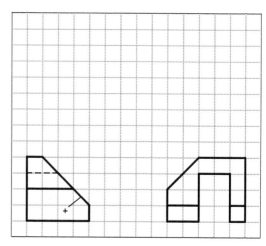

Figure P8-59

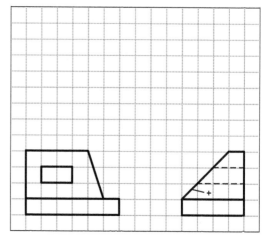

Figure P8-60

(A) Given the two views, sketch the <u>full</u> section view and an isometric view of the object.

(B) Given the two views, sketch the <u>half</u> section view and an isometric view of the object.

Drawing 8-1	Name	Date

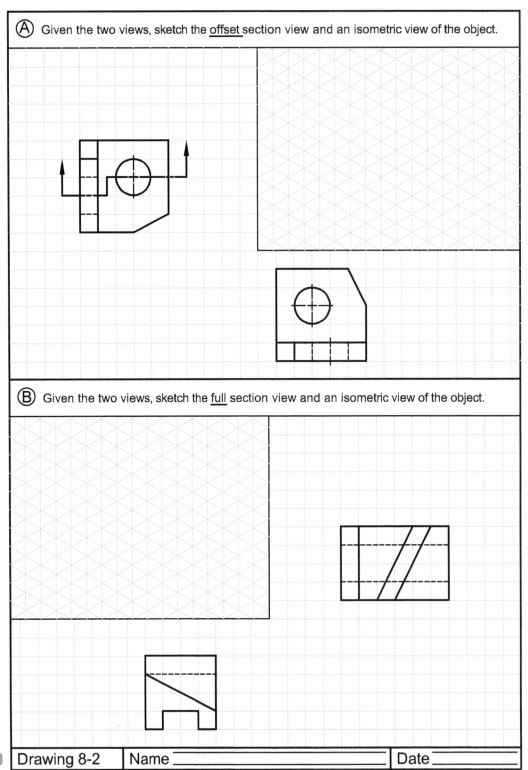

Ⓐ Given the two views, sketch the <u>offset</u> section view and an isometric view of the object.

Ⓑ Given the two views, sketch the <u>full</u> section view and an isometric view of the object.

SS

| Drawing 8-2 | Name | Date |

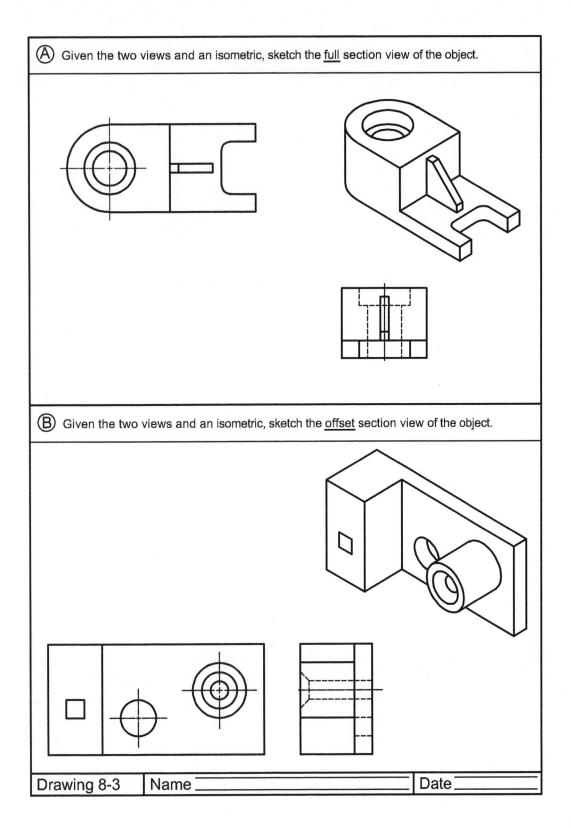

(A) Given the two views and an isometric, sketch the <u>full</u> section view of the object.

(B) Given the two views and an isometric, sketch the <u>offset</u> section view of the object.

Drawing 8-3 | Name | Date

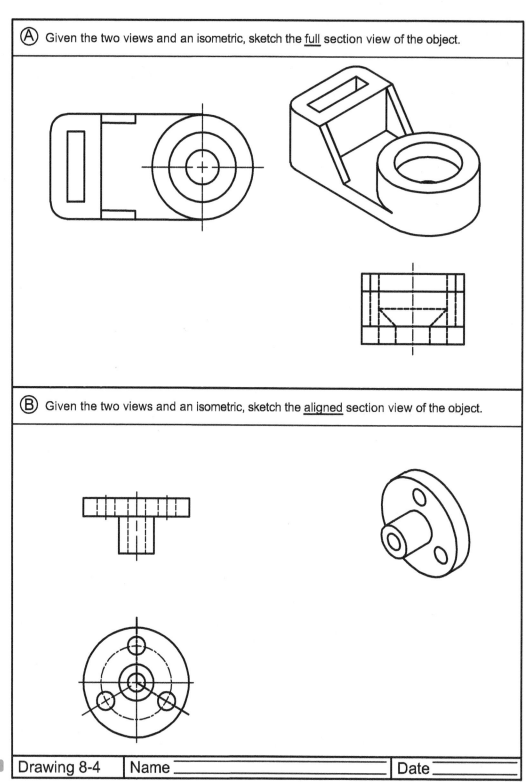

Ⓐ Given the two views and an isometric, sketch the <u>full</u> section view of the object.

Ⓑ Given the two views and an isometric, sketch the <u>aligned</u> section view of the object.

| Drawing 8-4 | Name | Date |

Ⓐ Given the two views, sketch the <u>half</u> section view of the object.

Ⓑ Given the two views, sketch the <u>full</u> section view of the object.

| Drawing 8-5 | Name | Date |

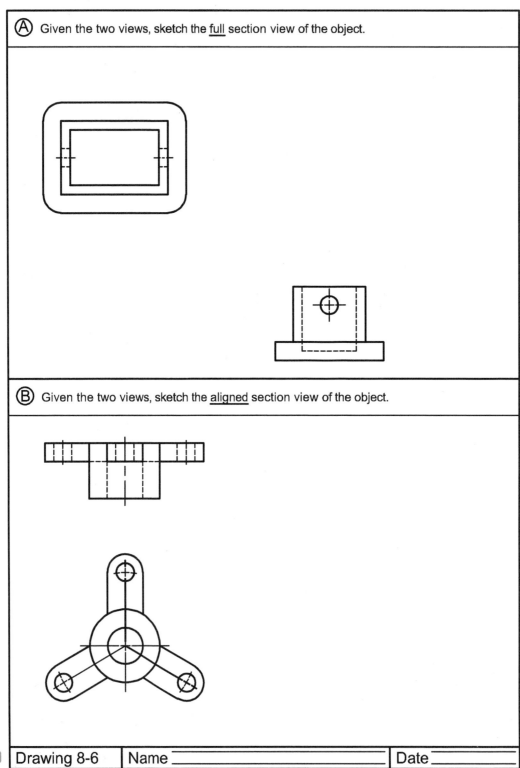

Ⓐ Given the two views, sketch the <u>full</u> section view of the object.

Ⓑ Given the two views, sketch the <u>aligned</u> section view of the object.

SS

| Drawing 8-6 | Name | Date |

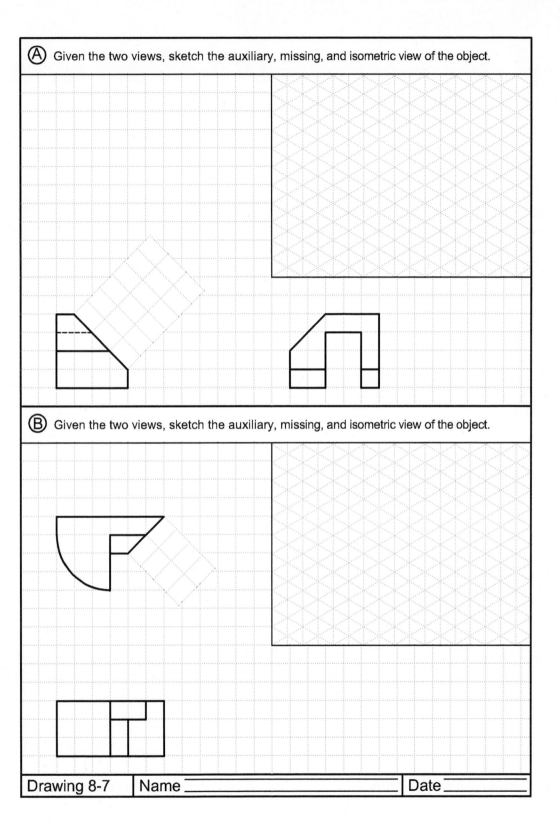

Ⓐ Given the two views, sketch the auxiliary, missing, and isometric view of the object.

Ⓑ Given the two views, sketch the auxiliary, missing, and isometric view of the object.

| Drawing 8-7 | Name _____ | Date _____ |

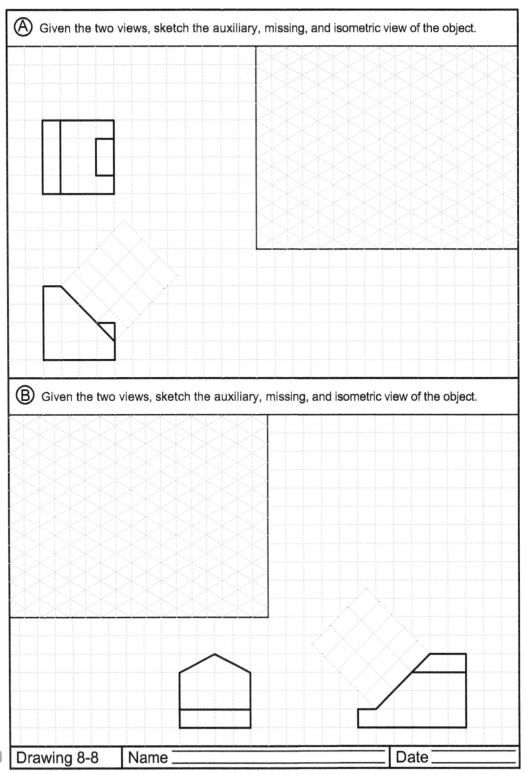

A Given the two views, sketch the auxiliary, missing, and isometric view of the object.

B Given the two views, sketch the auxiliary, missing, and isometric view of the object.

SS | Drawing 8-8 | Name _____ | Date _____

Ⓐ Given the two views, sketch the auxiliary, missing, and isometric view of the object.

Ⓑ Given the two views, sketch the auxiliary, missing, and isometric view of the object.

| Drawing 8-9 | Name _____ | Date _____ |

(A) Given the two views, sketch the auxiliary, missing, and isometric view of the object.

(B) Given the two views, sketch the auxiliary, missing, and isometric view of the object.

Drawing 8-10 | Name _____ | Date _____

CHAPTER 8 SECTION AND AUXILIARY VIEWS

SS

246

(A) Given the two views, sketch the auxiliary, missing, and isometric view of the object.

(B) Given the two views, sketch the auxiliary, missing, and isometric view of the object.

| Drawing 8-11 | Name | Date |

CHAPTER

9 DIMENSIONING AND TOLERANCING

▮ DIMENSIONING

Introduction

An engineering drawing, once submitted to production for manufacture or construction, must include all of the information needed to build the part, assembly, or system. To this end, technical drawings include dimensions and general notes describing the size and location of part features, as well as details related to the construction or manufacture of the part.

A ***dimension*** is a numerical value used to define the size, location, geometric characteristic, or surface texture of a part or feature. The main goals of dimensioning (as laid out in ANSI/ASME Y14.5M, "Dimensioning and Tolerancing for Engineering Drawings"), are the following:

1. Use only the dimensions needed to completely define the part, nothing more.
2. Select and arrange dimensions to support the function and mating relationship of the part. It is important that the dimensioned part not be subject to differing interpretations.
3. In general, do not specify the manufacturing methods to be used in building the part. This is done both to leave options open to manufacturing and to avoid potential legal problems.

4. Arrange the dimensions for optimum readability. Dimensions should appear in true profile views and refer to visible object edges.
5. Unless otherwise stated, assume angles to be 90 degrees

Units of Measurement

Drawings are typically dimensioned using either millimeters or decimal inches. Metric (Système Internationale) drawings normally employ millimeters specified as whole numbers, as shown in Figure 9-1a. In the English or Imperial system, the preferred units are inches expressed in decimal form, as seen in Figure 9-1b. In some disciplines,

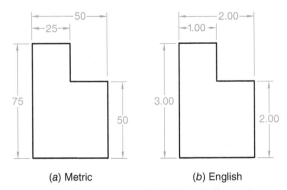

(a) Metric   (b) English

Figure 9-1 Metric versus English unit dimensions

notably architecture and construction, fractional inches are still employed, but decimals are preferred because of the easier arithmetic and the greater precision they provide. Drawings in inches are typically specified using two-decimal-place accuracy. In both metric and English drawings, there is no need to specify the units of individual dimensions. Rather, a general note similar to "Unless otherwise stated, all dimensions are in millimeters (or inches)" appears.

Application of Dimensions

Dimensions are applied to a drawing through the use of dimension lines, extension lines, and leaders from a feature to a dimension or note. General notes are used to convey additional information.

TERMINOLOGY

Dimension lines are thin, dark lines used to show the direction and extent of a dimension. See Figure 9-2. A *dimension value* indicating the number of units of a measurement is associated with the dimension line. The height of the dimension value is typically 3 mm. By preference, dimension lines are broken to allow for the insertion of the dimension value. There is, however, an alternative dimension style that places the dimension value above an unbroken dimension line. Dimension lines are terminated by *arrows*, where the length of the arrowhead is equal to the dimension text height.

The dimension line of an angle being dimensioned is an arc drawn with its center at the apex of the angle.

As shown in Figure 9-2, thin, dark *extension lines* are typically drawn perpendicular to the associated dimension line. Extension lines extend from the view of an object feature to which they refer. A short, visible gap (1.5 mm) is included between the extension line and the view for clarity. In addition, the extension line extends 3 mm beyond its outermost related dimension line.

Leader lines are drawn from a feature to a note, dimension, or symbol. As shown in Figure 9-2, leaders are inclined straight lines, except for a small horizontal shoulder that extends to the mid-height of the first or last letter (or digit) of

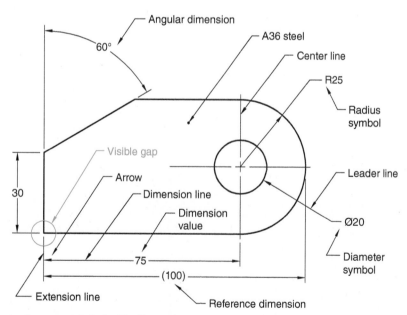

Figure 9-2 Terminology associated with dimensions

the note or dimension. These thin, dark lines start with an arrow at the feature being described. In certain circumstances, the feature is within an outline, in which case the arrow is replaced with a dot.

A **reference dimension** is used only for additional informational purposes. A reference dimension can be derived from other values shown on the drawing. It contains supplemental information and is not used for production or inspection purposes. Reference dimensions are easily identified because the associated dimensional value is placed in parentheses, as seen in Figure 9-2.

Thin, dark **center lines** also play a role in dimensioning, because they are used to locate the centers of cylindrical parts and holes.

READING DIRECTION FOR DIMENSIONAL VALUES

In **unidirectional dimensioning**, dimension values and text are oriented horizontally, as shown in Figure 9-3a. In an older dimensioning style called **aligned dimensioning**, dimensional values are oriented parallel to their dimension lines, as shown in Figure 9-3b. Aligned dimensioning is not recognized by ANSI.

ARRANGEMENT, PLACEMENT, AND SPACING OF DIMENSIONS

As was mentioned previously, dimensions are arranged for optimum readability. Several guidelines exist that govern the spacing, grouping, and staggering of parallel dimensions. There are also guidelines for dimensioning when space is limited.

A distance of at least 10 mm between the first dimension line and the part should be maintained. For succeeding parallel dimensions, this distance should be at least 6 mm (see Figure 9-4).

Parallel (i.e., either horizontal, vertical, or aligned) dimensions should be grouped and aligned, as shown in Figure 9-5, in order to present a uniform appearance.

The dimensional values of parallel dimensions should be staggered, as shown in Figure 9-6, to avoid crowding.

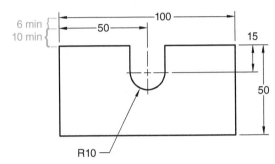

Figure 9-4 Spacing between parallel dimensions

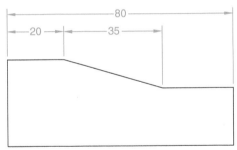

Figure 9-5 Grouping and alignment of parallel dimensions

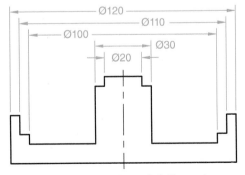

Figure 9-6 Staggering of parallel dimensions

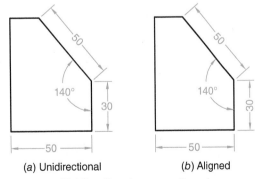

(a) Unidirectional   (b) Aligned

Figure 9-3 Unidirectional versus aligned dimensioning

By preference, the dimensional value and arrows should appear inside the extension lines. Depending on the available space, however, it may be necessary to leave only the dimensional value inside, only the arrows inside, or nothing inside. See Figure 9-7 for these possibilities. Note that this situation applies to horizontal, vertical, aligned, angular, and radial dimensions.

There are also guidelines for leaders that aim to improve the readability of a drawing. For instance, multiple leaders that are close to one another should be drawn parallel, as shown in Figure 9-8. Leader lines should not be overly long, and they should cross as few lines as possible. Leader lines should never cross one another. Finally, leaders directed to a circle or arc, as shown in Figure 9-9, should be radial (i.e., pass through the center, if extended) to the hole or arc.

Using Dimensions to Specify Size and Locate Features

Dimensions are used to specify the size and location of features. Features are sized and located with linear (horizontal, vertical, or aligned), radial, diametric, and angular dimensions. Figure 9-10

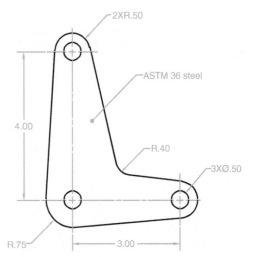

Figure 9-8 Multiple leaders in same vicinity should be parallel

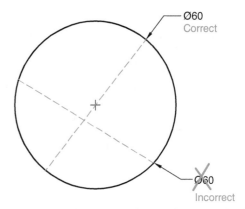

Figure 9-9 Leaders directed to circle or arc should pass through circle center when extended

provides a drawing with dimensions used to size the part features, and Figure 9-11, shows the same drawing, but with the dimensions used to locate part features now displayed.

Symbols, Abbreviations, and General Notes

A number of symbols are employed in association with dimensioning. Figure 9-12 shows several of these symbols, including radius, diameter, spherical radius, spherical diameter, counterbore, countersink, deep, and times.

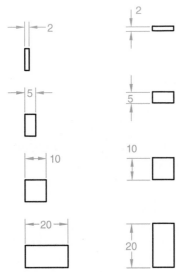

Figure 9-7 Placement of dimension text

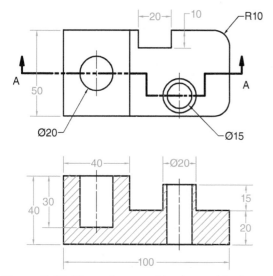

Figure 9-10 Dimensions used to size part features

Symbol Name	Symbol
Counterbore	⊔
Countersink	∨
Deep	�↧
Diameter	Ø
Square	□
Places, Times	X
Radius	R
Spherical radius	SR
Spherical diameter	SØ

Figure 9-12 Dimensioning symbols

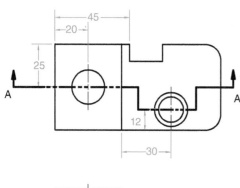

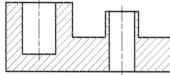

Figure 9-11 Dimensions used to locate part features

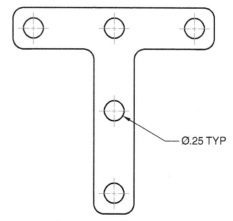

Figure 9-13 General notes and abbreviations used in dimensioning

Whenever several features (e.g., holes, fillets, rounds) of the same type and size appear in a drawing, either a general note or the abbreviation TYP (for typical) may be used, as shown in Figure 9-13. Also note that the *times* symbol (i.e., 2 × R10) is also used to dimension multiple features of the same size (e.g., Figure 9-8).

Dimensioning Rules and Guidelines

In this section several dimensioning rules or guidelines will be discussed. Note that on occasion these rules may be violated because of part complexity, lack of space, conflict with other rules, and the like. Rules concerning prismatic shapes will first be covered, followed by rules concerning cylinders and arcs.

PRISMS

1. *Do not repeat dimensions.* The depth of the object shown in Figure 9-14 is 30. Although this dimension appears in the top view, it could just as well have been placed on the

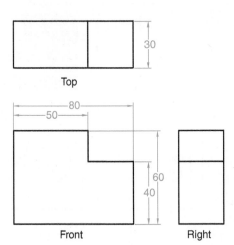

Figure 9-14 Rules and guidelines for dimensioning prisms

Figure 9-15 More rules and guidelines for dimensioning prisms

right view. In no case, however, should this same dimension appear in both views.

2. *Apply dimensions to a feature in its most descriptive view.* The object shown in Figure 9-14 contains a single (extruded "L" shape) feature. The most descriptive view of this feature is the front view. Note that four of the five dimensions needed to fully constrain the object appear on the front view.

3. *Dimension between views.* The object shown in Figure 9-14 contains two width dimensions (50 and 80) and two height dimensions (40 and 60). Because the width is shared by the front and top views, these width dimensions are placed between these views. Likewise, the height dimensions are placed between the front and right views.

4. *Omit one (intermediate) dimension.* The object in Figure 9-14 contains two intermediate width dimensions, 50 and 30 (since 80 – 50 = 30). Only one of them is shown. Likewise, there are two intermediate height dimensions, 40 and 20 (since 60 – 40 = 20). Only one of them is shown. This is done to avoid cluttering the drawing with unnecessary dimensions, and also to avoid ambiguity in

specifying tolerances (see the section on tolerance accumulation later in this chapter). As for which of the intermediate dimensions to include on the drawing, choose the one that is easiest to measure with calipers—in this case the width dimension of 50 and the height dimension of 40.

5. *Place smaller dimensions inside of larger dimensions.* Note that in Figure 9-14 the intermediate dimensions (50, 40) are placed closer to the view than the principal dimensions (80, 60). This practice helps to keep the drawing organized by avoiding the need for extension lines that cross dimension lines.

Two additional features, an extruded cut and a hole, have been added to the object depicted in Figure 9-14 and can be seen in Figure 9-15.

6. *Dimension to visible object lines, not to hidden lines.* The dimensions of the newly added cut, 30 and 10, are placed on the view that best describes this feature (i.e., the top view), and not in either the front (30) or the right (10) view, where it would have been necessary to apply a dimension to a hidden line.

DIMENSIONING

7. *Keep dimensions outside of the views.* The diameter of the through hole should be placed outside the view. If the drawing view is cluttered or the leader line needs to be extremely long to place the dimension outside the view, then this rule may be overridden.

8. *Extension lines may cross object lines and other extension lines.* In general it is desirable to avoid lines that cross; however, it is permissible for an extension line to cross an object line (e.g., extension lines for the dimensions of 20 and 15 locating the hole center) or another extension line (80 and 60).

As shown in Figure 9-16, calipers are used to directly measure the diameter of solid cylinders and round holes. It is for this practical reason that the diameter, rather than the radius, of circular features is specified on engineering drawings.

9. *Dimension the diameter of cylindrical parts in their rectangular view.* In Figure 9-17 the diameter of the boss (raised cylindrical) feature is dimensioned in its rectangular (i.e., front) view. Note that the diameter symbol Ø (the Greek letter phi), precedes the dimension.

10. *Dimension the diameter of cylindrical holes in their circular view.* In Figure 9-17 the diameter of the through hole is dimensioned in

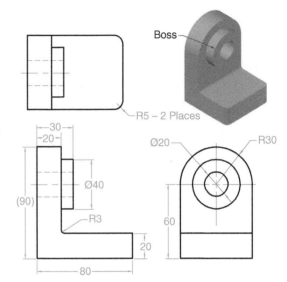

Figure 9-17 Rules and guidelines for dimensioning cylinders and arcs

the right view, where the hole appears as a circle.

11. *Dimension the radius of circular arcs in the view where their true shape is seen.* Note that the symbol R precedes the dimensional value of the radius. As seen in Figure 9-17, this rule applies to fillets (R3), rounds (R5), and other circular arcs (R30). For arcs less than or equal to 180 degrees, specify the radius. For arcs greater than 180 degrees, specify the diameter.

Some additional comments regarding Figure 9-17 are in order. First, note the use of the note Places (i.e., R5 – 2 Places) to eliminate the need for an additional R5 dimension. Similarly, the abbreviation TYP is common. Also note that the R30 radial dimension eliminates the need to dimension the overall depth (60) of the object. Similarly, the overall height (90) is not needed, although it is provided in the form of a reference dimension.

Finally, note that the overall width dimension (80) appears at the bottom of the front view, in apparent violation of Guideline 3, by which the width dimensions should appear between the front and top views. This is due to yet another guideline:

Figure 9-16 Use of calipers to measure the diameter of a cylinder

12. *Avoid overly long extension and leader lines.* Placing the 80 width dimension at the bottom, rather than the top, of the front view significantly reduces the length of the associated extension line.

Finish Marks

Parts formed by casting have rough external surfaces. When these cast parts are used in an assembly, surfaces in contact with other parts are machined or finished, in order to provide a smooth mating surface, reduce friction, and so on. In an engineering drawing, a finish mark symbol (√) is used to indicate that a surface is to be machined. As seen in Figure 9-18, finish marks are applied to all edge views, whether visible or hidden, of finished part surfaces.

▮ TOLERANCING

Introduction

In manufacturing, the same process is typically employed to mass produce a single part. These parts are then combined with other, similarly mass produced parts to create commercial products. Clearly these mass-produced parts must be interchangeable. However, when inspecting a batch of parts produced by the same manufacturing process, we would not find two parts that are exactly the same. In even the most precise manufacturing process, slight variations in part size are found.

Tolerancing is a dimensioning technique used to ensure part interchangeability by controlling the variance that exists in manufactured parts. This is accomplished by specifying a range within which a dimension is allowed to vary. As long as the size and location of part features fall within this tolerance zone, the part should function properly within an assembly.

Tolerancing is critical to the success of manufacturing. Beyond ensuring the interchangeability of parts, tolerancing directly influences both the cost and the quality of manufactured parts. Parts that are made to high accuracy are expensive. Depending on the type of product, extremely accurate parts may not be warranted. For example, the parts used to make a plastic toy do not need to be as accurate as automotive parts. As a general rule, tolerances should be stated as generously as possible, while still ensuring that the part will function properly. Doing so allows for the possibility of using a wider variety of processes to manufacture the part, and consequently it helps keep part costs low.

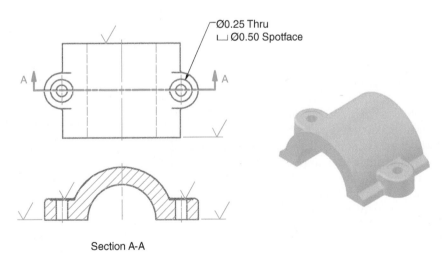

Ø0.25 Thru
⌴ Ø0.50 Spotface

Section A-A

Figure 9-18 Finish marks

Manufacturing quality is primarily a function of part accuracy. High-quality parts exhibit small variations in size and shape. By specifying tight tolerance zones and then controlling part variability using techniques like *statistical process control*, the designer can maintain and even improve upon product quality.

Definitions

A *tolerance* is the total permissible variation of a size, or the difference between the maximum and minimum *limits of size*. The tolerance 3.25 ± 0.03 indicates that the actual part size can range anywhere between 3.22 and 3.28 and still function properly. In this example the value 3.28 is the maximum limit of size, 3.22 is the minimum limit of size, and the tolerance is 0.06.

Whereas the *actual size* is the measured size of a finished part, the *basic size* is the theoretical size from which a tolerance is assigned. In the example cited above, 3.25 is the basic size, and the actual size will fall between 3.22 and 3.28, assuming that it is within tolerance.

Tolerance Declaration

Tolerances may be expressed in different ways, including:

1. Direct tolerancing methods
2. General tolerance notes
3. Geometric tolerances

Direct tolerancing methods include (1) limit dimensioning and (2) plus-and-minus tolerancing. In limit dimensioning, the limits of size are directly represented as part of the dimension. The upper limit (maximum value) is placed above the lower limit (minimum value). When this is expressed in a single line, the lower limit precedes the upper limit. Figure 9-19 provides some examples of limit dimensioning.

In plus-and-minus tolerancing, the basic dimension is given first, followed by a plus-and-minus expression of the tolerance. Plus-and-minus tolerances may be either *unilateral* or *bilateral*. A unilateral tolerance is permitted

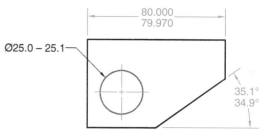

Figure 9-19 Limit dimensioning

to vary in only one direction from the basic size. Bilateral tolerances, on the other hand, may vary in either direction from the basic size. The variance of bilateral dimensions may be either equal or unequal. Figure 9-20 provides some examples of plus-and-minus tolerance dimensioning.

General notes such as "ALL DIMENSIONS HELD TO ±0.05" are sometimes used on engineering drawings. Such a note indicates that all dimensions that appear on the drawing should fall within 0.05 inch of the basic size.

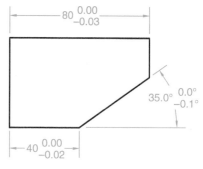

(*a*) Unilateral tolerancing

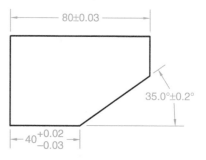

(*b*) Bilateral tolerancing

Figure 9-20 Plus-and-minus dimensioning

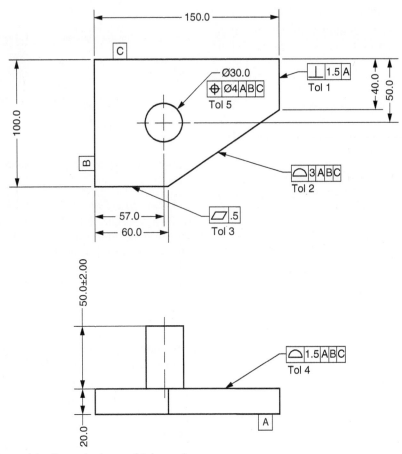

Figure 9-21 Geometric dimensioning and tolerancing

Figure 9-21 provides an example of an object that has been dimensioned using geometric dimensioning and tolerancing (GD&T) techniques. GD&T is not covered in this text.

Tolerance Accumulation

When the location of a feature depends on more than one tolerance value, these tolerances will be cumulative. In the *chain dimensioning* technique employed in Figure 9-22a, for example, the tolerance accumulation between surfaces X and Y is ±0.03. The tendency for tolerances to stack up, or accumulate, can be reduced by using *base line dimensioning*, where all dimensions of a given type are specified from the same datum, as

shown in Figure 9-22b. In this case, the tolerance variation between surfaces X and Y is reduced to ±0.02. If necessary, the tendency for tolerances to accumulate can be further controlled by *direct dimensioning*. The maximum variation between the directly dimensioned surfaces X and Y in Figure 9-22c is ±0.01.

Mated Parts

The tolerance of a single standalone part is of little importance. When a part is mated with other parts in an assembly, the true value of tolerancing becomes apparent. Mated parts must be toleranced as a system to fit within a

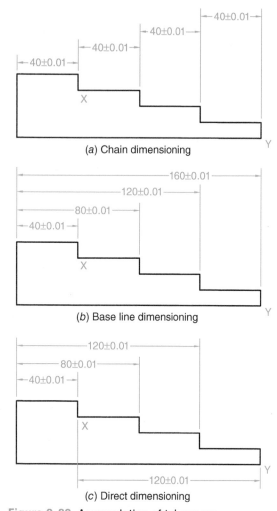

(a) Chain dimensioning

(b) Base line dimensioning

(c) Direct dimensioning

Figure 9-22 Accumulation of tolerances

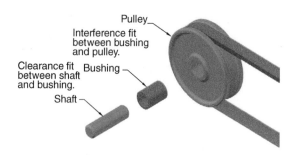

Figure 9-23 Pulley assembly tolerance fits

prescribed degree of accuracy. Figure 9-23, for example, shows a shaft, bushing, and pulley assembly. The shaft must be able to turn freely within the bushing, while the bushing is force-fit into the pulley.

Fit refers to the degree of tightness or looseness between two mating parts. In a ***clearance fit***, the internal member (e.g., shaft) is always smaller than the external member (e.g., hole). The shaft and bushing in Figure 9-23 have a clearance fit. The shaft is free to turn inside the hole.

In an ***interference fit***, the internal member is always larger than the external member. An interference fit requires that the two parts be forced together, as is the case with the bushing and pulley shown in Figure 9-23. Note that an interference fit fastens two parts together without using adhesive or mechanical fasteners.

A ***transition fit*** ranges between a pure clearance fit and a pure interference fit. In a transition fit, either the internal shaft or the external hole may be larger, so that parts either slide or are forced together. If an assembly calls for a transition fit, the two sets (hole, shaft) of components can be measured and sorted into groups according to size (e.g., small, medium, large). The components are then assembled, with components from one group being mated with corresponding components from the matching group. This method, known as *selective assembly*, is a relatively inexpensive way to manufacture tight clearance or interference fits.

In a ***line fit***, one of the limits on both the hole and the shaft are equal, which means that the shaft and hole may have the same size.

Figure 9-24 shows examples of clearance, interference, transition, and line fits, along with their upper and lower limits. Later in this chapter we will see that each of these classes of fit can be further categorized into subclasses.

The *allowance* is the tightest possible fit between two mated parts. It is the difference between the smallest hole size and the largest shaft size. For a clearance fit, the allowance is positive and represents the minimum clearance between

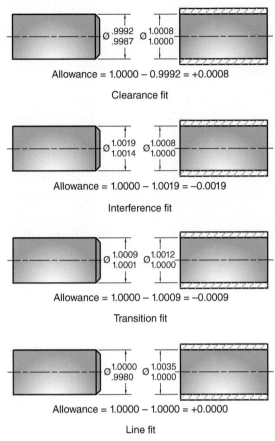

Allowance = 1.0000 − 0.9992 = +0.0008

Clearance fit

Allowance = 1.0000 − 1.0019 = −0.0019

Interference fit

Allowance = 1.0000 − 1.0009 = −0.0009

Transition fit

Allowance = 1.0000 − 1.0000 = +0.0000

Line fit

Figure 9-24 Comparison of different types of fits

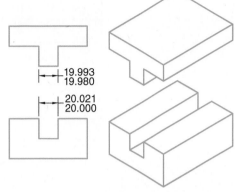

Figure 9-25 Fit of mated parts with parallel surfaces

the two parts. For an interference fit, the allowance is negative and represents the maximum interference between the two parts. The allowance is calculated for the four cases shown in Figure 9-24.

Although these different types of fit typically refer to cylindrical features like shafts and holes, they also apply to parts with parallel surfaces that fit inside one another, as depicted in Figure 9-25.

With an understanding of tolerance, allowance, basic size, and types of fit, we can assign tolerances to a system of mated parts in order to achieve a particular type of fit (i.e., clearance, interference). All that is needed is a reference system, or method of calculation, for relating the tolerances and allowance to the basic size. In the

following sections, two reference systems (hole and shaft) are discussed. Both English and metric units, though calculated differently, employ hole and shaft methods of calculation.

Basic Hole System: English Units

In the basic hole system, the minimum (i.e., lower limit) hole size is taken as the basic size. The allowance between the hole and shaft is then determined, and the tolerances are applied. The basic hole system is widely used because of the ready availability of tools (e.g., drills, reamers) capable of producing standard-size holes with precision. In effect, when using the basic hole system, we are choosing a standard drill size to create the hole and then turning down the shaft to fit that hole. Figure 9-26 illustrates the basic hole system for both a clearance fit and an interference fit.

Basic Shaft System: English Units

Less common than the basic hole system is the basic shaft system. In the basic shaft system, the maximum (i.e., upper limit) shaft is taken as the basic size. When using the basic shaft system, one selects a stock shaft and creates the mating hole to suit the shaft. The basic shaft system should be used only when there is a specific reason for using it, such as when a shaft cannot be easily

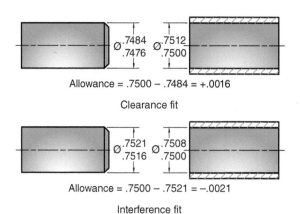

Allowance = .7500 − .7484 = +.0016

Clearance fit

Allowance = .7500 − .7521 = −.0021

Interference fit

Figure 9-26 Basic hole system for clearance and interference fits

Step-by-step tolerance calculation of a clearance fit using the basic hole system (see Figure 9-27)

Given:

- Basic size is .5000
- Allowance is +.0020
- Hole tolerance is .0016
- Shaft tolerance is .0010

1. Hole minimum = basic size = .5000
2. To find the upper limit on the shaft:

    Since

    Allowance = Hole minimum − Shaft maximum,

    Shaft maximum = Hole minimum − Allowance

    = .5000 − .0020 = .4980

3. To find the upper limit on the hole:

    Hole maximum = Hole minimum + Hole tolerance

    = .5000 + .0016 = .5016

4. To find the lower limit on the shaft:

    Shaft minimum = Shaft maximum − Shaft tolerance

    = .4980 − .0010 = .4970

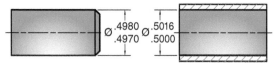

Figure 9-27 Step-by-step tolerance calculation: clearance, basic hole

Given:

- Basic size is 4.0000
- Allowance is –.0049
- Hole tolerance is .0014
- Shaft tolerance is .0009

1. Hole minimum = basic size = 4.0000
2. To find the upper limit on the shaft:

    Shaft maximum = Hole minimum – Allowance

    = 4.0000 – (–.0049) = 4.0049

    (because Allowance = Hole minimum – Shaft maximum)
3. To find the upper limit on the hole:

    Hole maximum = Hole minimum + Hole tolerance

    = 4.0000 + .0014 = 4.0014
4. To find the lower limit on the shaft:

    Shaft minimum = Shaft maximum – Shaft tolerance

    = 4.0049 – .0009 = 4.0040

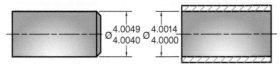

Figure 9-28 Step-by-step tolerance calculation; interference, basic hole

machined to size, or when several parts requiring different fits must be mated to the same shaft. Figure 9-29 illustrates the basic shaft system for clearance and interference fits.

Preferred English Limits and Fits

In order to simplify the process of tolerancing mated parts, ANSI standards and accompanying tables have been developed for both English and metric units. The English unit standards are described in B4.1-1967 (R1994), "Preferred Limits and Fits for Cylindrical Parts." Although these standards are intended for holes, cylinders, and shafts, they can also be used for fits between

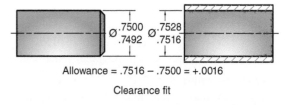

Allowance = .7516 – .7500 = +.0016

Clearance fit

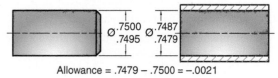

Allowance = .7479 – .7500 = –.0021

Interference fit

Figure 9-29 Basic shaft system for clearance and interference fits

parallel surfaces (see Figure 9-25). The tables taken from B4.1 appear in Appendix A. To use these tables, the user provides the basic size and the type of fit. These tables employ the basic hole system.

B4.1 recognizes five different types of fit, as well as different classes within each type of fit. For any one class of fit (e.g., RC5), the fit produced between the mated parts results in the same fit characteristics, regardless of the basic size of the part features. The characteristics of the different types and classes of fit are described below.

RUNNING OR SLIDING CLEARANCE FIT (RC)

Clearance fits provide a similar running performance throughout a range of sizes. The clearances of the first two classes (RC1, RC2) are intended for use as slide fits, and the other classes (RC3 through RC9) are for free-running operation.

RC1 and RC2 clearances increase more slowly with diameter than the other classes to maintain accurate location at the expense of free relative motion. The other (free-running) classes range from precision (RC3) to loose (RC9).

LOCATIONAL CLEARANCE FIT (LC)

Tighter clearance fits than RC, intended for parts that are normally stationary but can be freely assembled and disassembled. They range from snug line fits for parts requiring accuracy of location, to looser fastener fits where freedom of assembly is of prime importance.

TRANSITION CLEARANCE OR INTERFERENCE FIT (LT)

Compromise between clearance and interference fits, for application where accuracy of location is important, but a small amount of either clearance or interference is permissible.

Step-by-step tolerance calculation of a clearance fit using the basic hole system (see Figure 9-30)

Given:

- Basic size is 8.0000
- Allowance is +.0150
- Hole tolerance is .0120
- Shaft tolerance is .0070

1. Shaft maximum = basic size = 8.0000
2. To find the lower limit on the hole:

        Hole minimum = Shaft maximum + Allowance

               = 8.0000 + .0150 = 8.0150

    (because Allowance = Hole minimum – Shaft maximum)

3. To find the upper limit on the hole:

    Hole maximum = Hole minimum + Hole tolerance

            = 8.0150 + .0120 = 8.0270

4. To find the lower limit on the shaft:

    Shaft minimum = Shaft maximum – Shaft tolerance

            = 8.0000 – .0070 = 7.9930

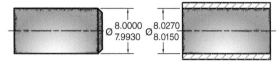

Figure 9-30 Step-by-step tolerance calculation: clearance, basic shaft

LOCATIONAL INTERFERENCE FIT (LN)

Force fit used where accuracy of location is of prime importance and for parts requiring rigidity and alignment with no special requirements for bore pressure. Not intended for parts designed to transmit frictional loads from one part to another based on tightness of fit.

FORCE OR SHRINK FIT (FN)

Force fit characterized by maintenance of constant bore pressures throughout the range of sizes. The interference varies almost directly with diameter, with the difference between its maximum and minimum values kept small to maintain the resulting pressures within reasonable limits.

Step-by-step tolerance calculation of a clearance fit using the basic hole system (see Figure 9-31)

Given:

- Basic size = 2.0000
- Fit type is RC8
- Calculation method is basic hole

1. Using Appendix A, Running and Sliding Fits, with a basic size of 2.0000 and an RC8 fit type, the following information is extracted:

Nominal Size Range (Inches)	Class RC8		
Over To	Limits of Clearance	Standard Limits	
		Hole H10	Shaft c9
1.97 – 3.15	6	+4.5	– 6.0
	13.5	0	– 9.0

A note at the top of the table indicates that the limits are in thousandths of an inch. Thus the upper and lower limits on the hole, +4.5 and 0, are actually .0045 and 0, and the limits on the shaft, −6.0 and −9.0, are −.0060 and −.0090. These standard limits are added algebraically to the basic size to determine the actual tolerance limits. The limits of clearance give the tightest (0 − (−.0060) = +.0060) and the loosest (+.0045 − (−.0090) = .0135) possible fits. Recall that the tightest fit is the allowance.

2. Tolerance limits on the hole:

   Upper limit = 2.0000 + .0045 = 2.0045

   Lower limit = 2.0000 (= basic size)

3. Tolerance limits on the shaft:

   Upper limit = 2.0000 − .0060 = 1.9940

   Lower limit = 2.0000 − .0090 = 1.9910

4. Allowance = Hole minimum − Shaft maximum

   $\qquad$ = 2.0000 − 1.9940 = .0060

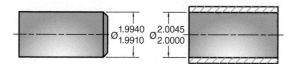

Figure 9-31 Step-by-step tolerance calculation: LC3, basic hole, preferred English-unit fit tables

Given:

- Basic size is 1.0000
- Fit type is LN2
- Calculation method is basic shaft

ANSI Standard B4.1 provides preferred limits and fits tables using only the Basic Hole System. To calculate tolerances of mated parts using the Basic Shaft System, the standard limits must be converted from basic hole to basic shaft.

1. Using Appendix A, Locational Interference Fits, with a basic size of 1.0000 and an LN2 fit type, the following information is extracted:

Nominal Size Range (Inches)		Class LN2		
Over To	Limits of Clearance	Standard Limits		
			Hole H7	Shaft c9
0.71 – 1.19	0	+0.8	+1.3	
	1.3	–0	+0.8	

2. These standard limits are using the Basic Hole System. In the Basic Shaft System, the upper limit on the shaft is taken as the basic size, meaning that the upper limit on the shaft should be 0, not +1.3. We can therefore convert the standard limits from basic hole to basic shaft by subtracting +1.3 from all of the standard limits. This has been done below:

	Hole	Shaft
Upper limit	+0.8 – 1.3 = –0.5	+1.3 – 1.3 = 0
Lower limit	–0 – 1.3 = –1.3	+0.8 – 1.3 = –0.5

3. Tolerance limits on the hole:

   Upper limit = 1.0000 – .0005 = .9995
   Lower limit = 1.0000 – .0013 = .9987

4. Tolerance limits on the shaft:

   Upper limit = 1.0000 (= basic size)
   Lower limit = 1.0000 – .0005 = .9995

5. Allowance = Hole minimum – Shaft maximum
   $$= .9987 – 1.0000 = – .0013$$

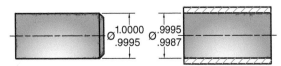

Figure 9-32 Step-by-step tolerance calculation: LN2, basic shaft, preferred English-unit fit tables

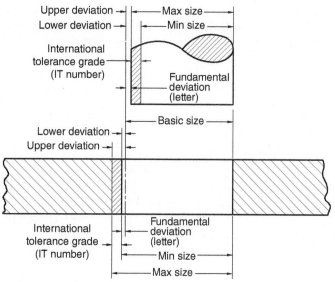

Figure 9-33 Illustration of definitions for metric limits and fits

Preferred Metric Limits and Fits

ANSI B4.2 – 1978 (1994), "Preferred Metric Limits and Fits," provides standards and tables for tolerancing fitted parts using metric units. Using this standard, a tolerance is specified using a special designation, for example 40H7. ANSI B4.2 begins with a series of definitions and an accompanying illustration similar to the one shown in Figure 9-33. These definitions include:

Basic size—The size to which limits or deviations are assigned. It is designated by the number 40 in 40H7.

Deviation—The algebraic difference between a size and the corresponding basic size.

Upper deviation—The algebraic difference between the maximum limit of size and the corresponding basic size.

Lower deviation—The algebraic difference between the minimum limit of size and the corresponding basic size.

Fundamental deviation—The deviation, upper or lower, that is closest to the basic size. It is designated by the letter H in 40H7.

Tolerance—The difference between the maximum and minimum size limits on a part.

Tolerance zone—A zone representing the tolerance and its position in relation to the basic size.

International tolerance grade (IT)—A group of tolerances that vary depending on the basic size, but that provide the same relative accuracy within a given grade. It is designated by the number 7 in 40H7 (IT7).

Hole basis—The system of fits where the minimum hole size is equal to the basic size. The fundamental deviation for a hole basis system is "H."

Shaft basis—The system of fits where the maximum shaft size is equal to the basic size. The fundamental deviation for a shaft basis system is "h."

Figure 9-34 shows a toleranced size, along with the associated terminology. The International Tolerance grade establishes the magnitude of the tolerance zone (i.e., the amount of variation in part size that is allowed) for both internal (hole) and external (shaft) dimensions. It is expressed in grade numbers (e.g., IT7), with smaller grade numbers indicating a smaller tolerance zone.

The fundamental deviation establishes the position of the tolerance zone with respect to the basic size. It is expressed by "tolerance position

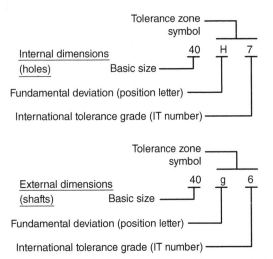

Figure 9-34 Metric-toleranced size with associated terminology

letters"; upper-case letters (e.g., H) being used for hole dimensions, and lower-case letters (e.g., h) used for shaft dimensions.

A tolerance symbol (e.g., H7) is formed by combining the IT grade number and the tolerance position letter. The tolerance symbol identifies the actual maximum and minimum limits of the part. Toleranced sizes (e.g., 40H7) are determined by the basic size followed by a tolerance symbol.

A fit (e.g., 40 H8/f7) between mated parts is indicated by the basic size common to both components, followed by a symbol corresponding to each component, with the symbol for the internal (hole) part preceding the symbol for the external (shaft) part. Figure 9-35 shows a fit as designated using B4.2.

Standard, or preferred, sizes of round metal parts should be used whenever possible. Table 9-1

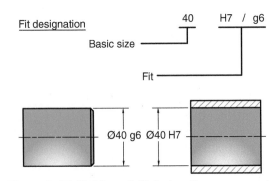

Figure 9-35 Metric-unit fit designation

shows these preferred basic sizes. The basic size of mating parts should, when possible, be chosen from the first choice sizes listed in this table.

As with English fits, metric preferred fits are based on either hole or shaft parts. Preferred fits to relative scale are shown in Figure 9-36 for hole basis and in Figure 9-37 for shaft basis fits. Hole basis fits have a fundamental deviation of "H" on the hole, whereas shaft basis fits have a fundamental deviation of "h" on the shaft. Hole basis is the preferred system in most cases, but shaft basis should be used when a common shaft mates with different holes.

Figure 9-38 provides a description of hole basis and shaft basis fits that have the same relative fit condition. The limits and fits of clearance, transition, and interference fits are provided in the table appearing in Appendix A. Appendix B uses the fit types described in Figure 9-38 and the preferred sizes appearing in Table 9-1.

Table 9-1 **Preferred basic sizes**

Basic Size (mm)		Basic Size (mm)		Basic Size (mm)	
First Choice	Second Choice	First Choice	Second Choice	First Choice	Second Choice
1		1		100	
	1.1		11		110
1.2		12		120	
	1.4		14		140
1.6		16		160	
	1.8		18		180
2		20		200	
	2.2		22		220
2.5		25		250	
	2.8		28		280
3		30		300	
	3.5		35		350
4		40		400	
	4.5		45		450
5		50		500	
	5.5		55		550
6		60		600	
	7		70		700
8		80		800	
	9		90		900
				100	

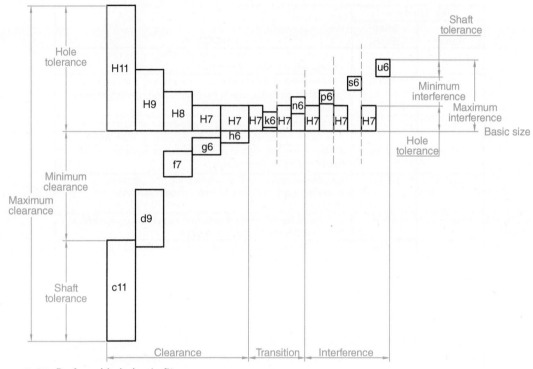

Figure 9-36 Preferred hole basis fits

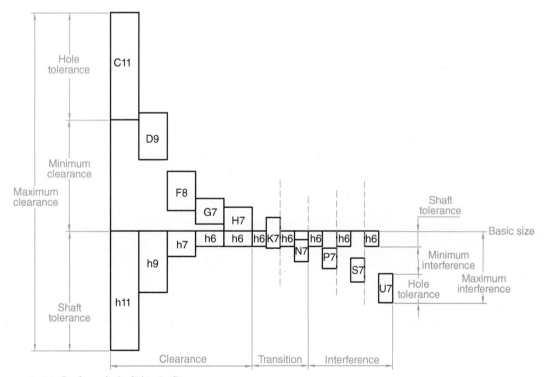

Figure 9-37 Preferred shaft basis fits

ISO Symbol		Description
Hole Basis	**Shaft Basis**	**Description**
H11/c11	C11/h11	*Loose-running fit* for wide commercial tolerances or allowances on external members.
H9/d9	D9/h9	*Free-running fit* not for use where accuracy is essential, but good for large temperature variations, high running speeds, or heavy journal pressures.
H8/f7	F8/h7	*Close-running fit* for running on accurate machines and for accurate location at moderate speeds and journal pressures.
H7/g6	G7/h6	*Sliding fit* not intended to run freely, but to move and turn freely and locate accurately.
H7/h6	H7/h6	*Locational clearance fit* provides snug fit for locating stationary parts, but can be freely assembled and disassembled
H7/k6	K7/h6	*Locational transition fit* for accurate location, a compromise between clearance and interference.
H7/n6	N7/h6	*Locational transition fit* for more accurate location where greater interference is permissible.
H7/p6	P7/h6	*Locational interference fit* for parts requiring rigidity and alignment with prime accuracy of location but without special bore pressure requirements.
H7/s6	S7/h6	*Medium drive fit* for ordinary steel parts or shrink fits on light sections, the tightest fit usable with cast iron.
H7/u6	U7/h6	*Force fit* suitable for parts that can be highly stressed or for shrink fits where the heavy pressing forces required are impractical.

(Left margin brackets: Clearance fits, Transition fits, Interference fits. Right margin brackets: More clearance, More interference)

Figure 9-38 Description of preferred fits

Step-by-step tolerance calculation using metric-unit fit tables, hole basis (see Figure 9-39)

Given:

- Basic size is 50
- Fit type is free running, H9/d9
- Calculation method is hole basis

1. Entering the table in Appendix B, Preferred Hole Basis Clearance Fits, with a basic size of 50 and a free-running H9/d9 fit type, the following information is extracted:

BASIC SIZE		FREE RUNNING		
		Hole H9	Shaft d9	Fit
50	MAX	50.062	49.920	0.204
	MIN	50.000	49.858	0.080

2. Tolerance limits on the hole:

Upper limit = 50.062

Lower limit = 50.000 (= basic size)

3. Tolerance limits on the shaft:
   Upper limit = 49.920
   Lower limit = 49.858

4. Allowance = Hole min − Shaft max
   $$= 50.000 - 49.920 = +0.080$$

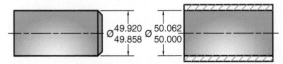

Figure 9-39 Step-by-step tolerance calculation: free-running, hole basis, preferred metric-unit fit tables

---

Step-by-step tolerance calculation using metric-unit fit tables, hole basis (see Figure 9-40)

Given:

- Basic size is 30
- Fit type is medium drive, S7/h6
- Calculation method is shaft basis

1. Entering the table in Appendix B, Preferred Shaft Basis Interference Fits, with a basic size of 30 and a medium drive S7/h6 fit type, the following information is extracted:

BASIC SIZE		MEDIUM DRIVE		
		Hole S7	Shaft h6	Fit
30	MAX	29.973	30.000	−0.014
	MIN	29.952	29.987	−0.048

2. Tolerance limits on the hole:
   Upper limit = 29.973
   Lower limit − 29.952

3. Tolerance limits on the shaft:
   Upper limit = 30.000 (= basic size)
   Lower limit = 29.987

4. Allowance = Hole min − Shaft max = 29.952 − 30.000 = − 0.048

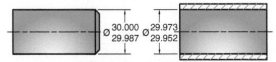

Figure 9-40 Step-by-step tolerance calculation: medium drive, shaft basis, preferred metric-unit fit tables

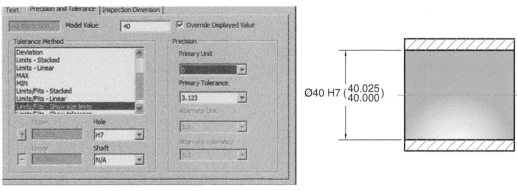

Figure 9-41 Tolerance specification using CAD (Courtesy of Autodesk, Inc.)

Tolerancing in CAD

CAD programs typically provide a variety of tools for the specification of toleranced dimensions within a drawing. The dialog box shown on the left in Figure 9-41, for example, allows the designer to display tolerances in several ways. In Figure 9-41 the dialog box settings on the left result in the toleranced dimension shown on the right.

▌ QUESTIONS

1. Select the dimensions A-Z, which represent 〔SS〕 good dimensioning (Figures P9-1 and P9-2). When finished, the object should be fully dimensioned. In cases where two dimensions locate or size the same feature, accept the one that best meets good dimensioning practice guidelines, while rejecting the other.

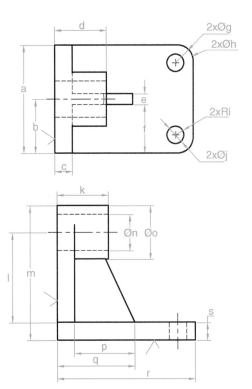

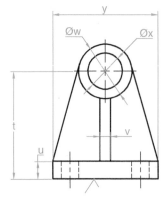

Figure P9-1

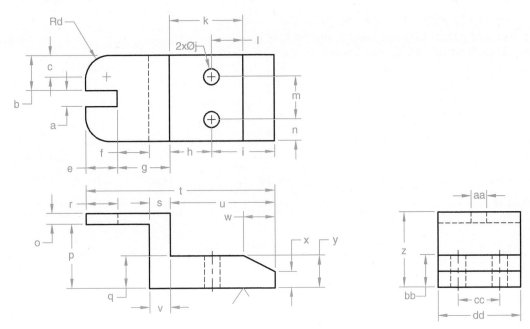

Figure P9-2

TOLERANCING

2. Given the method of calculation, basic size, fit (clearance, interference, or transition), tolerances, and the allowance, determine the limits on the hole and shaft.

	a.	b.	c.	d.	e.
Basic Size	10.000	1.5000	0.5000	30.000	16.000
Class of Fit	Clearance	Interference	Clearance	Interference	Transition
Method of Calculation	Shaft	Hole	Hole	Shaft	Hole
Limits of Hole Size					
Limits of Shaft Size					
Tolerance on Hole	0.200	0.0015	0.0010	0.016	0.025
Tolerance on Shaft	0.100	0.0014	0.0007	0.014	0.024
Allowance	0.150	−0.0030	0.0017	−0.032	−0.016

3. Using the appropriate limit dimensioning tables and either the basic hole or the basic shaft method of calculation, complete the table.

	a.	b.	c.	d.
Basic Size	2.5000	1.5000	4.2500	10.0000
Class of Fit	RC1	LT4	LN1	FN2
Method of Calculation	Basic Hole	Basic Shaft	Basic Hole	Basic Hole
Limits of Hole Size				
Limits of Shaft Size				
Tolerance on Hole				
Tolerance on Shaft				
Allowance				

	e.	f.	g.	h.
Basic Size				3.7500
Class of Fit	RC7	LC8	LN3	FN5
Method of Calculation	Basic Hole	Basic Hole	Basic Shaft	
Limits of Hole Size	0.7520	0.5028		3.7522
	0.7500	0.5000		3.7500
Limits of Shaft Size	0.7475		5.0000	
	0.7463		4.9990	
Tolerance on Hole				
Tolerance on Shaft		0.0016		0.0014
Allowance			−0.0035	

4. Using the appropriate limit dimensioning tables and either the basic hole or the basic shaft method of calculation, complete the table.

	a.	b.	c.	d.
Basic Size	10	250	25	4
Class of Fit	Loose Running H11/c11	Close Running H8/f 7	Locational Clearance H7/h6	Medium Drive H7/s6
Method of Calculation	Preferred Hole Basis	Preferred Hole Basis	Preferred Shaft Basis	Preferred Hole Basis
Limits of Hole Size				
Limits of Shaft Size				
Tolerance on Hole				
Tolerance on Shaft				
Allowance				

	e.	f.	g.	h.
Basic Size	6			50
Class of Fit	Sliding H7/g6	Locational Transition N7/h6	Locational Interference H7/p6	Force U7/h6
Method of Calculation		Preferred Shaft Basis	Preferred Hole Basis	
Limits of Hole Size	6.012		80.030	
	6.000		80.000	
Limits of Shaft Size		160.000		50.000
		159.975		49.984
Tolerance on Hole		0.040		
Tolerance on Shaft	0.008		0.019	0.016
Allowance		−0.052		

Construct all extension lines, dimension lines, leaders, and arrowheads that are necessary to fully dimension the objects.

Ⓐ

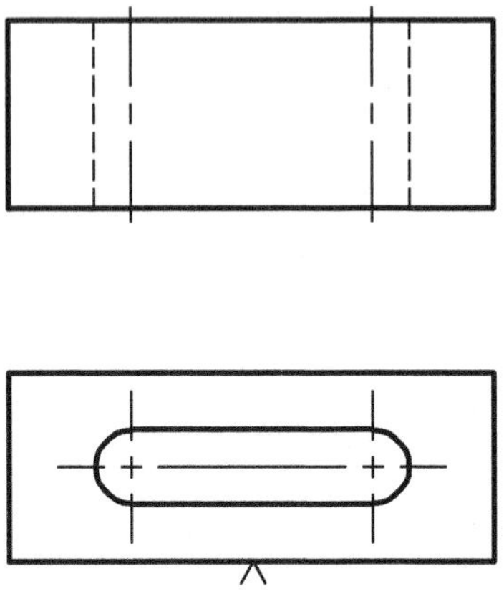

Ⓑ

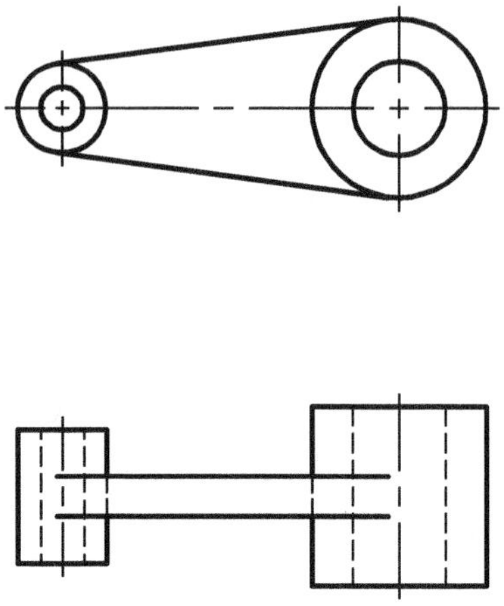

Ⓒ

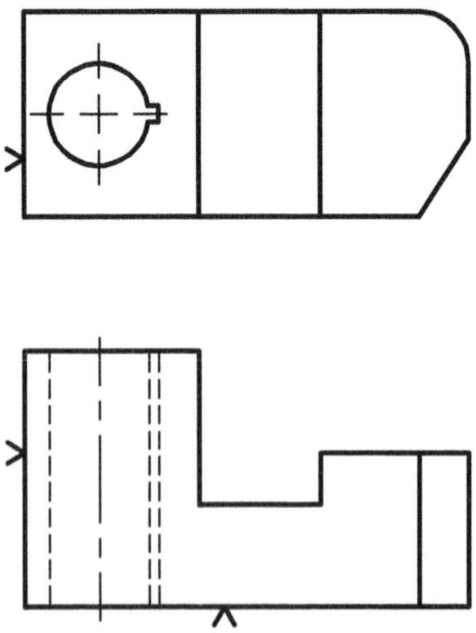

Ⓓ

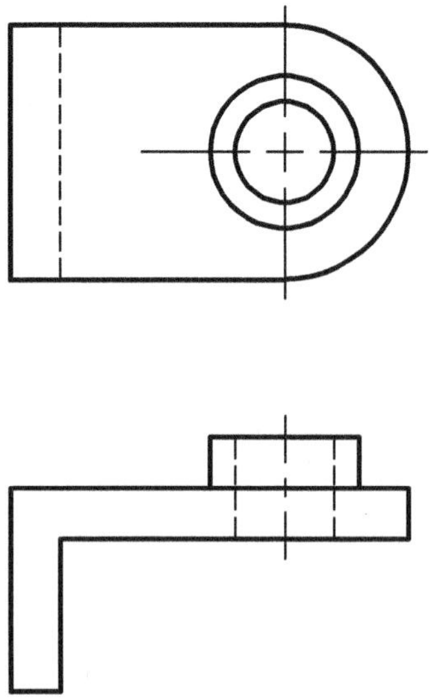

| Drawing 9-1 | Name | Date |

Construct all extension lines, dimension lines, leaders, and arrowheads that are necessary to fully dimension the objects.

10

CAD: SOLID MODELING

▌ INTRODUCTION

Computer-Aided Design

Computer-aided design (CAD) is a technology concerned with the use of computer-based tools employed by engineers, architects, and other design professionals in their design activities. CAD is used in the design of such artifacts as consumer products, tools, machinery, buildings, and infrastructure. Current applications include 2D drafting, solid modeling (both parametric and direct), surface modeling (both NURBS and freeform) and building information modeling (BIM). Although 2D drawing remains popular, CAD capabilities have developed well beyond the ability to generate drawings. Today's model-based, object-oriented CAD programs provide designers, engineers, and architects with the ability to digitally capture a product's or building's definition, and to integrate this definition into the knowledge base of the entire enterprise. The resulting digital product (or building) model can then be used to simulate the behavior of the design under varying conditions.

Categories of CAD Systems

In this section, 2D CAD, solid, parametric, direct, surface, and freeform modeling are briefly discussed. Later in this chapter parametric solid modeling is discussed in greater detail. The chapter concludes with a discussion of cloud-based CAD. NURBS surface and freeform modeling are discussed in Chapter 11.

COMPUTER-AIDED DRAWING

In 1982 Autodesk launched AutoCAD®, the first commercially successful 2D vector-based drafting program. Vector graphics employs geometric elements like points, lines, curves, and polygons to represent images. Since these elements are defined mathematically, they can be stored in a database and later manipulated (e.g., copied, moved, rotated, scaled, arrayed). 2D CAD is used by civil engineers, architects, land developers, interior designers, and other design professionals. The principal output of 2D CAD programs are the drawings themselves, rather than a model from which drawings can be extracted. Figure 10-1 shows a 2D CAD drawing created by a student in a first-year engineering graphics course. Figures 1-13, 1-15, and 1-16 in Chapter 1 and Figures 12-1, 12-2, and 12-4 in Chapter 12 are professionally developed 2D CAD drawings.

Many computer-aided drawing programs also have 3D capabilities. A 3D *wireframe* drawing, like the one shown in Figure 10-2, is created using the same geometric elements (e.g., line, circle, arc, polyline) as those used to create a 2D drawing. Although they contain only edge and vertex information, wireframe drawings provide

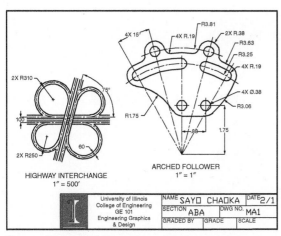

Figure 10-1 2D CAD student drawing (Courtesy of Sayo Chaoka)

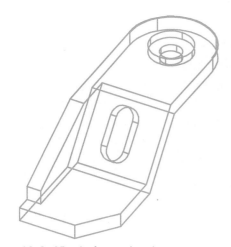

Figure 10-2 3D wireframe drawing

a relatively simple and fast means to convey the three-dimensional form of an object.

SOLID MODELING

Figure 10-3 shows an example of a solid model created in AutoCAD® by a student in a first-year engineering graphics course. Note that the solid model is composed of recognizable solid features such as boxes and cylinders.

Solid models are built up from primitives and sweeps. *Primitives* are solid entities similar to the 2D line, circle, and arc entities used in wireframe modeling. Examples of primitives include

Figure 10-3 Solid model created by a student

box, sphere, cylinder, cone, wedge, and torus (see Figure 10-4).

Sweep operations include extruding, revolving, and sweeping. An *extrusion*[1] is a modeling technique that creates a 3D shape by translating a 2D closed profile along a linear path. A *revolution* is a modeling process that creates a 3D shape by revolving a 2D closed profile about an axis. A *sweep* is created by moving a 2D closed profile along a path. Examples of extruded, revolved, and swept solids are shown in Figure 10-5. A *profile* is simply a 2D outline or shape that does not cross back over on itself. Most solid modeling operations also require that the profile be closed. Note that all primitives can be created with a single sweeping operation.

The Boolean operations union, subtraction, and intersection are used to combine solid primitives and sweeps. These combined solids are called *composites*. The results of the different Boolean operations for two different solids are shown in Figure 10-6.

The two most popular representation schemes, or data structures, employed for describing solid models are *constructive solid geometry (CSG)*[2] and *boundary representation (B-rep)*.[3] The CSG representation stores the model data in terms of the solid primitives and the Boolean operations that are used to combine them. The model history is stored in a *binary tree structure*,[4] with the

[1] In manufacturing, an extrusion is a process that involves forcing a material through a shaped opening.

[2] See Bezier Award sidebar in Chapter 11.

[3] See Bezier Award sidebar in Chapter 11.

[4] https://en.wikipedia.org/wiki/Binary_tree.

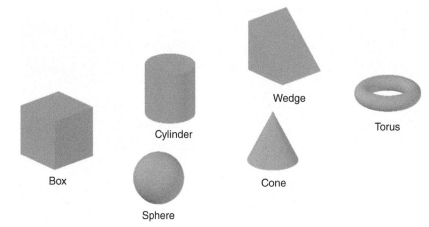

Figure 10-4 Solid primitives

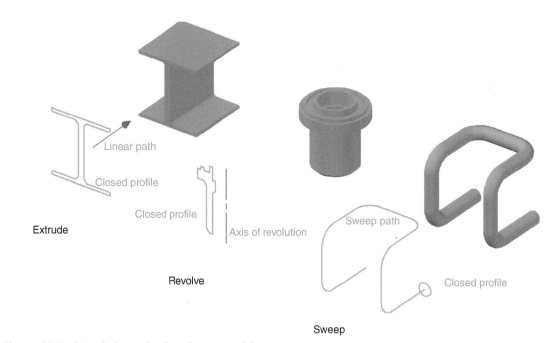

Figure 10-5 Extruded, revolved, and swept solids

Figure 10-6 Boolean operations

solid primitives serving as *leaves* and the Boolean operators as *branches*. Figure 10-7 shows an example of a CSG tree structure.

Because they organize data solely in terms of primitives and Boolean operations, CSG data structures are very simple and compact. On the other hand, because data are stored only in terms of primitives and Boolean operations, information regarding the faces, edges, and vertices of a solid is not readily available. This is a significant shortcoming because this information is frequently required in working with solid models. For example, when rendering a solid object, information regarding the surfaces of the object is required. Surface information is also required in preparation for machining a part, in order to calculate numerical control tool paths.

Boundary representation, on the other hand, stores the boundaries of the solid (e.g., vertices, edges, faces) in the database, along with information regarding how these entities are connected. Figure 10-8 shows a tetrahedron with labeled faces, edges, and vertices. The figure also includes tables that show how the geometry is stored. In B-rep, the outside of a face is determined by the order of its bounding edges. When a face is viewed from the outside, the bounding edges of the face are stored in the database in

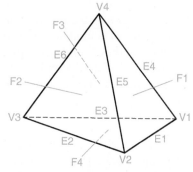

Face Table	
Face	Edges
F1	E1, E4, E5
F2	E2, E5, E6
F3	E3, E6, E4
F4	E1, E2, E3

Edge Table	
Edge	Vertices
E1	V1, V2
E2	V2, V3
E3	V3, V1
E4	V1, V4
E5	V2, V4
E6	V3, V4

Vertex Table	
Vertex	Coordinates
V1	x1, y1, z1
V2	x2, y2, z2
V3	x3, y3, z3
V4	x4, y4, z4

Figure 10-8 Boundary representation (B-rep) of a simple polyhedron

a counterclockwise direction around the face. For example, the edges of face F1 are ordered as follows: E1, E4, and E5. Because the face, edge, and vertex information of a solid is directly available when using boundary representation, shading, hidden-line removal, and other types of display algorithms can be directly evaluated.

Most commercial solid modelers employ boundary representation or a combination of both CSG and B-rep for storing models. Adopting this hybrid approach retains the strengths of both systems. The drawback to storing models using both representations is, predictably, increased file size.

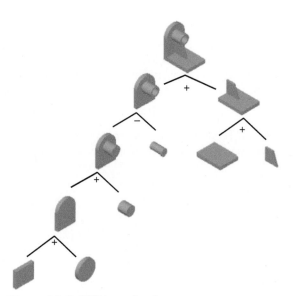

Figure 10-7 CSG tree structure

With solid models like the ones described thus far, it is possible to determine the mass and other material properties of a modeled part. It is also possible to display these solids in various ways, including wireframe, with hidden lines removed, shaded, and rendered. There are, however, a number of capabilities that are not available. For instance:

- It is difficult to change geometry. Existing geometry can be added to, subtracted from, or intersected with, or the model can be deleted entirely and started over. It is not, however, easy to modify the dimensions of an existing solid.

- Features contain only geometric information; for example, a hole is created by subtracting a cylinder. Associated manufacturing and other information is not included in the model database.

- Dimensions are nonparametric; they are added after the primitive is created. The dimensions are not associative with the geometry; they do not *drive* the geometry.

- The model history is unavailable; that is, the order of operations is unavailable to the operator.

- There is no true assembly modeling environment; only parts can be modeled. Without an assembly environment, the motion of moving parts cannot be investigated, and interferences between parts cannot be detected.

In order to emphasize these shortcomings, the models discussed in this section are sometimes referred to in industry as *dumb* solids.

Topology

A solid consists of both geometric and topological data. Geometry refers to the actual dimensions of the solid, whereas topology refers to how the different geometric elements (i.e., face, edge, and vertex) of the solid are connected. Without both geometric and topological information, a solid model database is incomplete and ambiguous.

Topology is a branch of mathematics that is concerned with the properties of certain shapes that are preserved under transformations or deformations that stretch, compress, bend, or twist the shape, but without tearing, puncturing, or introducing self-intersections. Some geometric problems do not depend upon the exact shape of the objects involved, but rather on how they are put together. Topology has to do with dimensional continuity and connectivity.

Topology is sometimes referred to as "rubber sheet geometry." Two objects are considered to be *topologically equivalent* if they can be continuously deformed into one another using operations like stretching, but not tearing. For example, an ellipse is topologically equivalent to a circle. A rectangle—imagine a rubber band stretched around four square pegs—is also topologically equivalent to a circle. In three dimensions, an ellipsoid is topologically equivalent to a sphere. A cube (or any parallelepiped) is equivalent to a sphere.

Euler's Polyhedron Formula

A *polyhedron* is a solid bounded by a set of flat polygons. A *simple polyhedron* is a polyhedron that is topologically equivalent to a sphere; if inflated it would take the shape of a sphere. Euler's formula applies to simple polyhedra, and states that:

$$V - E + F = 2$$

Where V, E, and F are the number of vertices, edges, and faces, respectively. Figure 10-9 shows some examples:

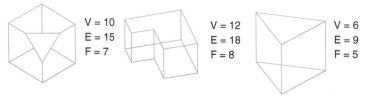

Figure 10-9 Simple polyhedra with V, E, and F indicated

Nonsimple polyhedra are topologically equivalents of any solid object with holes in it. For example, a rectangular parallelepiped with a through hole is topologically equivalent to a torus. Euler's polyhedron formula has been extended to address nonsimple polyhedra but is beyond the scope of this text.

Manifold versus Nonmanifold Solids

In mathematics a *manifold* is defined as a topological space that is connected and locally Euclidian, but for our purposes a manifold solid is one that has a closed volume, that is, one that is manufacturable.[5] Nonmanifold models, on the other hand, are not manufacturable. Figure 10-10 shows some examples of nonmanifold geometries.

Note that nonmanifold models allow for the combination of one-, two-, and three-dimensional entities (i.e., lines, surfaces, and solids). Early solid modeling systems could only represent manifold solids. At the time, this was viewed as a positive feature because all models would then be guaranteed to be manufacturable. For design and analysis purposes, however, it is useful to be able to create abstract (i.e., nonmanifold) models. Today's solid modeling systems can represent both manifold and nonmanifold geometries. Note that nonmanifold geometries are a common source of errors when 3D printing.

Seven Bridges of Konigsberg

In 1736, Leonhard Euler laid the foundations of *graph theory*, and prefigured the idea of topology, with his seven bridges of Konigsberg problem. Figure 10-11*a* shows a map of Konigsberg and its seven bridges, Figure 10-11*b* is an abstraction of the map, and Figure 10-11*c* is a *graph* of the map. In the graph, each vertex represents a different land mass, while each edge represents a different bridge. From a

[5]For every point on the boundary of a manifold solid, a small enough sphere around the point can be divided into two pieces, one inside and one outside the solid.

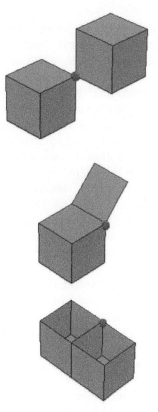

Figure 10-10 Nonmanifold geometries

topological perspective, the metrical shape of the graph on the right does not matter, only the connectivity information (4 vertices, 7 edges, connected as shown in Figure 10-11*c*) is important. Another graph with a different dimensional layout is *topologically equivalent* to Figure 10-11*c*, as long as the connectivity is the same.

In Chapter 11, we will see that the principal difference between NURBS and both T-splines and subdivision modeling is their topology.

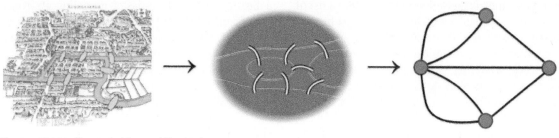

Figure 10-11 Seven bridges of Konigsberg

3D parametric solid modeling addresses all of the shortcomings of dumb solids listed in the previous section. That is, when a parametric modeler is used, geometry is easy to modify, features contain manufacturing as well as geometric information, dimensions are parametric, the model history is available to the operator, and there is an assembly environment. Parametric modeling is so significant that it will be discussed in detail later in this chapter.

The four principal **MCAD (mechanical CAD)** companies—Autodesk, Dassault Systemes, Parametric Technology Corporation (PTC), and Siemens—all include two parametric modelers in their product lines. These products include, from Autodesk, Inventor and Fusion 360, from Dassault, SolidWorks and CATIA, from PTC, Creo and Onshape, and from Siemens, NX and Solid Edge. Onshape is the first pure **Software-as-a-Service (SaaS)**, cloud-native CAD solution. Fusion is also cloud-based, although it does require the user to install the application locally. Cloud-based CAD is discussed at the end of this chapter. All of the other applications run exclusively on the Windows platform.

DIRECT MODELING

History-based parametric modeling has dominated the MCAD market since the 1980s. This technology has been successful in part because its history tree brings order and intelligence to the model-building process. Parametric modeling is especially good at building complex models, large assemblies, and part families. It is characterized by a feature tree and parent–child relationships between features, where recently added features depend on earlier ones. History-based parametric modeling functions like a list of instructions that, like software code, must be successfully executed to produce a result. Changes to early instructions can cause later ones to fail.

Direct modeling, sometimes called explicit modeling, is an alternative technique that has been around for at least as long as history-based modeling. In the past decade, direct modeling has experienced a resurgence in popularity, thanks to advances in both solid modeling technology and high-performance computing. Unlike parametric modelers, direct modelers do not use instructions; rather, they offer direct interaction with the geometry. Pure direct modelers are highly interactive tools that enable users to model without worrying about model history.

Owing to the complexity of its feature tree, history-based model construction requires careful planning to ensure that the model behaves as intended. Without this forethought, other users may find it difficult to edit the history-based model. Users, for this reason, must be well trained to be successful. Conceptual modeling is also somewhat constrained by the rigidity of the history-based structure, as is the ability to collaborate with computer-aided manufacturing (CAM), computer-aided engineering (CAE), and other CAD systems.

In contrast, direct modelers are relatively easy to learn and use. History-free, explicit modelers appeal to non-CAD specialists and to designers and engineers who are new to CAD. For these users, the power of traditional parametric tools may be unnecessary. Direct modelers are well suited for design conceptualization and for preparing CAD models for analysis. Because of their ability to work with multiple CAD systems, direct modelers provide a solution to industry interoperability problems.

Most standalone direct modelers have been acquired by other CAD/CAM/CAE companies in recent years. One exception is KeyCreator from Kubotek3D. CoCreate, originally a standalone explicit modeler, was purchased by the Parametric Technology Corporation in 2007.[6] Another standalone direct modeler, SpaceClaim, was purchased by Ansys in 2014. Today all of the major MCAD software companies have some sort of direct modeling presence, including some hybrid approaches. These include Siemens Synchronous Technology, a history-free, feature-based modeling approach used in both NX and Solid Edge. Another hybrid, Autodesk's Fusion technology, unites direct and parametric workflows within a single digital model. SolidWorks does not offer a separate product, but it regularly adds direct modeling functionality to its parametric tool.

[6]After this acquisition, PTC changed the name of their flagship product from Pro/Engineer to Creo Elements/Pro, and now simply Creo.

Figure 10-12 provides a glimpse of the direct modeling workflow. Starting with a block, step 1 shows the Push/Pull tool used to fillet an edge. In step 2, the Move/Rotate tool is used to move a face. In steps 3 and 4, the same Move/Rotate tool is used to chamfer the same face, first about one axis and then about another. Finally, in step 5, the Push/Pull tool is again used to fillet an edge.

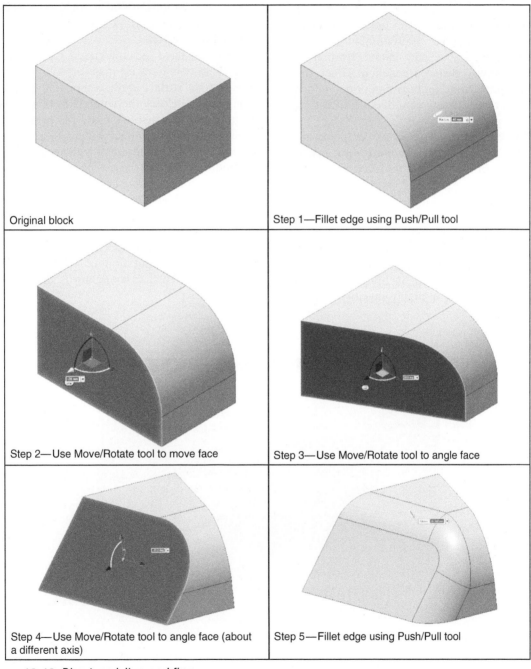

Original block	Step 1—Fillet edge using Push/Pull tool
Step 2—Use Move/Rotate tool to move face	Step 3—Use Move/Rotate tool to angle face
Step 4—Use Move/Rotate tool to angle face (about a different axis)	Step 5—Fillet edge using Push/Pull tool

Figure 10-12 Direct modeling workflow

Surface modeling is used extensively in a number of industries requiring smooth, organic shapes. Surface modeling applications are used by industrial designers, design engineers, and architects, among others. Industrial sectors that employ surface modeling software include automotive, aerospace, ship building, consumer product design, and entertainment animation. Best-in-class design and manufacturing companies employ surface modeling to set themselves apart from their competition, allowing them to create attractive products that customers want to own.

Most computer graphics and computer-aided design applications use *nonuniform rational B-splines,* or *NURBS*, to model curves and surfaces. NURBS provide a mathematically precise representation of both standard analytic *curves* such as straight lines, circles, and other conic sections, as well as freeform shapes. Numerically stable and accurate computer algorithms are available to quickly evaluate NURBS. Despite their technical complexity, NURBS modeling is highly intuitive, allowing the user to operate without knowledge of the underlying math.

Commercial NURBS-based *computer-aided industrial design* software includes standalone applications like Rhinoceros, as well as other products owned by the major CAD companies. These applications/modules include Autodesk Alias, Autodesk Fusion, CATIA Design/Styling, and NX for Design.

Surfaces have no thickness. Unlike solids, surface models have no mass or volume, unless a watertight surface quilt is modeled so that it completely encloses the volume. In this case the surface becomes a solid, and a 3D print can be produced from the model. Figure 10-13 shows a rendered surface model of a Volkswagen Beetle created by an industrial design and engineering student team.

The term *freeform modeling* is used here to describe CAD tools based on either subdivision modeling or T-splines. Freeform modeling is an outgrowth of surface modeling that improves upon some of the limitations of NURBS. NURBS and freeform surface modeling is discussed in detail in Chapter 11.

Figure 10-13 Volkswagen Beetle rendered surface model (Courtesy of Ian Bradley and Yunjin Kim)

Building Information Modeling (BIM) is an intelligent 3D model–based process that provides insight to help in the planning, designing, and management of buildings and infrastructure. A *building model* is a digital database of a building that contains information about its objects. This information might include geometry (defined by parametric rules), its performance, its planning, its construction, and its operation. Building models are the next-generation replacement for construction or architectural drawings. BIM models contain coordinated, consistent, and computable information about a building project. Shown below are examples of student design team projects. Figures 10-14, 10-15, and 10-16 are images taken from campus building projects at the University of Illinois Urbana-Champaign. Figures 10-17 and 10-18 are images for residential neighborhoods.

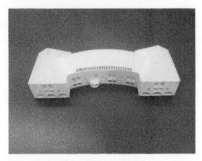

Figure 10-14 UIUC Grainger Engineering Library, rendered image on left, 3D print on right

Figure 10-15 UIUC Armory Building by student design team (Courtesy of Lauren Schissler, Robert Wiggins, Kweku Osei, and Ariana Nevarez)

Figure 10-16 UIUC ACES Library by student design team (Courtesy of Dale Robbennolt, Bryar Lindenmeyer, Max McAvoy, and Gabriel Piechnik)

Figure 10-17 Example of neighborhood BIM by student design team (Courtesy of Maddie Dearborn, Aaron Morita, Ryan Hammond, Ahmed Baig, and Jacob Quintana, Spring 2021)

Figure 10-18 Example of neighborhood BIM by student design team (Courtesy of Michael Cano, Felipe Maganhoto, Sedona Ivy)

CAD Viewing and Display

CAD viewing tools like pan, zoom, and orbit are used to control the position from which an object is viewed. When using a viewing tool, it is important to keep in mind that it is the viewer that moves, not the object or scene.

The term ***display*** refers to the way in which objects appear on the computer screen. CAD display types include wireframe, hidden-line/hidden-surface, shaded, and rendered, as seen in Figure 10-19. Using wireframe display, model edges and contours are visible. Wireframe is simpler and faster than either hidden or shaded display; it is frequently used when editing, because all model edges and vertices are available for selection. Solid, surface, and wireframe models can all be displayed in wireframe mode.

Hidden-line or hidden-surface display refers to a class of computer graphics algorithms used to determine which edges, surfaces, and volumes in a model are visible from a particular viewpoint. Object faces are opaque when displayed using hidden-line removal. For this reason, any portion of a face or edge that is obscured by another face from a given viewpoint is not visible. Both solid and surface models can be displayed using hidden-line removal.

Shaded display is a technique that applies flat colors to visible surfaces. Both solid and surface models can be displayed in shaded mode.

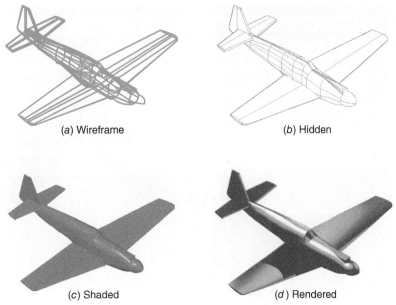

(a) Wireframe

(b) Hidden

(c) Shaded

(d) Rendered

Figure 10-19 Wireframe, hidden, shaded, and rendered displays (Courtesy of Tim Lingner)

A rendered image is a snapshot of a 3D model, to which color, texture, and lighting have been applied to create a single photorealistic image. Both surface and solid models can be rendered. Wireframe, hidden, and shaded displays can all be orbited in real time, but rendered views are still images, unless an entire scene is rendered frame-by-frame to create an animation.

■ PARAMETRIC MODELING

Introduction

Parametric solid-modeling software has the ability to reflect the way in which modern manufacturing companies develop their products. Owing to its object-oriented, parametric nature, parametric modeling has expanded the traditional role of CAD beyond geometry creation and into the realm of product realization.

Unlike primitives-based solid modelers such as the one found in early versions of AutoCAD®, a parametric modeler provides the operator with multiple work environments, each with its own file type. Work environments common to most parametric modelers include part, assembly,

and drawing.[7] The assembly environment, for example, allows the user to combine components in order to form a virtual product model. Figure 10-20 shows a typical assembly modeling environment.

Most commercial products may be thought of as an assembly composed of different components. A moderately complex product like a bicycle will contain multiple subassemblies as well as individual parts. The rear derailleur, for example, is a subassembly containing many different parts. This natural product decomposition hierarchy is duplicated virtually in the assembly environment tree structure. Figure 10-21 shows the assembly structure for a commercial ball valve.

If the actual product has moving parts, it is possible to constrain the virtual assembly in order to simulate the motion of the part(s). The virtual product model can also be used to identify any static interference between parts, and to determine the range of allowable motion of the product's moving parts.

[7]Examples of other specialized work environments include sheet metal parts and welded assemblies.

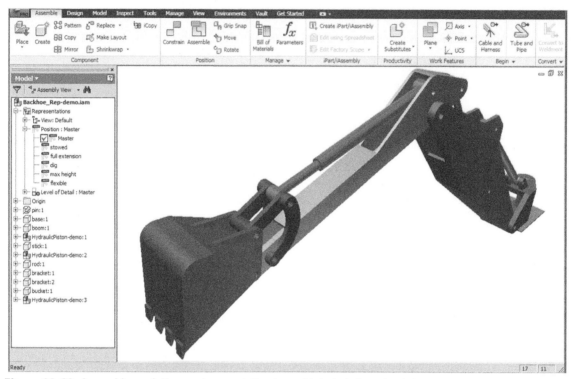

Figure 10-20 Assembly modeling environment (Courtesy of Autodesk, Inc.; Jacob Borgerson)

Within the part environment, this ability to mirror reality continues. Parts are modeled using feature-based techniques, with names like extrude, hole, draft, rib, shell, fillet, and chamfer. These parametric modeling features, as their naming suggests, capture manufacturing as well as geometric data.

Geometric constraints are used to control sketch geometry by limiting the number of necessary dimensions. Note that these geometric constraints have names like parallel, perpendicular, concentric, collinear, tangent, equal, and symmetric, which reflect the precise language employed by a machinist when fabricating parts.

When creating solid model features, designers often use work or construction geometry (planes, axes, and points) to locate the feature. These work features are analogous to the reference datums employed in manufacturing to fabricate parts.

In parametric modeling, dimensions are associatively linked to the model geometry that they describe. If a dimension changes, so does the associated geometry. This gives the user considerable

latitude to explore design alternatives. In addition, the parametric basis of model dimensions enables the designer to build intelligence into the model. Using equations, a parametric dimension can be linked to other parameters of the same feature, different part features, and even different parts within an assembly. This flexibility opens the way to the development of part and even product families, something clearly valued in today's customer-focused, option-driven market.

The concept of associative linking is deeply embedded in parametric modeling system design. Beyond the previously discussed link between a dimension and the size (or location) of a feature, associative links are maintained across the part, drawing, and assembly environments, between an assembly and its associated parts list, between a file's summary information and title block text fields, and so on.

Beyond this, CAD software programs are able to link associatively with the parametric model. As an example, a CAD model can be imported into a finite element analysis (FEA) program

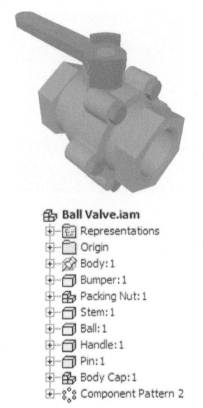

Ball Valve.iam
- ⊞ ▦ Representations
- ⊞ ▢ Origin
- ⊞ ▨ Body: 1
- ⊞ ▨ Bumper: 1
- ⊞ ▨ Packing Nut: 1
- ⊞ ▨ Stem: 1
- ⊞ ▨ Ball: 1
- ⊞ ▨ Handle: 1
- ⊞ ▨ Pin: 1
- ⊞ ▨ Body Cap: 1
- ⊞ ▨ Component Pattern 2

Figure 10-21 Assembly environment tree structure

for stress analysis. If results indicate an unsatisfactory level of stress, the CAD model can be modified, the geometry updated within the FEA program, and the analysis repeated. This ability to use the CAD model for analysis, manufacturing, and other downstream applications is a cornerstone of product lifecycle management; it is discussed in further detail later in this chapter.

Terminology

Parametric solid modeling employs parametric dimensions and geometric constraints to define part features, and to create relationships between these features in order to create intelligent part models. These parts can then be combined to form virtual assembly models. Drawing documents can then be extracted from both part and assembly models.

Features are the basic three-dimensional building blocks for creating parts. Model features available in a parametric modeler mimic actual design and manufacturing features (e.g., counterbore hole,

rib, draft angle). There are essentially two kinds of model features, sketched and placed. Sketched features (e.g., extrude, revolve) require that a 2D sketch be made before the feature can be created. Placed features (e.g., hole, chamfer, fillet, shell, face draft) can be created without sketch geometry. Because features are so important to the part-modeling process, parametric modeling is sometimes called *feature-based modeling*.

A *parameter* is a named quantity whose value can be changed; consider, for example, d0 = 10. Here d0 is the name of the parameter, and 10 is its value. Parametric dimensions control the size and position of features. If the value of a parameter is changed, the feature geometry changes as well. Because of this, we say that parametric models are dimensionally driven. In addition, because a parameter has a name as well as a value, parametric relationships can be formed across features, parts, and assemblies. Consider, for example, d0 = 2*d1. If d1 is 5, then d0 is 10. When d1 changes, so does d0, always being equal to twice the current value of d1.

Constraints are mathematical requirements imposed on the geometry of a 3D model. Regarding part models, there are two kinds of constraints, dimensional and geometric. Dimensional constraints (also called parametric dimensions) place limits on the size or position of a feature. Geometric constraints (e.g., parallelism, tangency, concentricity) place limits on the shape or position of a feature. In some software packages, geometric constraints are called relations. In addition to part constraints, there are also joints and/or assembly constraints. These joints/constraints determine how different parts are positioned with respect to one another. Examples of assembly constraints include mate and insert. Parametric modeling is sometimes referred to as *constraint-based modeling*.

Part Modeling

INTRODUCTION

The part creation process starts with a two-dimensional sketch. This sketch, normally a closed profile, is used to create the first feature, or *base feature*, of the part. The base feature is frequently either an extrusion or a revolved feature.

Additional features are then added until the part is complete. These features can either be based on a sketch, or directly placed on the model without the need to create additional geometry.

New part files contain three mutually perpendicular work planes, three work axes, and a common origin point, similar to that shown in Figure 10-22. One of the work planes is selected and used to sketch the profile for the base feature. The origin of the part file is defined at the common intersection of the three reference planes.

SKETCH MODE

As seen in the previous section, 2D sketches play an important role in the creation of 3D parametric parts. To support this crucial aspect of part creation, a sketch mode, sometimes called a sketcher, is included within the part-modeling environment. Figure 10-23 shows a typical sketching interface.

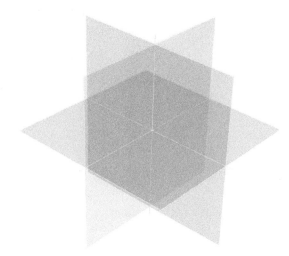

Figure 10-22 Default work geometry

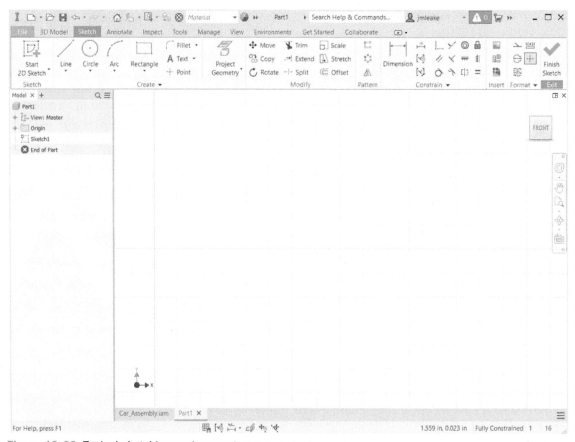

Figure 10-23 Typical sketching environment (Autodesk, Inc.)

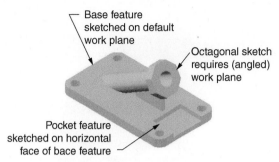

Figure 10-24 Sketch plane categories

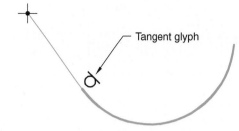

Figure 10-25 Inferred geometric constraints

Prior to entering the sketching environment, a sketch plane must be defined. This sketch plane can be (1) one of the default work planes, (2) a planar face on the existing part geometry, or (3) a new work plane. Figure 10-24 shows examples of these three cases.

Tools found in sketch mode are similar to those found in a 2D CAD program; they include line, circle, arc, trim, offset, and so on.

Depending on how the sketch is made, some geometric constraints may be inferred. Figure 10-25, for example, shows a line being drawn from an arc. When the line appears to be approximately tangent to the arc, a small symbol called a *glyph* appears. If the endpoint of the line is then selected, a tangent constraint is applied between the line and the arc.

After the rough sketch is complete, additional geometric constraints are added. Figure 10-26a shows a recently completed sketch. Figure 10-26b shows the same sketch after concentric, tangent, collinear, and symmetric constraints have been applied.

Once the sketch has been geometrically constrained, parametric dimensions are added in order to fully constrain the sketch. Figure 10-27 shows a fully constrained sketch.

FEATURE CREATION

Upon exiting sketch mode, the feature creation tools become available. The most commonly used sketched features are extrude, revolve, sweep, and loft (blend). Extrude and revolve both require a single sketch, whereas sweep and loft need more than one available sketch.

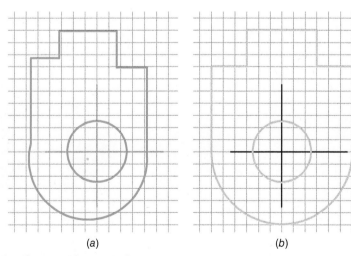

(a)   (b)

Figure 10-26 Additional geometric constraints

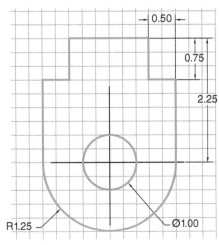

Figure 10-27 A fully constrained sketch

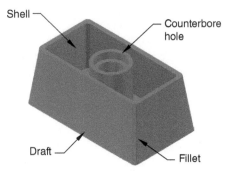

Figure 10-29 Placed features

The base feature geometry is always additive. Subsequent sketched features are added, subtracted, or intersected with the existing geometry, depending on which Boolean operator (join, cut, or intersect) is selected. Figure 10-28 shows the result of joining, cutting, and intersecting an extruded sketch with a base feature.

Placed features require that the user select existing geometry and supply specific parametric inputs; no sketches are needed. Figure 10-29 shows a part with several placed features, including draft, shell, counterbore hole, and fillet.

Work features (also called construction geometry) include work planes, work axes, and work points. Unlike sketched and placed features, work features do not have a direct effect on the model geometry. Rather, work features are employed to make possible the creation of

a subsequent geometric feature. For example, both the fuselage and the wings of the airplane shown in Figure 10-30 were created using a loft feature. A loft requires multiple sketches. These sketches, in turn, require sketch planes. To create each loft, multiple work plane features were created by offsetting from a datum plane. Sketches were then made on each of the work planes, and lofted solids were created that pass through the sketches.

The sidebar that follows provides a simple description of the part creation process.

PART EDITING

Once created, a part feature can be modified at any time, either by editing its **consumed sketch**[8] or by changing inputs and/or parametric values

---

[8]A sketch is said to be consumed once it has been used to create a feature. In some parametric modeling programs, the consumed sketch can be *shared*, which makes it available once again for creating additional features.

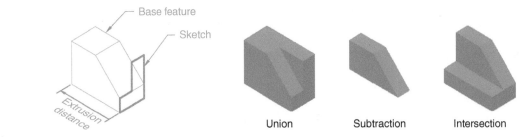

Figure 10-28 Boolean operations on a base feature

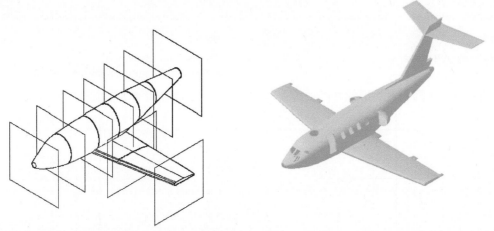

Figure 10-30 Use of work planes to create loft features (Courtesy of Michael Rybalko, Rob Wille, Paul D. Arendt, Stephanie Schachtrup, Dominic Menoni, Daniel J. Weidner)

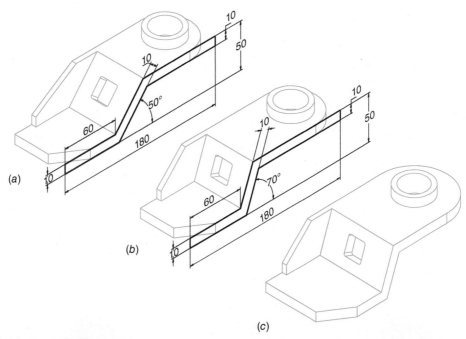

Figure 10-31 Feature editing

made at the time the feature was created. In Figure 10-31a, a part and its base feature sketch are shown. Figure 10-31b shows the same part after the base feature sketch has been modified by changing the angle from 50 to 70 degrees. Figure 10-31c shows the part after the extruded distance of the modified base feature has been increased from 60 to 80.

Step 1

Define a sketch plane.

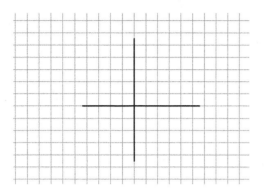

Step 2

Create a rough sketch:
• Some geometric constraints are inferred

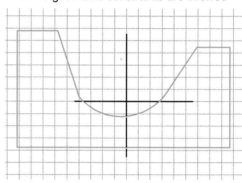

Step 3

Add additional geometric constraints:
• Coincident, tangent, symmetric

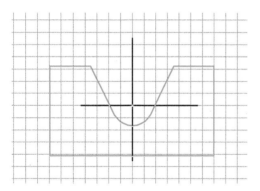

Step 4

Add dimensions to fully constrain sketch.

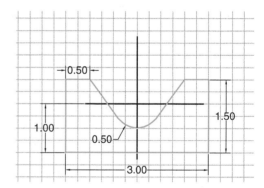

Step 5

Create the base feature.

Step 6

Add additional features:
• Extrude, hole, fillet, chamfer

Figure 10-32 Part creation process

Features may also be ***suppressed***, meaning that the effect of the feature operation on the part is not applied. Figure 10-33 shows the effect of suppressing hole pattern and fillet features on a garlic press handle. It may be desirable to suppress certain features before importing the part geometry into another CAD/CAM/CAE application. For example, small features may be suppressed before importing them into a finite element analysis package, since these features may unnecessarily increase the size of the analysis model.

Access to part features, whether to edit, suppress, rename, or otherwise manage them, is easily gained via the ***feature tree***. Figure 10-34 shows a simple part and its associated feature tree. Note that work features (i.e., Work Plane1) appear in the tree structure, as well as sketched and placed features, and unconsumed sketches. Expanding a sketched feature reveals its consumed sketch. As new features are added, they appear at the

bottom of the feature tree list. In this way the feature tree captures the model history, showing the order in which the model was created. The different parametric modeling packages have various tools that enable a user to review the way in which a model has been constructed.

At least in some modeling packages, a feature's position within the feature tree may be modified by dragging it up or down the tree structure, unless feature dependencies prevent this. These feature dependencies are called ***parent–child relationships***. A parent–child relationship refers to the way in which one feature is derived from, and is consequently dependent on, another feature. This dependency is created when a new child feature is positioned with respect to an older parent feature. It is possible to avoid many dependencies by positioning features with respect to the default work planes, rather than to other features. Note that a child feature must

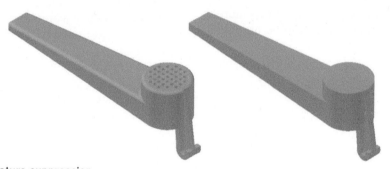

Figure 10-33 Feature suppression

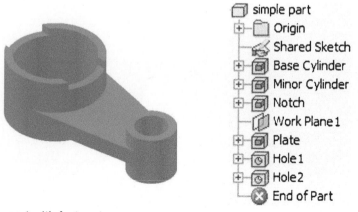

Figure 10-34 Simple part with feature tree

simple part
├─ Origin
├─ Shared Sketch
├─ Base Cylinder
├─ Minor Cylinder
├─ Notch
├─ Work Plane1
├─ Plate
├─ Hole1
├─ Hole2
└─ End of Part

always appear below its parent in the feature tree, and that the base feature will typically be a parent to all other features. If a parent feature is deleted, then any children will be deleted as well. The raised cylinder shown in Figure 10-35a, for example, is the parent to the hole, fillet, and chamfer features shown in Figure 10-35b.

Assembly Modeling

INTRODUCTION

Used extensively in the automotive and aerospace industries, assembly modeling allows the operator to combine components to create a 3D parametric assembly model. Assembly modeling is an essential tool for any work group engaged in the development of a product composed of multiple parts. The benefits of assembly modeling include:

- A fully associated bill of materials (BOM) can be extracted from the assembly.
- The weight, center of gravity, and other inertial properties of the assembly can be automatically tracked.
- Interferences between parts can be detected, avoiding embarrassing and costly design mistakes.
- Kinematic analysis of moving parts within the assembly can be conducted, including determination of the range of motion of parts, and the position, velocity, and acceleration of linkages.

- Exploded and assembly section views can be easily created (see Chapter 12 for further discussion).

The initial task performed within the assembly modeling environment is to bring the components into the assembly. Options for bringing a part into an assembly include (1) directly importing the component, (2) creating the part within the assembly environment, and (3) importing a standard part from an internal part library.

Once in the assembly environment, the components must be correctly positioned relative to one another in the assembly. The following section discuss how this is accomplished.

JOINTS

Solid parts and components are rigid bodies and consequently they have six degrees of freedom (DOF), three in translation and three in rotation. Parametric modelers use either joints or assembly constraints (called mates in SolidWorks) to position and define motion between components. A ***joint*** is a mechanical relationship that defines the relative position and motion between two components in an assembly. While joints focus on what components can do (e.g., rotate about an axis, translate along an axis), assembly constraints or mates lock down the available degrees of freedom to limit motion. Each joint type uses different degrees of freedom to define motion. Standard joints are shown in Table 10-1.

Good practice dictates that at least one component should be grounded. Grounding removes

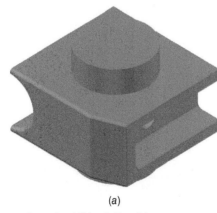 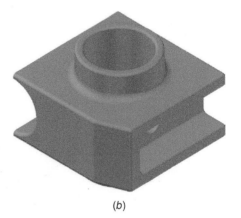

(a)

(b)

Figure 10-35 Parent–child relationships

Table 10-1	**Standard joint types**
Type	**Description**
Rigid	Components are locked together removes all degrees of freedom
Revolute	Component rotates about an axis
Slider	Component translates along an axis
Cylindrical	Component rotates about and translates along an axis
Pin-slot	Component rotates about an axis, and translates along a different axis
Planar	Component translates on a plane, and rotates about the normal to the plane
Ball	Component rotates about all three axes (spherical joint)

all component degrees of freedom and aligns the component coordinate system with the assembly coordinate system. Doing so provides a stable platform for simulating motion within an assembly.

CAD LIBRARIES

Most parametric modeling packages include a library of standard parts that is linked to the assembly environment. Standard parts include threaded fasteners, washers, and O-rings, as well as structural shapes. Figure 10-36 shows the interface for one of these CAD libraries. In addition to these internal libraries, external CAD part libraries can be accessed via the Internet.

Advanced Modeling Strategies

What is the optimal modeling approach, anticipating that the model will need to be modified in the future? This question captures the essence of *design intent*, a term frequently used in conjunction with parametric modeling. Whereas parametric features are easily adjusted, a parametric model built from these features may be either easy, difficult, or nearly impossible to modify. This is a result of the feature dependencies that tend to accumulate as a model is built.

The goal of design intent is to optimize the usability of the model. To achieve this, operators must carefully weigh the consequences of

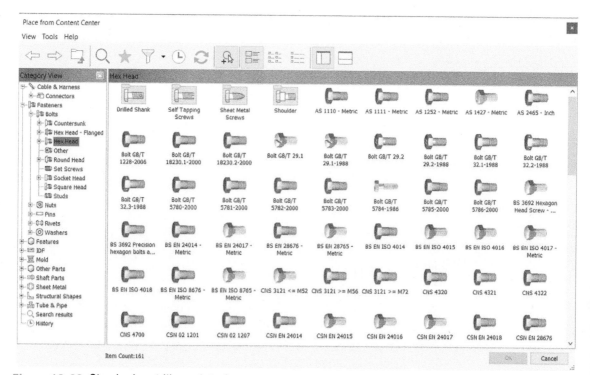

Figure 10-36 Standard part library interface (Autodesk, Inc.)

their modeling decisions. The easy solution today may unfortunately make the model unusable in the future. Careful planning is necessary if a parametric design is to function as intended.

As an example, assume that equally spaced holes are to be made in a flat bar. Given the following dimensions:

$$\text{Flat bar length (L)} = 48''$$

$$\text{Number of holes (N)} = 5$$

$$\text{Distance from flat bar ends to hole center nearest end (D)} = 2''$$

This model can be built in three steps:

1. Create the flat bar.

2. Create a hole at one end of the bar.

3. Use a rectangular array to pattern the other holes. Note that the hole spacing must be determined.

The result is shown in Figure 10-37a. Now assume that a manufacturing company uses a family of parts like the one described previously,

where L, N, and D are the driving parameters. In other words, the intent of the design is to be able to enter a flat bar length, the number of holes, and the distances from the ends of the flat bar to the nearest hole and arrive at the desired part. Figure 10-37b shows the same part, but with L = 36'', N = 8, and D = 1''. Figure 10-37c shows the parameters dialog box used to establish the relationships that allow the part to be updated as intended, by simply changing the three parameters.

Figure 10-38a provides another example. A manufacturing company uses wire shelving similar to that shown in the figure. Longitudinal and transverse lengths of wire are welded together to form the shelf. Relevant dimensions include the diameter and length of the wire, the number of wire lengths used, the spacing between the lengths of wire, and the overhang, for both the longitudinal and the transverse wires. It is fairly easy to create this model as a single part, for example:

1. Create a longitudinal wire.

2. Array.

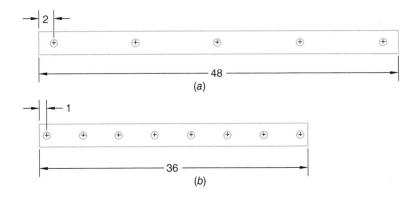

Parameter name	Units	Equation	Parameter value
L	in	36 in	36.00000
D	in	1 in	1.00000
N	–	8	8.00000
hole_spacing	in	$(L - 2 * D)/(N - 1)$	4.8571

(c)

Figure 10-37 Plate with holes part family

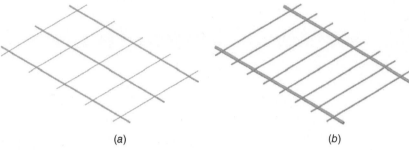

(a)   (b)

Figure 10-38 Wire shelf part family

3. Create a transverse wire, making sure that this wire just touches the longitudinals.

4. Array.

However, if the intent is to vary these parameters (diameter, length, count, spacing, and overhang) in order to generate a family of wire shelving configurations (see Figure 10-38b), then more planning is necessary.

Parametric modeling can also be used to create entire product families. Figure 10-39 shows a popular mixer blender design employed in the food processing industry. The mixer includes multiple subassemblies and contains several hundred parts. This mixer is available in a wide range of capacities. In addition, customers require that the product be customized to suit the particular

needs of their facilities, which can impact, for example, the parameters shown in Figure 10-40.

A senior project design team developed a product family model of the mixer blender for an industrial sponsor. The team modeled a specific-capacity mixer, and then used it as the basis for a mixer blender product family. As suggested in Figure 10-41, the design team's solution was to link the model assembly parameters via multiple spreadsheets. By specifying the driving parameters in a master spreadsheet, a mixer assembly model of the specified capacity can be generated.

Cloud-Based CAD

In recent years *cloud computing* has emerged to compete with the traditional model of a single-user, on-premises computer containing

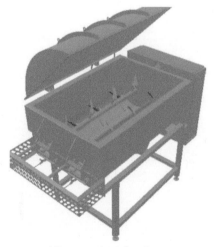

Figure 10-39 Mixer product family (Courtesy of Cozzini, Inc.; Adam R. Andrea, Katie Kopren, Philip Kunz)

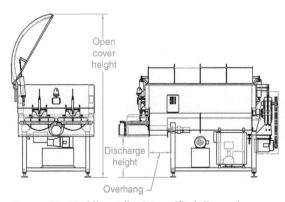

Figure 10-40 Mixer client-specified dimensions (Courtesy of Cozzini, Inc.; Adam R. Andrea, Katie Kopren, Philip Kunz)

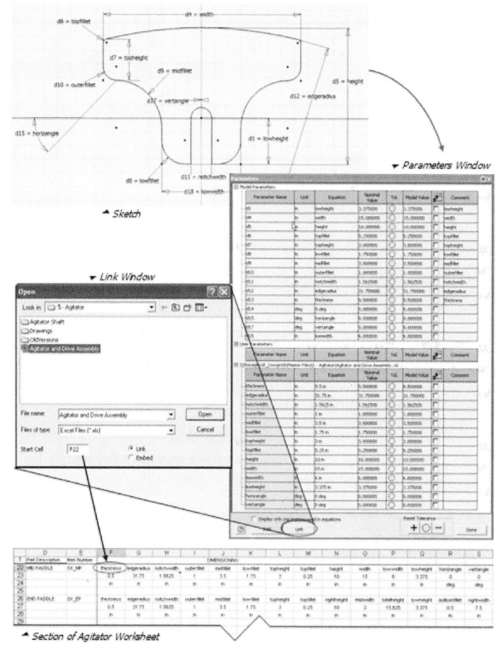

Figure 10-41 Mixer product parameter roadmap (Courtesy of Cozzini, Inc.; Adam R. Andrea, Katie Kopren, Philip Kunz)

locally installed application software and local data storage. Cloud computing refers to the on-demand availability of computer system resources, accessed via the internet on remote servers. These computing resources include data storage (i.e., cloud storage) and computing power. Examples of established cloud storage and file-sharing services include Dropbox, Box, Microsoft OneDrive, and Google Drive, whereas familiar Software-as-a-Service (SaaS) cloud computing applications include Google Docs (word processing), Google Sheets (spreadsheets), and Google Slides (presentations).

Traditionally CAD software is installed locally on a standalone computer. The software is licensed on a perpetual basis and can be updated on a schedule, typically one year. SaaS cloud-based CAD, on the other hand, runs either in a browser (e.g., Onshape) or provides cloud services through an application (e.g., Autodesk Fusion 360), which is regularly updated on a remote server, and acquired by paying a subscription fee (monthly, annual).

The SaaS CAD model offers significant advantages over traditional CAD, including flexibility, maintenance, collaboration, and versioning. Traditionally CAD operators are limited to using computers where the application is installed locally. They are also limited to using the Windows PC platform. Each new update must be downloaded and installed. New releases are typically available once per year. Collaboration is difficult, because users can only modify their own copy of a design file while another user modifies theirs. Modified files must be manually swapped back and forth to collaborate. Users must manually keep track of file versions.

Contrast this with working in the cloud, where CAD operators can use any device at any time, on any platform. If connected to the internet, you can access the application.[9] Updates are seamlessly rolled out as available with no interruption to the user. There is no need to wait a year for the latest improvements and tools. Users can collaborate simultaneously on the same document without worrying about different file versions. There is a single source of truth regarding the current version of a file. Data can be rolled back indefinitely to capture previous versions of a document, as shown in Figure 10-42.

An important drawback to using cloud-based CAD is security. Initially there was considerable resistance to storing intellectual property on external servers in the cloud, although this concern has significantly abated in recent years. Another disadvantage is the need to maintain a continuous connection to the internet to work.

Cloud-based CAD first emerged in 2012 with the release of Autodesk Fusion 360. Fusion

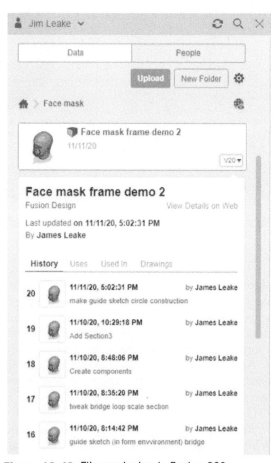

Figure 10-42 File versioning in Fusion 360 (Autodesk, Inc.)

[9]In the case of a hybrid on-premises/cloud-based solution like Fusion, it is also possible to work offline.

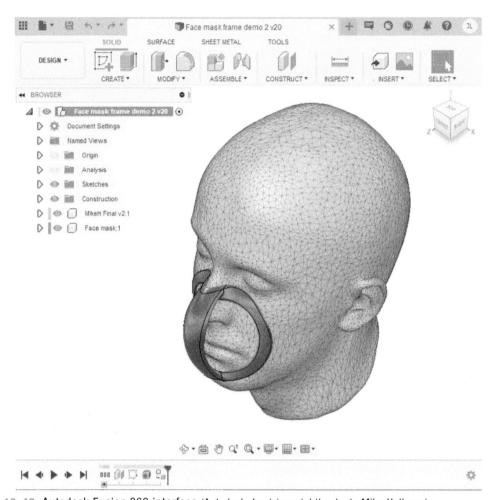

Figure 10-43 Autodesk Fusion 360 interface (Autodesk, Inc.) (special thanks to Mike Halloran)

requires that a small application be installed locally, after which the application is seamlessly updated from a remote server. Design files are stored on the external server. Figure 10-43 shows a screen capture of the Fusion 360 interface.

Another cloud-based CAD application, Onshape, is currently the only CAD software that runs entirely in the cloud. Released in 2015, Onshape runs within a web browser (see Figure 10-44). In 2019 PTC acquired Onshape. With Onshape there is no software to install, no files to save, and no licenses to keep track of.

Today almost every major CAD vendor has a cloud service available or in development.

Another important benefit of cloud CAD is that most computationally intensive tasks can be performed remotely on an external server. Examples of such tasks include rendering, finite element analysis, topology optimization, generative design, and computational fluid dynamics. Figure 10-45 shows a dialog box where two simulation studies, static stress and shape optimization, are about to be solved on the cloud.

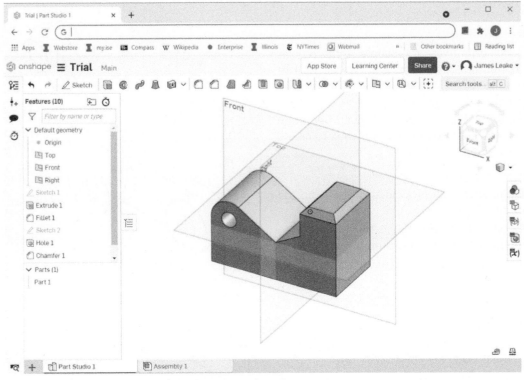

Figure 10-44 Onshape web browser interface (Onshape Inc.)

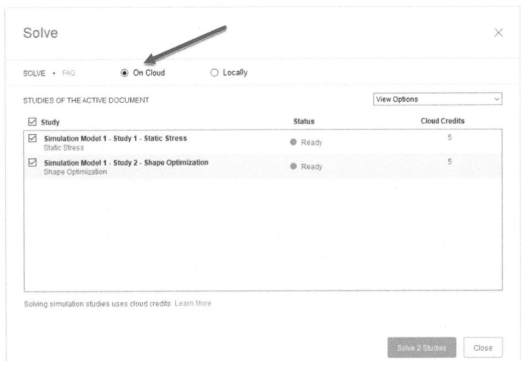

Figure 10-45 Fusion Solve dialog box (Autodesk, Inc.)

■ QUESTIONS

SHORT ANSWER

1. Solid primitives are simple solid objects created directly within a CAD system. List (and sketch) three to six examples of solid primitives.

2. Review Figure 10-5. What operation (extrude, revolve, sweep) was used to create each of the three solids?

TRUE OR FALSE

3. Parametric solid modeling involves the use of dumb solids.

4. Geometry is relatively straightforward to modify when using a parametric modeler.

5. A building information model is the same thing as an architectural drawing.

MODELING

SS
6. Given the dimensioned isometric view (P10-1 through P10-24), create a solid model of the object.

7. Given the dimensioned assembly view (P12-1 through P12-5 in Chapter 12), create a solid assembly model.

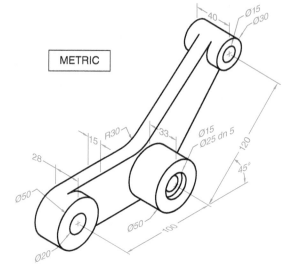

Figure P10-2

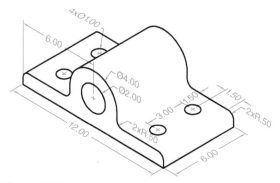

Figure P10-3

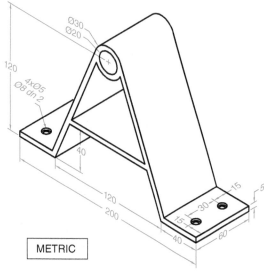

Figure P10-1

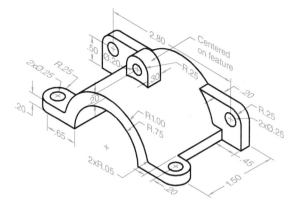

Figure P10-4

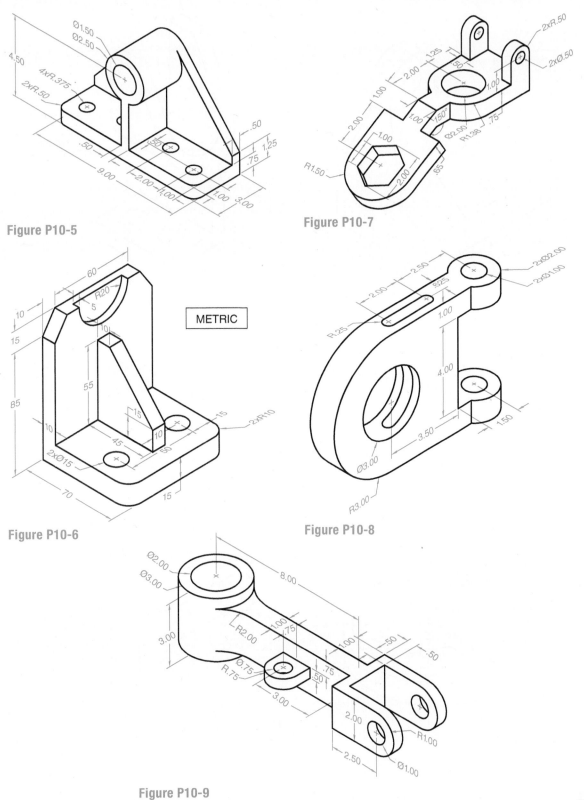

Figure P10-5

Figure P10-7

Figure P10-6

METRIC

Figure P10-8

Figure P10-9

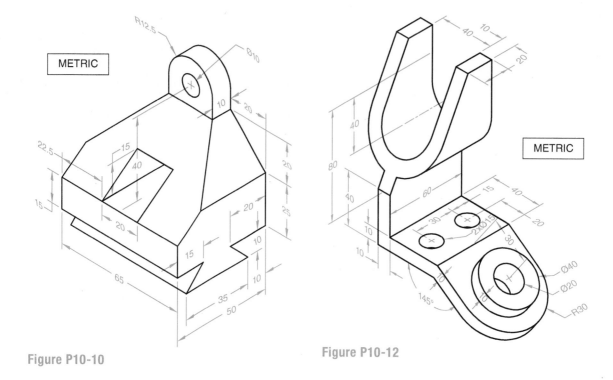

Figure P10-10

METRIC

Figure P10-12

METRIC

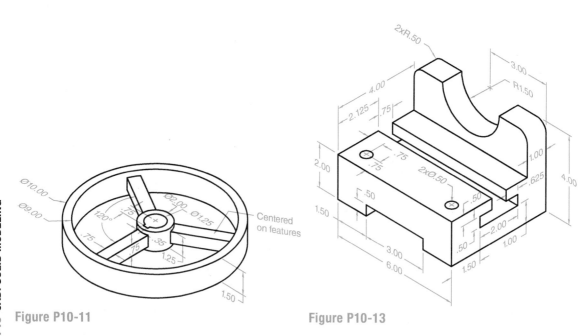

Figure P10-11

Centered on features

Figure P10-13

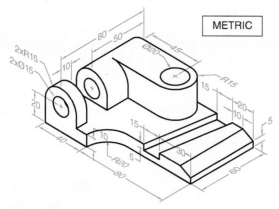

METRIC

Figure P10-14

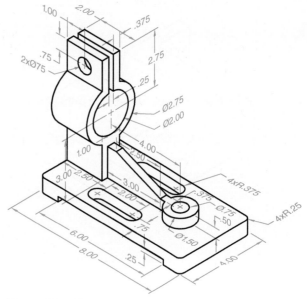

Figure P10-16

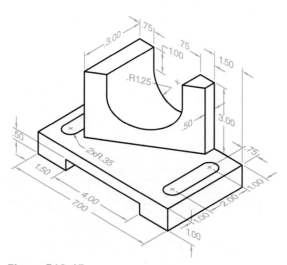

Figure P10-15

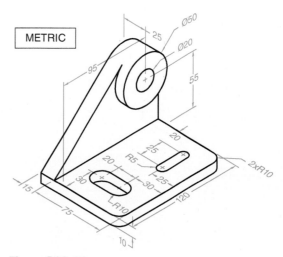

METRIC

Figure P10-17

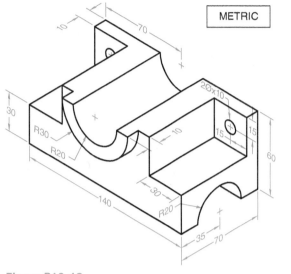

Figure P10-18

METRIC

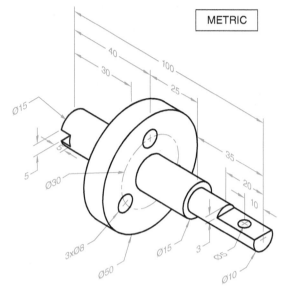

Figure P10-20

METRIC

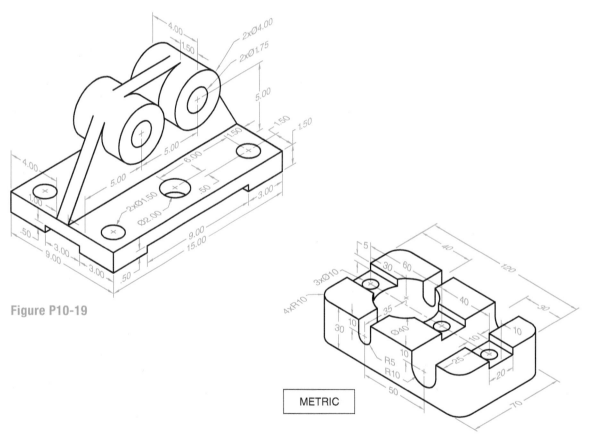

Figure P10-19

Figure P10-21

METRIC

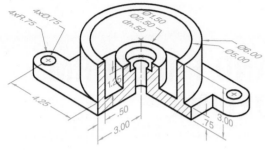

Figure P10-22

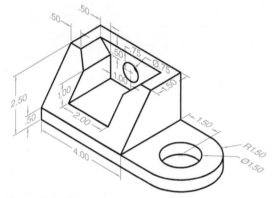

Figure P10-23

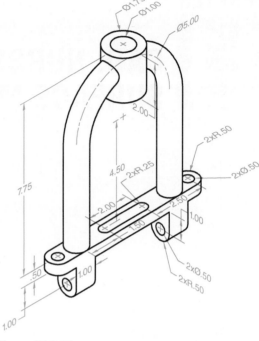

Figure P10-24

CHAPTER 11

CAD: NURBS AND FREEFORM SURFACE MODELING

■ NURBS SURFACE MODELING

Introduction

Nonuniform rational B-spline (NURBS) curves and surfaces are used to model ship hulls, car bodies, consumer products, and even the animated characters seen in 3D digital games and feature films. The most common approach to creating a NURBS surface model is first to build an underlying framework of curves and then to create surfaces using these construction curves. Figure 11-1 shows on the left the underlying curves, and on the right the surface patches, for a Mini Cooper modeled by an industrial design and engineering student team. A second modeling technique is to directly manipulate the surfaces.[1] On the left in Figure 11-2, a revolved surface representing a joystick handle is created. On the right, the surface is sculpted to improve the handle's ergonomics.

The historical development of NURBS is linked with the emergence of computer technology. Prior to the development of personal computers, 2D freeform curves were drawn with physical *splines*. These long, smooth, and flexible strips of wood or plastic were used on shipyard lofting-room floors and in the offices of naval architects to produce the curves that define the shape of a ship's hull. Lead weights called *ducks* were used to maintain the shape of the spline as it passed through fixed data points. The edge of the spline was used to pass a smooth interpolating curve in either pencil or ink along the spline edge. Figure 11-3 shows several ducks and a spline positioned to draw a portion of a curve. Traditionally, splines would have been used to create the *lines plan* for the boat shown in Figure 11-4. Today, however, these drawings are created using computer software.

In the 1940s, mathematicians began to study the traditional spline with the aim of modeling its behavior, in order to make the spline creation process reproducible. The physical spline behaves like a thin, elastic beam, with bending deflections forming a smooth curve. It can be shown[2] that the shape of the physical spline is mathematically described by cubic polynomials.

Parametric Curves and Cubic Splines

A mathematical spline may be described as a "piecewise *parametric* polynomial curve." Although curves can be represented in either nonparametric or parametric form, the parametric representation of a curve is most useful for the purposes of computer graphics and CAD. For

---

[1]Veteran Alias users refer to this as "pulling CVs."

[2]David F. Rogers and J. Alan Adams, *Mathematical Elements for Computer Graphics*, 2d ed., McGraw-Hill, 1990, page 252.

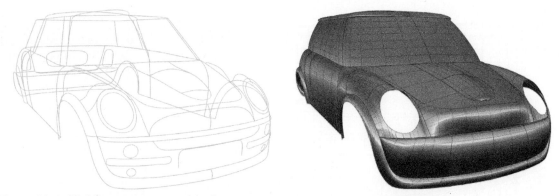

Figure 11-1 Mini Cooper curves and surfaces (Courtesy of Todd Cao and William Bergen)

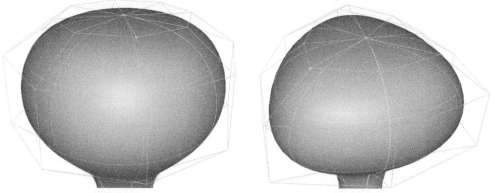

Figure 11-2 Surface manipulation

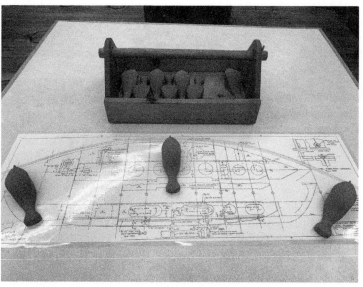

Figure 11-3 Spline with ducks

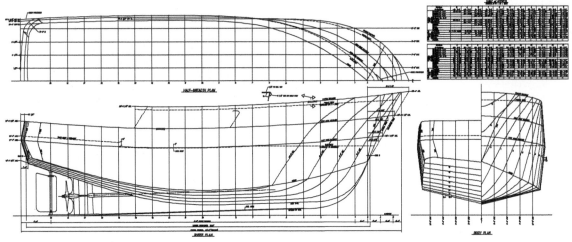

Figure 11-4 Lines plan (Courtesy of Jensen Maritime Consultants, Inc.)

curves, only a single parameter, typically u, is needed. For surfaces, two parameters, u and v, are needed. These parameters typically range from 0 to 1, although this is not strictly necessary. The mathematics of parametric equations is the basis for understanding all of the synthetic curves discussed here, including cubic splines, Bézier curves, B-splines, NURBS, and T-splines. The parametric form is also used to represent analytic curves (lines, conic sections, and so on).

A ***cubic spline*** is modeled after a physical spline. A cubic spline can be defined by four points, with the spline passing through, or interpolating, the points. These points are called ***boundary conditions***, and they are used to define the spline. If these points are not all in the same plane, then the curve is three dimensional and is called a ***space curve***. If all points do lie in the same plane, then the curve can have an inflection point, as shown in Figure 11-5.

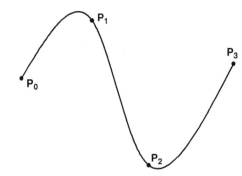

Figure 11-5 Cubic spline with inflection point

Parametric Representation of a Curve

In the parametric representation of a curve, each point on the curve is defined by a position vector $\mathbf{P}$ (see Figure 11-6), where the vector components are:

$$x = x(u), y = y(u), z = z(u)$$
$$\mathbf{P}(u) = [x(u)\, y(u)\, z(u)], 0 \le u \le 1$$

where

$x, y,$ and z are polynomials and u is the parameter

For example, the coordinates defining the 2D curve shown in Figure 11-7 are generated from the following equations, with the parameter u varying between 0 and 1:

$$x(u) = 3u^2$$
$$y(u) = 2u$$
$$z(u) = 0$$

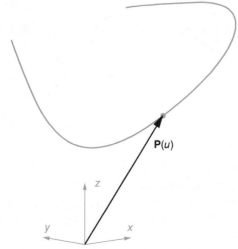

Figure 11-6 Parametric representation of a curve

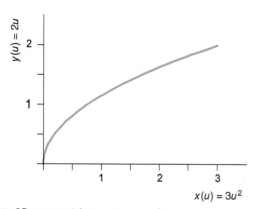

u	$x(u) = 3u^2$	$y(u) = 2u$
0	0	0
0.1	0.03	0.2
0.2	0.12	0.4
0.3	0.27	0.6
0.4	0.48	0.8
0.5	0.75	1.00
0.6	1.08	1.20
0.7	1.47	1.40
0.8	1.92	1.60
0.9	2.43	1.80
1	3.00	2.00

Figure 11-7 2D parametric curve example

The shape of a traditional spline is described by a cubic polynomial, but quadratic and linear splines are also possible. A quadratic spline is a second-degree polynomial. The quadratic spline interpolates, or fits, all three data points used to define it. The quadratic spline is a *plane curve* because it must lie in a plane. A quadratic spline cannot have an inflection point. Figure 11-8 shows a quadratic spline defined by three points. A linear spline is a first-degree polynomial, defined by two points that the curve passes through.

Rather than using four points on the curve as the boundary conditions defining a cubic spline, a *Hermite cubic spline* is defined by the two end points of the curve and the tangent vectors at

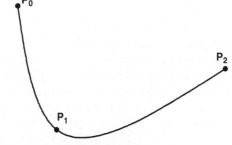

Figure 11-8 Quadratic spline

these end points. Figure 11-9 shows a Hermite cubic spline and its boundary conditions, the two end points and the tangent vectors at these points.

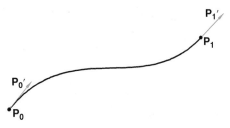

Figure 11-9 Hermite cubic spline

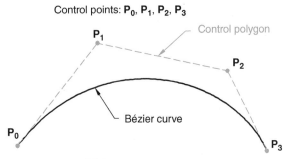

Figure 11-10 Bézier curve

The shape of a mathematical spline is determined by the multiplication of a **blending function** matrix and a boundary condition matrix. The blending functions are polynomial expressions in terms of the parameter u, where $0 <= u <= 1$. The polynomial expressions depend on the boundary conditions used to define the curve.

Bézier Curves

Bézier curves and surfaces are named for Pierre Bézier (1910–1999). Bézier was a French engineer who, in the early 1960s, used and further developed these parametric curves and surfaces to design automobile bodies. See the section on the Bézier Award later in this chapter for more information on Bézier.

Although the cubic spline is relatively simple, it is difficult to control. The Bézier curve, on the other hand, provides a more intuitive way to manipulate a curve. This is largely due to the fact that cubic splines interpolate all data points, whereas the Bézier curve passes through only the first and last points, while approximating the interior ones.

Figure 11-10 shows a cubic Bézier curve. **Control points**, sometimes called control vertices, are used to manipulate the curve. A Bézier curve always interpolates the first and last control points, while approximating all interior control points. The curve is pulled toward the interior control points, without passing through them. The tangent to the curve at $\mathbf{P}_0$ is given by $\mathbf{P}_1 - \mathbf{P}_0$, and the tangent to the curve at $\mathbf{P}_n$ is given by $\mathbf{P}_n - \mathbf{P}_{n-1}$. The **control polygon** (not a polygon really, more like a polyline) is formed by connecting the control points in sequence. Note how the shape of the curve is suggested by its control polygon.

The degree of a Bézier curve is always one less than the number of control points. In this case, four control points means that the curve is of degree 3—that is, a cubic polynomial.

The **convex hull** of any polygon is the convex polygon formed by stretching a rubber band over the vertex points of the polygon. Figure 11-11 shows a Bézier curve and its control polygon, together with the convex hull of the polygon. Note that the curve lies entirely within the convex hull of the polygon. Since Bézier curves cannot intersect if their convex hulls do not overlap, the intersection of the convex hulls is calculated first. If there is no intersection, clearly the curves do not intersect. If, however, the convex hulls intersect, then a second calculation is made to determine whether the curves themselves intersect.

As mentioned previously, for Bézier curves, the number of control points and the degree of the curve are coupled, the degree of the curve being

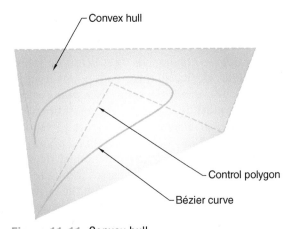

Figure 11-11 Convex hull

Table 11-1 **Common Bézier curves**		
Curve	**Degree**	**Control Points**
Linear	1	2
Quadratic	2	3
Cubic	3	4
Quartic	4	5
Quintic	5	6

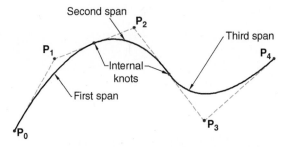

Figure 11-12 B-spline with 3 spans, degree = 2

one less than the number of control points. This means that if several control points are needed to form a curve, the Bézier curve will necessarily be of high degree. Curves of high degree are difficult to control because they tend to oscillate, as well as being difficult to compute. The most commonly used Bézier curves are quadratic, cubic, and quintic. A second problem with Bézier curves is that there is no local control. If one control point is moved, the entire curve is affected. Table 11-1 shows the name, degree, and number of control points for Bézier curves.

B-Splines

B-splines address two shortcomings of Bézier curves: (1) The number ($n + 1$) of control points determines the degree (n) of the polynomial, and (2) because of the global nature of Bézier curves, moving one control point changes the entire curve. B-splines, on the other hand, are generally nonglobal, and the degree of the curve can be changed without changing the number of control points.

A B-spline is actually a composite of several curves, where the different curve *spans* are joined at *knots*. This is accomplished by requiring that, regardless of the total number of control points, only a certain number of them, k, influence the curve at any one time. Also, the degree of the curve is one less than the number of control points, $k - 1$. The first k points are used to generate the first span. The next span uses the next k points, with the first control point being dropped. This continues until all of the control points have been used.

The curve in Figure 11-12 has five control points, with $k = 3$, so the degree of the curve is 2, quadratic. There are three spans separated by two

internal knots. The curve shown in Figure 11-13 uses the same five control points as those used in Figure 11-12, but now $k = 4$, so the degree of the curve is 3, cubic. There are two spans separated by a single internal knot. Figure 11-14 again uses the same five control points, but now $k = 2$, so the degree is 1, linear. There are four spans separated by four knots. The curve is reduced to its own control polygon.

Finally, in Figure 11-15, $k = 5$ and the degree of the curve is 4. Note that k is equal to the total number of control points. In this case, all of the control points are used all at once. There is a single span without internal knots. If the degree of the curve is one less than the number of control

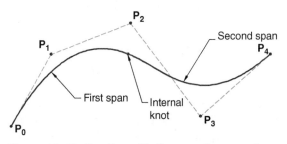

Figure 11-13 B-spline with 2 spans, degree = 3

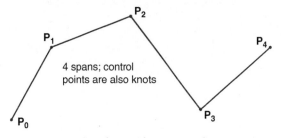

Figure 11-14 B-spline with 4 spans, degree = 1

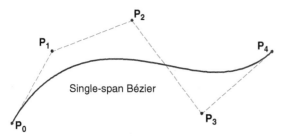

Figure 11-15 B-spline with 1 span, degree = 4 (Bézier)

points, then a B-spline is identical to a Bézier. A Bézier curve is a special case of the B-spline.

Figure 11-16 shows all four curves in a single figure. Note that for all of the curves, the number of spans plus the degree of the curve equals 5, the number of control points.

All of the curves described above have **uniform** knot spacing, meaning that there is a uniform gap between neighboring knots. The general case is for the knot spacing to be **nonuniform**. Nonuniform knots arise in the process of editing a curve—for example, by adding or deleting a knot. Note that this is the "nonuniform" appearing in the name *nonuniform rational B-spline*.

To summarize, the polynomial degree of a B-spline is independent of the number of control points, so a given control point influences the curve only locally. Also, the degree of the curve can be changed without changing the number of control points.

NURBS

So far the most general form discussed has been the nonuniform B-spline. To complete the

definition of the nonuniform rational B-spline, only the **rational** element remains to be discussed. Unlike the B-spline definition, the NURBS definition is a rational formula involving **homogeneous coordinates**, a topic beyond the scope of this text. The key point is that in the NURBS formulation, the control points can be weighted. These **weights** are used to control the shape of the curve.

For example, increasing the weight of a control point pulls the curve toward the point, even to the extent that the curve interpolates the control point. Figure 11-17 shows how a weighted control point affects the shape of the curve.

Most important, rational B-splines provide a single precise mathematical form capable of representing not only freeform shapes, but analytic ones as well (e.g., lines, planes, conic sections, including circles). Note that NURBS are the general form, capable of representing B-splines and Bézier curves alike.

Surfaces

A parametric surface requires two independent parametric variables, rather than the single variable used with curves. Normally u and v are chosen to represent the parameters. Typically both parameters are limited to a unit interval ranging between 0 and 1. Setting either u or v equal to a constant, for example $u = 0.5$, generates a curve on the surface called an **isoparametric** curve. Figure 11-18 shows several *isoparms* (short for *isoparametric curves*) on a surface.

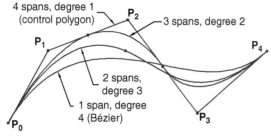

Figure 11-16 Four B-splines with the same control points

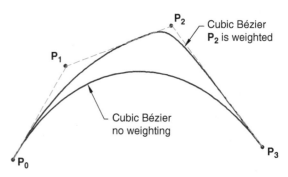

Figure 11-17 NURBS curve with weighted control point

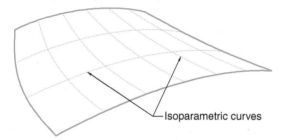

Figure 11-18 Isoparametric curves

A **surface patch** is a piecewise parametric surface. A more complex surface quilt will be built up from surface patches. Figure 11-19, for example, shows a car body composed of different surface patches.

Figure 11-20 shows a 4 × 4 bicubic Bézier surface patch. The shape of the surface is controlled by the *polygon net*, a grid of control points analogous to a curve's control polygon. Only the corner points of the polygon net and the surface

Figure 11-19 Surface patches on a new Volkswagen Beetle

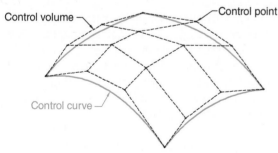

Figure 11-20 Bézier surface patch (4 × 4)

are coincident. The surface is contained within the convex hull of the defining polygon net. The four curves bounding the Bézier surface are themselves Bézier curves.

Note that, because of the rectangular topology of the polygon net that defines these surfaces, surface patches are four-sided. Nonquadrilateral surfaces require special treatment, for example, by trimming the surface. Before discussing continuity requirements used to connect different surfaces patches into a quilt, we will discuss some curvature tools.

Curvature

It is useful to be able to evaluate the curvature of a curve or surface. Curvature can intuitively be defined as the rate of deviation of a curve (or surface) from a straight line (or plane) tangent to it. The curvature of a circle is defined as the reciprocal of the circle's radius. In this way, a small circle has large curvature, and a large circle has smaller curvature. Using an **osculating circle**, this same definition can be extended to plane curves. An osculating circle, seen in Figure 11-21, is the circle on the inner normal line of a curve at a point that best fits the curve. The radius of an osculating circle at a point is called the **radius of curvature** of the curve at that point.

Another curvature analysis tool commonly available in surface modeling applications is the **curvature comb**, as shown in Figure 11-22. The curvature comb is used to help visualize the curvature and overall smoothness of a spline. The length of a *quill* represents the curvature of the spline at that point. The longer the quill is, the greater the curvature at that point.

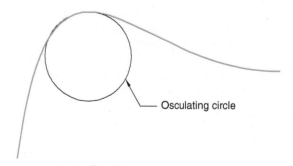

Figure 11-21 Osculating circle

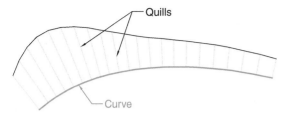

Figure 11-22 Curvature comb

Continuity

The ability to control the ***continuity*** of curves and surfaces is important to the appearance of these objects. All commercial surface-modeling applications provide tools to control continuity. If a single curve is not sufficiently complex, then multiple curves can be joined together. To create a smooth transition between curve segments, make the tangent vectors of adjoining curves collinear. Surface patches are used to build up a complex design, and it is the relationship between these surface patches that gives the design its special character.

Curves and surfaces can have different levels of continuity, referred to as ***geometric (G^n) continuity***, with higher values of n indicating increasing smoothness. Three types of continuity between curves are shown in Figure 11-23. They are:

- ***Positional (G^0) continuity***, where two curves share a common end point
- ***Tangent (G^1) continuity***, where the tangent vector directions of both curves at a common point are the same
- ***Curvature (G^2) continuity***, where the tangent vector directions and radius of curvature at a common point are the same

Note that geometric continuity implies that the directions of the tangent vector at a common point are equal, but not the magnitudes. There is another, stronger type of continuity, called ***parametric (C^n) continuity***, which implies that both the magnitude and the direction of the tangent vectors at a common point are equal. Geometric (G^n) continuity is typically used in commercial surface modeling applications.

With regard to surfaces, positional continuity implies that two surfaces meet at an angle,

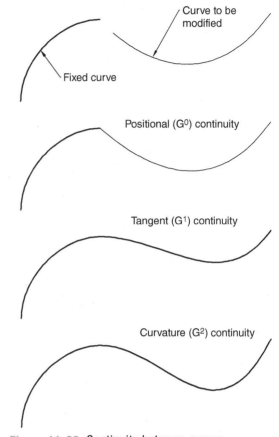

Figure 11-23 Continuity between curves

sometimes called a "knuckle." A minimum of positional continuity is essential if a watertight model is required, either to convert it to a solid or to make a 3D print of the object. Tangent continuity between surfaces, seen for example in a fillet, creates a smooth transition between the surfaces, but visually one can tell when one surface ends and the other begins. With curvature continuity, the join between the surfaces is smooth, and it is hard to tell when one surface ends and the other begins. Figure 11-24 shows examples of surfaces exhibiting these three kinds of continuity.

Understanding how continuity influences control points is useful when evaluating the smoothness of curves and surfaces. In positional continuity (Figure 11-25 at the top), the end control point of the aligned curve is in the same position as the end control point of the curve being aligned to. Tangent continuity (Figure 11-25 in

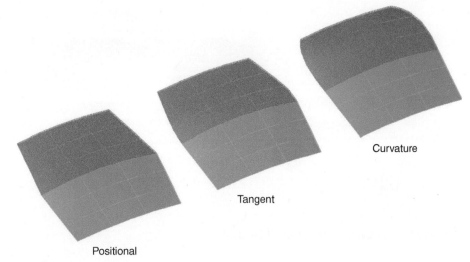

Curvature

Tangent

Positional

Figure 11-24 Continuity between surfaces

the middle) requires that there be a straight-line relationship through the first and second control points of both curves, in addition to the positional continuity requirements. Curvature continuity (Figure 11-25 at the bottom) requires that, in addition to the tangent continuity requirements,

the third control point of the aligned curve be in a constrained relationship with the third control point of the curve being aligned to.

The curvature comb is a useful tool when evaluating continuity, as shown in Figure 11-26. With

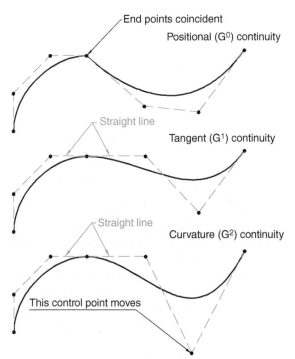

End points coincident

Positional (G⁰) continuity

Straight line

Tangent (G¹) continuity

Straight line

Curvature (G²) continuity

This control point moves

Figure 11-25 Continuity and control points

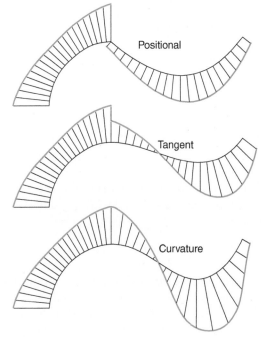

Positional

Tangent

Curvature

Figure 11-26 Evaluating continuity with the curvature comb

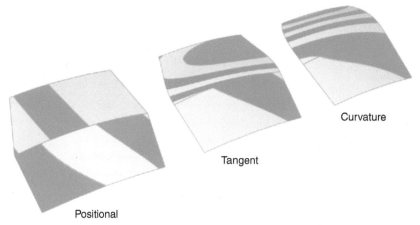

Curvature

Tangent

Positional

Figure 11-27 Evaluating continuity with Zebra stripe shading

positional continuity, although the end points meet exactly, the direction of each curve at that point is different. With tangent continuity, the end tangents of the curves match, meaning that the curves have the same direction, but the distinct jump in the length of the comb quills means that the curvatures are different. With curvature continuity, the curve directions match and there is a smooth transition in the lengths of the quills, indicating that the curvatures are equal at the join point between the curves.

Zebra stripe shading, a tool used to evaluate surface smoothness, is commonly available in surface-modeling packages. This shader mimics the zebra stripe analysis used in automotive design, where a model is placed under strip lights in order to evaluate the continuity of the model's surfaces. As seen in Figure 11-27, the zebra stripes are disjointed between positionally continuous surfaces. On tangent surfaces, the zebra stripes match up but are not smooth. The zebra stripes are smooth across curvature-continuous surfaces.

Class A Surfaces

The term *class A surface* is used in automotive design to describe a set of high-quality, high-efficiency surfaces. The term tends to be more qualitative than quantitative, although G^2 (and even G^3) curvature continuity may well be the case. Class A surfaces are perfectly smooth,

pleasing to the eye, and pleasant to the touch. Styling intent is also implied by the term.

With regard to efficiency, a single span (that is, Bézier) is smoother than multispan curves and surfaces. In automotive design, considerable effort is taken to break the surfaces up into single spans to ensure smoothness and controllability. Class A surfaces consequently imply that single-span Bézier surfaces are used.

▮ FREEFORM SURFACE MODELING

Introduction

A freeform surface is a smooth, nonanalytic (e.g., not spherical, conical, cylindrical, toroidal) surface whose shape can be controlled by a designer. Figure 11-28 provides an example. Freeform surfaces are also referred to as sculpted or organic surfaces, and they are often associated with concept design. In the twenty-first century, design has become an important way to differentiate products, and one approach used by product designers to accomplish this is to create products with a pleasing freeform shape. Common metaphors used to describe working with freeform surfaces include fabric sewing and clay sculpting.

There are two ways to produce freeform shapes: nonuniform rational B-splines (NURBS) and subdivision surfaces. We have already considered NURBS, and we will discuss subdivision surfaces and their cousin, T-splines, later in this

Figure 11-28 "Cloud Gate" sculpture in Millennium Park in Chicago

chapter. Both NURBS and subdivision surfaces were first described in the late 1970s. Both representations use a control frame or grid, with control points connected together in a mesh (see Figure 11-29), and both are constructed using B-spline mathematics.

Each representation has its own strength. While NURBS offer control over surface parametrization and smoothness, subdivision surfaces provide freedom from topological constraints. These different strengths led to each representation being preferred in different industries. While NURBS are the dominant standard in engineering and CAD, subdivision surfaces dominate in the younger media and entertainment industry, for applications like film and video games. Subdivision surfaces were not widely

used until the 1990s, by which time NURBS was the standard in design and manufacturing. For purposes like character animation, subdivision surfaces provide a significant advantage over NURBS because there is no need to stitch together separate surface patches.

In product design, the need for the flexibility afforded by subdivision surfaces has grown in recent years, as products have taken on smoother, more freeform shapes. Issues preventing the adoption of subdivision surfaces for CAD include the following:

- Almost all existing CAD data, dating back to 1970s, employs NURBS to represent freeform surfaces, and until just recently subdivision surfaces were not compatible with NURBS

- Subdivision does not provide the quality and precision required for engineering applications; subdivision surfaces do not have a closed-form representation, meaning that most subdivision schemes cannot be parameterized and evaluated exactly

In the 2010s, significant progress was made on these fronts.[3] The discussion will now turn to polygon meshes and polygonal modeling, subdivision surfaces, limitations of NURBS for freeform modeling, and finally, T-splines.

Polygon Meshes and Polygonal Modeling

Before discussing subdivision surfaces, polygon meshes and polygonal models will first be described. A polygon mesh is a collection of vertices, edges, and faces that defines the shape of a polyhedral object. The faces usually consist of triangles, quadrilaterals, or other simple convex polygons, since this simplifies rendering. Polygon meshes have been around since early days of computer graphics, and their characteristics include being easy to compute, to render, and to store.

Polygonal modeling is an older approach for modeling objects by representing or

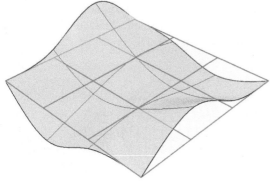

Figure 11-29 Freeform surface with control frame

[3]For example, see Thomas J. Cashman, *NURBS-compatible subdivision surfaces*, 2010.

approximating their surfaces using polygons. The main advantage of polygonal modeling is that it is faster than other representations. Polygon meshes are easy to edit, and even to deform in a clay-like fashion. The main disadvantage of polygonal modeling is that the objects created are faceted.

Subdivision Surfaces

Subdivision algorithms are used in computer graphics to define a smooth curve or surface as the limit of a sequence of successive refinements. One of these algorithms is Catmull-Clark subdivision, named for Edwin Catmull, the former president of Pixar, and Jim Clark, who founded several notable Silicon Valley technology companies, including Silicon Graphics and Netscape. Subdivision surface modeling was first used in Pixar's computer-animated short film, *Geri's Game*, in 1997. The following year the film won the Academy Award for Best Animated Short Film. Today, Pixar uses subdivision to model almost all of their characters, as well as other objects.

The Catmull-Clark algorithm was first described by Catmull and Clark in their short 1978 paper, *Recursively generated B-spline surfaces on arbitrary topological meshes*. The title of this paper provides several insights into the nature of subdivision modeling. Subdivision (1) is recursive, (2) employs B-splines, (3) works with arbitrary topology, and (4) uses (polygon) meshes.

Subdivision modeling is a recursive procedure for generating smooth surfaces of arbitrary topology from a grid of control points, that is, a polygon mesh. A refinement scheme is successively applied to the mesh, subdividing it to produce new vertices, edges, and faces.

See Figure 11-30. The positions of the new vertices are calculated based on the positions of the nearby given vertices. This produces a more refined mesh that is then used to generate an even denser and smoother mesh. The limit subdivision surface is the surface produced from this process being iteratively applied infinitely many times, although in reality the algorithm is only applied a limited number of times. We might say that subdivision modeling uses a polygon mesh as a scaffold, over which a smooth surface is draped.

Subdivision surfaces can be categorized according to the type of mesh (i.e., quadrilateral, triangular), the refinement approach (i.e., Doo-Sabin, Catmull-Clark), and the fairing scheme (i.e., approximating or interpolating).

In the control mesh for a subdivision surface, an arbitrary number of edges can be incident at a given vertex. The number of edges incident at a vertex is referred to as the ***valence*** of the vertex. A vertex is considered to be *extraordinary* if its valence differs from that of the regular configuration. Catmull-Clark subdivision surfaces, for example, employ a quadrilateral mesh, so the expected valence is four. Irregular behavior of subdivision surfaces will be found in the neighborhood of ***extraordinary points***, that is, points where the valence is extraordinary. Catmull-Clark subdivision surfaces are curvature continuous (C^2), except near extraordinary points. Later in our discussion of T-splines, we will see that star points (for T-splines) and extraordinary points (for subdivision surfaces), refer to the same regions where the topology is allowed to deviate from the anticipated configuration.

CAD solutions that provide freeform modeling capabilities based on subdivision modeling include CATIA's Image and Shape (Dassault), NX Shape Studio (Siemens), and Creo Interactive Surface

Figure 11-30 Catmull-Clark subdivision acting on a cube

Design Extension (PTC). Both CATIA and NX are high-end CAD solutions, and consequently are unlikely to be used by smaller design firms.

NURBS Limitations

All NURBS surfaces have an underlying rectangular topology, as seen in Figure 11-31. The limitations of NURBS surfaces stem from this rectangular topology restriction. The control mesh used to define a NURBS surface has four edges, and two parametric directions, u and v. Isoparametric curves (isoparms) lying on the surface are defined by setting either u or v equal to a constant value. In a NURBS surface, every interior control point must be connected to four other control points. This means that, in order to add localized detail to a NURBS surface, additional isoparms must be added to the entire surface, as shown in Figure 11-32. This significantly increases the number of control points required to define the surface, even though many of these control points are superfluous. They do not contain any significant geometric information and are only needed to satisfy the topological constraint.

Additionally, this underlying rectangular topology implies that multiple NURBS surfaces must be employed in order to build up a complex surface or a watertight body. Figure 11-33 shows a car body that is built up from multiple surface patches. In Figure 11-34 of the device housing, we can see that it is constructed from several primary (top, bottom, and sides) surfaces, which are then joined by additional secondary (blend, fillet) surfaces. As a consequence, it is common for small gaps to exist in a NURBS surface quilt built up from multiple surface patches, as seen in Figure 11-35. In transitioning a design to analysis and manufacturing, these gaps must

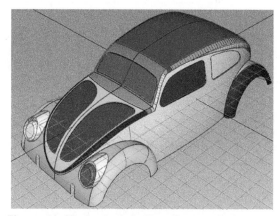

Figure 11-33 Car body built up from multiple surface patches

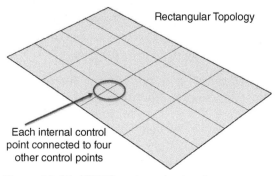

Figure 11-31 NURBS rectangular topology

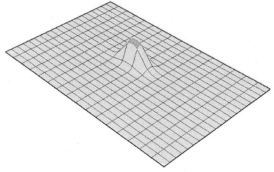

Figure 11-32 Need to add additional isoparms to add small, local features

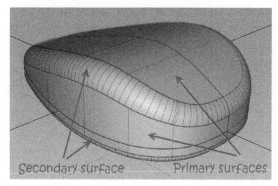

Figure 11-34 Device housing created from primary and secondary surfaces

FREEFORM SURFACE MODELING

323

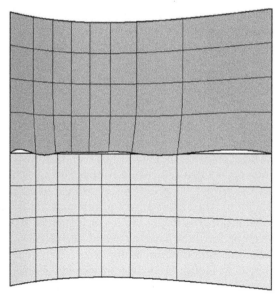

Figure 11-35 Small gaps between surface patches

be eliminated, requiring time-consuming and expensive surface repair routines.

Table 11-2 provides a comparison of NURBS and subdivision surfaces.

T-Splines

T-splines were first defined by Thomas W. Sederberg (see Bézier Award sidebar) in 2003.[4] T-Splines, Inc. was founded in 2004 to commercialize the technology. A U.S. patent for technologies related to T-splines was awarded in 2007. In 2011, the

company was acquired by Autodesk. T-splines are used in Autodesk Fusion 360.

In the previous section, the limitations of NURBS surfaces were discussed. These limitations include:

- A uniform grid of *uv* curves, forcing a uniform level of detail across the entire surface
- Rectangular topology, forcing the user to break up model geometry into many individual surface patches
- Characteristic and unavoidable small gaps between patches in trimmed NURBS surfaces, forcing time-consuming and expensive surface repair operations

T-splines break free from the limitations of a rectangular topology. A T-spline surface:

- Allows *local refinement*, thus requiring fewer control points. Modeling surfaces with T-splines can significantly reduce the number of control points in comparison to NURBS surfaces.[5]
- Can produce a *single, unified surface*, which is watertight by definition. T-splines are smooth and continuous and are easy to thicken or shell.
- Is *gap-free*, with no need for repair.

A **T-spline** is a nonuniform B-spline surface that permits T-junctions. Unlike NURBS surfaces, a row of control points in a T-spline can terminate without traversing the entire surface. These partial rows of control points

Table 11-2 Comparison of NURBS and subdivision surfaces

Freeform surfaces	NURBS	Subdivision surfaces
Year	1970s	1990s
Industry	Engineering, geometric modeling, CAD	Media & entertainment
B-splines	Yes	Yes
Control mesh	Yes	Yes
Key strength	Surface parametrization, smoothness	Freedom from topological constraints
Topology	Rectangular	Arbitrary
Surface patches	Yes	No (single, unitary surface)
Watertight	No	Yes

[4]T.W. Sederberg, J. Zheng, A. Bakenov, and A. Nasri, T-Splines and T-NURCCS, *ACM Transactions on Graphics* 22, no. 3 (July 2003): 477–484.

[5]In comparison to NURBS surfaces, modeling surfaces with T-splines can reduce the number of control points. T-splines typically have 50–70% fewer control points than the equivalent NURBS surface (Tsplines.com).

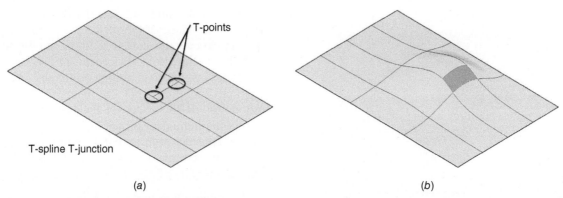

T-points

T-spline T-junction

(a)　　　　　　　　　　　　(b)

Figure 11-36　T-spline surface with T-point

terminate in **T points**, hence the name T-splines. See Figure 11-36. T points allow a surface to be locally refinable, so that different levels of detail and complexity can be added to a single surface, all while using fewer control points. This makes it easier to model complex shapes, and to create smooth watertight models. The surface around a T point will stay smooth and will not stretch or kink. Mathematically speaking, T points are curvature continuous (C^2).

In addition to T points, T-splines also have **star points**. Star points exist when three, five, or more edges come together at a point. Star points occur naturally in the box and quadball primitives, as seen in Figure 11-37a and 11-37b. More commonly, star points are created when extruding a face, as shown in Figure 11-38. Star points, in effect, provide the means for a T-spline to change from a rectangular to an arbitrary topology.

As we have seen, using NURBS to construct a complex shape with varying detail, curvature, and smoothness requires many individual rectangular patches, and maintaining continuity and smoothness across these patch surfaces requires significant effort. Star points allow T-splines to overcome this fundamental limitation in NURBS modeling. Star points enable modeling techniques such as extrusion, face deletion, and merging of surfaces that greatly increase design freedom for the user.

It is difficult to control the shape of a T-spline at star points, so they should only be used where necessary. Good practice dictates that star points should be placed on the flatter parts of a surface in order to obtain an aesthetically pleasing model more predictably. Poor locations for star point placement include the sharper portions of a model like creased edges where the curvature changes

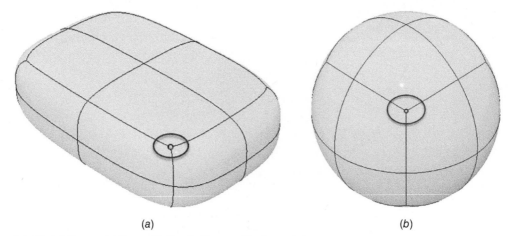

(a)　　　　　　　　　　　　(b)

Figure 11-37　(a) Box and (b) quadball primitives with star points

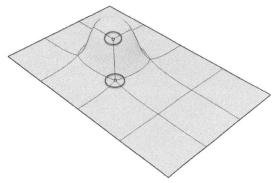

Figure 11-38 Star points created upon extruding a face

significantly, or on the edge of an open surface. Star points are also used in subdivision modeling, where they are called extraordinary points.

Another important consideration is that T-splines can be converted to untrimmed NURBS surfaces, and vice-versa, without loss or change to the shape of the surface. T-splines can, in theory, do everything that NURBS can do. As Dr. Richard Riesenfeld, the creator of B-splines, says, "The T-Spline technology addresses some important limitations that are inherent in conventional NURBS surfaces. T-Splines are based on solid mathematical principles. An important practical consideration is that T-Splines are forward and backward compatible with NURBS."

In summary, T-splines allow designers to:

- Add detail only where necessary
- Model geometries with arbitrary (i.e., nonrectangular) topology
- Easily edit complex freeform models
- Maintain NURBS compatibility

T-splines can be displayed in smooth, control frame, or box mode. See Figure 11-39. Note

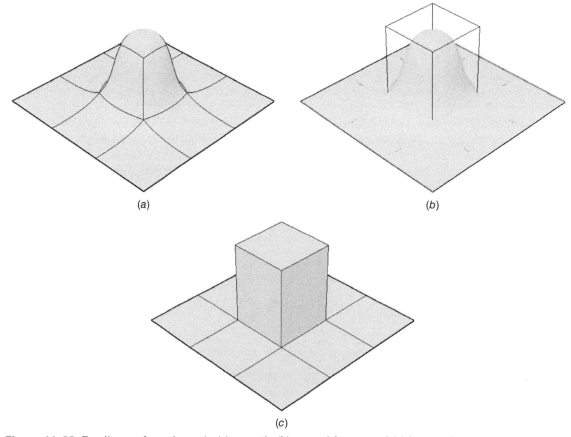

(a)

(b)

(c)

Figure 11-39 T-spline surface shown in (a) smooth, (b) control frame, and (c) box mode

Table 11-3 Comparison of subdivision surfaces and T-splines

Freeform surfaces	Subdivision surfaces	T-splines
Year	1990s	2000s
Industry	Media & entertainment	CAD & engineering
B-splines	Yes	Yes
Control mesh	Yes	Yes
Arbitrary topology	Yes	Yes
Single, unitary surface (i.e., no surface patches)	Yes	Yes
Watertight	Yes	Yes
T-junctions	No	Yes (local refinement)
Patented	No	Yes
NURBS compatible	No	Yes
Points where valence differs from regular configuration	Extraordinary	Star

that in box mode, only the underlying control frame, or control mesh, is displayed. Experienced users frequently model in box mode, in order to ensure that the control polygons do not cross or intersect. Intersecting polygons in box mode indicate self-intersecting surfaces in smooth mode, meaning that the object is nonmanifold.

To conclude the chapter, Table 11-3 provides a comparison of subdivision surfaces and T-splines.

The Bézier Award

The Solid Modeling Association (SMA) serves research, development, and user communities in CAD/CAM, and in the broader field of solid modeling and its applications. The SMA is an outgrowth of the Association of Computing Machinery's (ACM) Symposium on solid modeling and applications. In 2007, the SMA established the Bézier Award to recognize individuals or teams who have made long-lasting contributions in solid, geometric, or physical modeling, or in their applications. The Bézier Award is modeled after the ACM's A.M. Turing Award for computing.

The Bézier award is named for Pierre Bézier (see Figure 11-40) who, in the 1950s and 1960s, worked for the French car manufacturer, Renault. Bézier is one of the founders of the fields of solid, geometric, and physical modeling. Bézier developed one of the first CAD/CAM systems (UNISURF) at Renault. He is best known for his work in representing curves. The *Bézier curve* is named for him.

In 2007, the Bézier Award recipients (see Figure 11-41) were Aristides Requicha and Herbert Voelcker, who worked at the University of Rochester from the early 1970s to the mid-1980s. Here they developed early CAD systems utilizing *constructive solid geometry* (CSG) data structures.

In 2008, Ian Braid, Alan Grayer, and Charles Lang received the Award for their work at

Figure 11-40 Photo of Pierre Bézier (Courtesy of the Bézier family)

Cambridge University in the late 1960s through the 1980s. The three developed early *boundary representation* (B-rep) solid modelers. In the 1980s, the three were involved in the development of such *solid modeling kernels* as Parasolid in 1985 and ACIS in 1989. Their work has had a profound

Figure 11-41 List of Bézier Award recipients

influence on today's commercial solid modeling systems.

Richard Riesenfeld and Elaine Cohen received the Award in 2009. In his PhD dissertation (Syracuse University, 1973), Riesenfeld introduced **B-splines**.

The 2012 recipient was Paul de Faget de Casteljau. At around the same time that Bézier worked at Renault, de Casteljau worked at Citroen, another French car company. Paul de Casteljau made many significant contributions in the area of surface modeling but, owing to policies at Citroen, he was prevented from publishing his work until 1974.

In 2013, Thomas W. Sederberg, a professor at Brigham Young University, was the Bézier Award recipient. In a 2003 paper, Sederberg introduced **T-splines**. In 2004, he co-founded a company to commercialize T-splines, which was then acquired by Autodesk in December 2011.

▮ QUESTIONS

MULTIPLE CHOICE

1. If a B-spline has five control points and two spans, what is the degree of the curve?
   a. One
   b. Two
   c. Three
   d. Four
   e. Five
   d. Six

SS 2. If the degree of a Bezier curve is four, then how many control points must it have?
   a. One
   b. Two
   c. Three
   d. Four
   e Five
   f. Six

3. NURBS and T-Splines have which of the following in common:
   a. Control points
   b. T-points
   c. Star points
   d. Rectangular topology
   e. All of the above
   f. None of the above

4. The natural shape of a NURBS surface is: SS
   a. Triangular
   b. Quadrilateral
   c. Pentagonal
   d. Hexagonal

5. Which of the following modeling methods employs a recursive technique to arrive at a smooth surface?
   a. Subdivision surfaces
   b. Polygon meshes
   c. NURBS surfaces
   d. T-spline surfaces

6. In the figure of the T-spline surface shown SS here, how many star points are visible?
   a. One
   b. Two
   c. Three
   d. Four
   e. Cannot be determined

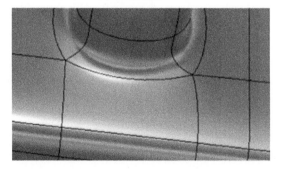

CHAPTER

12

PRODUCT DOCUMENTATION

▐ WORKING DRAWINGS

At the beginning of any project, there is no guarantee that a design will actually be executed. If a project is put out to bid, for example, only one of several competing preliminary designs will be selected. Even then, funding for the project may fall through. Similarly, for a company engaged in both research and development, management may decide to abandon the development of a new product.

Once the decision is made to build, however, the existing preliminary design must be further developed and detailed for production. The term *working drawing* is used to describe the complete set of drawing information needed for the manufacture and assembly of a product based on its design.

As discussed in a previous chapter, commercial products are almost always assemblies comprising several different parts. Perhaps the most recognizable element of a working drawing set is the *assembly drawing*. The purpose of the assembly drawing is to show how the different components fit together to form the product. An example of an assembly drawing is shown in Figure 12-1.

An essential element of a working drawing is the parts list, or *bill of materials* (abbreviated BOM). The purpose of the BOM is to identify all parts, both standard and nonstandard, used in an assembly. The BOM for the assembly shown in

Figure 12-1 is located in the upper-right corner of the drawing.

In addition to the assembly drawing, a set of working drawings also includes *detail drawings* of all nonstandard parts. As seen in Figure 12-2, these individual part drawings contain multiple views, dimensions, notes, tolerance information, material specifications, and any other information necessary to manufacture the part. Working drawings are typically accompanied by written instructions called *specifications*, which serve to further clarify the details for manufacturing the product. An excerpt from the specifications describing the construction of a tugboat appears in Figure 1-14 in Chapter 1.

In addition to describing the details of a product's design and manufacture, working drawings and written specifications also serve as legal contracts. In the event that a problem arises with a design, working drawings may be called upon to help establish liability. Figure 12-3 shows a professional engineer's stamp, taken from an engineering drawing. In stamping and signing the drawing, the engineer takes responsibility for the accuracy of the drawing's contents.

▐ MODEL-BASED DEFINITION

2D engineering drawings, frequently derived from 3D solid models, have historically been used to provide product documentation. This

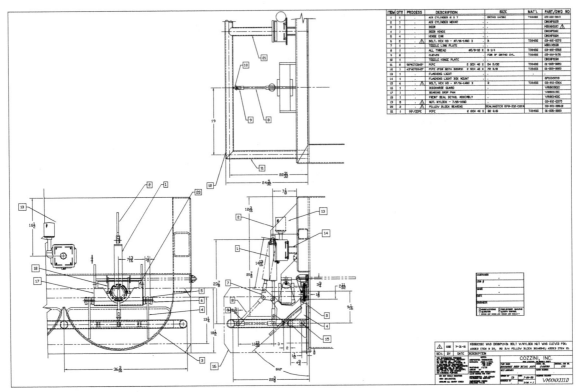

Figure 12-1 Typical assembly drawing (Courtesy of Cozzini, Inc.)

documentation contains sufficient information to manufacture and inspect the product. This product information includes dimensions, geometric dimensioning and tolerancing (GD&T), notes,

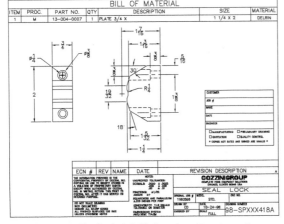

Figure 12-2 Typical detail drawing (Courtesy of Cozzini, Inc.)

EXPIRES   07/18/2008

NOT VALID WITHOUT SIGNATURE.

Figure 12-3 Professional engineer's stamp (Courtesy of Jensen Maritime Consultants, Inc.)

specifications, process and inspection information, component materials, assembly level bills of materials, etc.

Still, for decades now, the goal[1] has been to eliminate the need for 2D working drawings, to go "paperless." **Model-Based Definition (MBD)** is the practice of using 3D CAD models to fully define (that is, provide specifications for) product components and assemblies. Model-based definition also refers to the 3D annotated model with associated data attributes that fully captures the product definition so that it can be effectively used by downstream stakeholders, without the use of traditional drawings.

The ASME first published a standard for developing MBD's in 2003. This document, ASME Y14.41-2003 Digital Product Definition Data Practices,[2] was revised in 2012 as ASME Y14.41-2012. The Y14-41 standard supports two alternative methods for the development of the product definition data: (1) model only in digital format or (2) model and drawing in digital format.

The term **Model-Based Enterprise (MBE)** is used to describe the collaborative environment within an organization that has (replaced 2D drawings and) adopted model-based product definition as the single, authoritative information source used for communication (between people, machines, hardware and software systems) across the enterprise.

The cornerstone of this digital transformation is the MBD. Additionally, because CAD models are stored in proprietary data formats, and the organization must protect their intellectual property, the MBD's must be translated into a suitable compatible format. Examples of these lightweight standard 3D formats include 3D PDF, JT, STEP AP 242, and ANSI QIF.

Modern manufacturing companies choose to focus on 3D in order to avoid duplication in 2D of what they already have in 3D. Relying both on a 3D CAD database and 2D drawings runs the risk of the two getting out of sync, causing inevitable mistakes. Plus a 3D model is easier to visualize and understand and is more accurate. Another factor driving this trend is the fact that today's manufacturing equipment can now

work directly with 3D, which again helps to minimize mistakes. Most importantly, 3D geometric product definition is critical for quality, collaboration, and efficiency. A change in the model is immediately felt throughout the enterprise. This is referred to as the **digital thread**.

▌DETAIL DRAWINGS

A set of working drawings includes detail drawings of all nonstandard parts included in the product or assembly. Standard parts, either vendor-purchased or developed within the company, do not require individual part drawings. A detail drawing is a fully dimensioned multiview drawing that contains all of the information necessary to manufacture the part. Figure 12-4 shows a detail drawing of a shaft part. Part drawings are either developed directly in 2D or extracted from a 3D model. A typical part drawing includes the following information:

- Drawing views
- Dimensions
- Tolerances
- Material designation
- Surface finish
- Notes
- Title block
- Revision block

Although many parts are simple enough to be included together with other parts on the same drawing, there are good reasons to place each part on its own drawing. Doing so makes it possible to use a single identification number to identify both the part and the drawing. This same number can also be used to name the CAD file that documents the part. This practice considerably simplifies the task of keeping track of the ever-increasing number of parts, drawings, and files that even a small company must manage. It also makes it easy for the company to reuse parts in different assemblies.

In many companies, parts are identified by a unique code number. This code or identification number contains information about the part. Formal classification and coding techniques have been developed and are used to group similar parts

---

[1]Especially in the military, aerospace, and automotive industries.

[2]A *digital product definition data set* is another term for model-based definition.

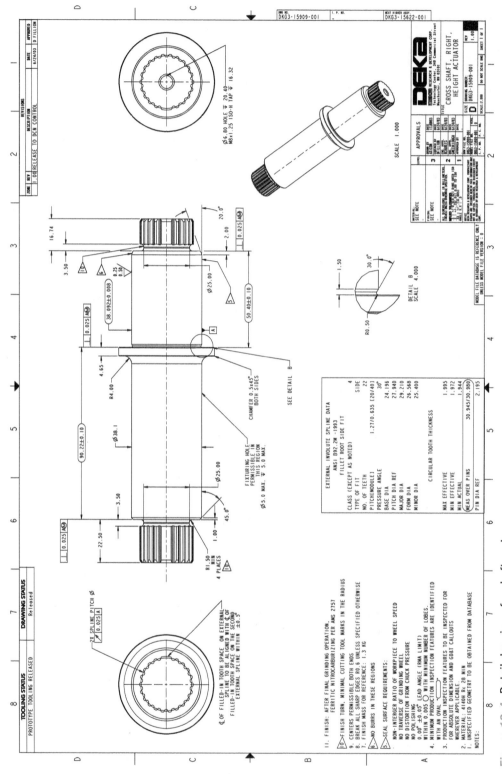

Figure 12-4 Detail drawing of a shaft part

on the basis of such characteristics as material, size, shape, function, process, or other information.

ASSEMBLY DRAWING VIEWS

The main purpose of an assembly drawing is to show all of the different parts in the assembly, and how these parts fit together to create the mechanism, device, component, or product. Using parametric modeling software, it is relatively easy to generate the views needed to accomplish this task. Two of the most commonly used assembly views are the section and the exploded view.

A section view of an assembly is used when the relationship between the different parts may not be apparent from an external view, as is the case with the ball valve shown in Figure 12-5. In addition to the full section, half section and broken-out assembly views are also used, as seen in Figures 12-6 and 12-7, respectively, for a globe valve.

Certain conventions regarding section lines are to be followed when creating assembly section views:

- The section lining used for adjacent parts through which the cutting plane passes should be different. Either use different section line patterns for each material (as shown in Figure 12-5) or use the general-purpose (cast

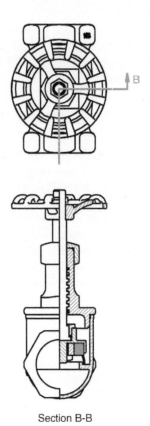

Section B-B

Figure 12-6 Half section view of a globe valve

iron) section lining and vary the angle to help distinguish different parts (the approach used in Figures 12-6 and 12-7).

- Standard parts and solid parts with no internal detail are not sectioned. These parts include

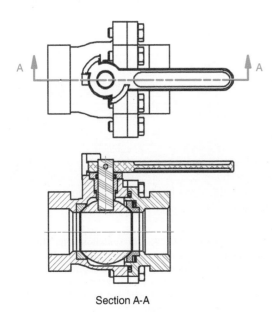

Section A-A

Figure 12-5 Assembly section view of a ball valve

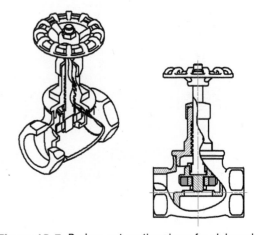

Figure 12-7 Broken-out section view of a globe valve

nuts, bolts, screws, shafts, pins, keys, bearings, spokes, and ribs. The shaft and nuts on the globe valve appearing in Figures 12-6 and 12-7 are unsectioned, for example.

- Extremely thin parts, such as gaskets and sheet metal parts, are shown solid, rather than sectioned.

An exploded view, like the one of the ball valve shown in Figure 12-8, is also fairly easy to extract from a parametric assembly model. Because exploded views are easy to visualize, clearly showing the different parts and how they fit together to form an assembly, these views are commonly found in installation manuals and part catalogs.

When creating an assembly drawing, use the minimum number of views necessary to describe the part relationships. Often a single view is sufficient. It is not necessary to show individual parts in detail, because this is accomplished in the detail drawings. Standard parts are shown in the assembly views. Dimensions are not shown in an assembly drawing unless they pertain directly to the assembly as a whole. Hidden lines are typically not shown in an assembly drawing.

In addition to the assembly view showing all of the parts, the other main elements of an assembly drawing include a parts list providing information about these parts, and balloons used to identify the parts and relate them to the parts list. The parts list and balloons are discussed in the following section.

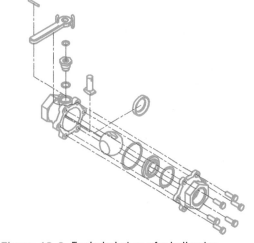

Figure 12-8 Exploded view of a ball valve

▌ BILL OF MATERIALS AND BALLOONS

The BOM (bill of materials), or parts list,[3] provides a tabular list of information about the individual parts in the assembly. As shown in Figure 12-9, a parts list includes an item number, a unique part identification number, a brief verbal description of the part, the quantity found in the assembly, the material, and possibly other information, such as weight or stock size. Note that standard parts are also included in the parts list.

The parts list generally appears on the right-hand side of the assembly drawing. Parts are listed in order of importance and/or size. If the column headings appear at the top of the parts list, then the most important parts appear at the top of the list. On some drawings the column headings appear at the bottom, in which case the most important parts appear at the bottom of the list. This practice makes it possible to add new, presumably less important parts without affecting the item numbering.

Balloons, like those shown in Figure 12-10, are used to identify parts and relate them to the parts list. Each balloon consists of a circled number and a leader line. The leader line is drawn to a specific part. The circled number is the item number (or find number) for the part; it is used to locate

ITEM	QTY	PROCESS	DESCRIPTION	SIZE	MAT'L	PART/DWG NO
1	1	.	AIR CYLINDER 2 X 7	ORTHO U45BC	T304SS	05-001-0015
2	1	.	AIR CYLINDER MOUNT	.		DXXXNP112B
3	1	.	DOOR	.		MBXXN0310C ⚠
4	1	.	DOOR HINGE	.		DXXXNP116C
5	4	.	HINGE EAR	.		DXXXNP118A
6	2	⚠	BOLT, HEX HD - Ø7/16-14NC X	3	T304SS	09-012-0070
7	1	.	TOGGLE LINK PLATE	.		MBXLX0133B
8	1	.	ALL THREAD	#5/8-18 X 8 1/4	T304SS	09-012-0351B
9	2	.	CLEVIS	FOR 2" ORTHO CYL.	T304SS	05-001-0150
10	1	.	TOGGLE HINGE PLATE	.		DXXXNP119A
11	2	90"MITER45"	PIPE	2 SCH 40 X 24 5/32	T304SS	01-005-0028
12	1	45"MITER45"	PIPE GTDR BOTH DOORS	2 SCH 40 X 72 3/8	T304SS	01-005-0029
13	1	.	FLASHING LIGHT	.	.	.
14	1	.	FLASHING LIGHT BOX MOUNT	.		SPXXX919B
15	6	⚠	BOLT, HEX HD - Ø7/16-14NC X	2	T304SS	09-012-0066
16	1	.	DISCHARGE GUARD	.		VXXXN0901C
17	1	.	BEARING DRIP PAN	.		VXXXK0118C
18	1	.	FRONT SEAL DETAIL ASSEMBLY	.		VXXXK0400C
19	8	⚠	NUT, NYLOCK - 7/16-14NC	.		09-012-0375
20	2	⚠	PILLOW BLOCK BEARING	SEALMASTER RPB-212-C2CR		03-001-0084B
21	1	90"/COPE	PIPE	2 SCH 40 X 22 3/8	T304SS	01-005-0026

Figure 12-9 Parts list for a specific product (Courtesy of Cozzini, Inc.)

[3]A **parts list** is a drawing file representation of the Bill of Materials information. The parts list is configured to provide the information required for the manufacturing drawing. A **BOM (Bill of Materials)** is a table that represents database information about the components in an assembly. This table includes quantities, names, costs, vendors, and other information someone building an assembly requires.

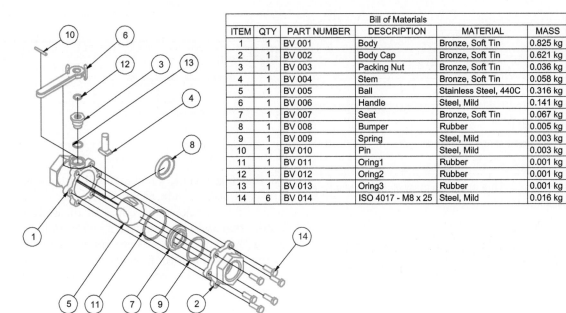

Bill of Materials					
ITEM	QTY	PART NUMBER	DESCRIPTION	MATERIAL	MASS
1	1	BV 001	Body	Bronze, Soft Tin	0.825 kg
2	1	BV 002	Body Cap	Bronze, Soft Tin	0.621 kg
3	1	BV 003	Packing Nut	Bronze, Soft Tin	0.036 kg
4	1	BV 004	Stem	Bronze, Soft Tin	0.058 kg
5	1	BV 005	Ball	Stainless Steel, 440C	0.316 kg
6	1	BV 006	Handle	Steel, Mild	0.141 kg
7	1	BV 007	Seat	Bronze, Soft Tin	0.067 kg
8	1	BV 008	Bumper	Rubber	0.005 kg
9	1	BV 009	Spring	Steel, Mild	0.003 kg
10	1	BV 010	Pin	Steel, Mild	0.003 kg
11	1	BV 011	Oring1	Rubber	0.001 kg
12	1	BV 012	Oring2	Rubber	0.001 kg
13	1	BV 013	Oring3	Rubber	0.001 kg
14	6	BV 014	ISO 4017 - M8 x 25	Steel, Mild	0.016 kg

Figure 12-10 BOM, balloons used to identify parts

the part in the parts list. The balloons should be neatly organized, preferably in horizontal or vertical rows. Balloon leaders should not cross, and adjacent leaders should be parallel.

Once an assembly model is available, a parts list can be automatically generated and customized using parametric modeling software. A balloon tool is also generally available.

SHEET SIZES

Standard drawing sheet sizes for both metric and English units are shown in Table 12-1. These and other engineering graphics standards and conventions are maintained in publications by the American Society of Mechanical Engineers (ASME) and are approved by the American National Standards Institute (ANSI).

TITLE BLOCKS

Every drawing contains a *title block* similar to the one shown in Figure 12-11. The purpose of the title block is to organize the information needed to identify the drawing. Additional information that is not given directly on the drawing is also included in the title block. Standard title block information includes the following:

Table 12-1 Standard sheet sizes

Metric (mm)		English (inches)	
A4	210 × 297	A	8.5 × 11.0
A3	297 × 420	B	11.0 × 17.0
A2	420 × 594	C	17.0 × 22.0
A1	594 × 841	D	22.0 × 34.0
A0	841 × 1189	E	34.0 × 44.0
ASME Y14.1M-1995		ASME Y14.1-1995	

- Company name and address
- Drawing title
- Drawing number
- Names, dates, and release signatures of designer, checker, and supervisor
- Revision number
- Sheet size
- Sheet number
- Scale of drawing
- Other information: project name, client name, material, general tolerance information, heat treatment, surface finish, hardness, weight estimate

/	OUTBOARD PROFILE		103025-101-1
NO.		TITLE	DRAWING NO.

REFERENCES

PROJECT 96 FT TRACTOR TUG – HULL 798

CLIENT EASTERN SHIPBUILDING GROUP, PANAMA CITY, FLORIDA

TITLE

GENERAL ARRANGEMENTS

JMC
Jensen Maritime Consultants Inc.
NAVAL ARCHITECTS   MARINE ENGINEERS
4241 21st Ave West, Suite 404
Seattle, Washington 98199
www.jensenmaritime.com   Ph 206 284 1274
Fax 206 284 2556

DWN	JGP/CSW	DATE	5/13/03
CKD		DATE	
APP		DATE	
SCALE	1/4" = 1'- 0"		
DWG NO.			REV
103025–101–3			–
SHEET	1 OF 1		

Figure 12-11 Title block (Courtesy of Jensen Maritime Consultants, Inc.)

The company logo is also frequently found on the title block. On most drawings the title block is located in the lower right-hand corner of the drawing.

ANSI standard title blocks are available for both metric and English units in various sheet sizes. Companies tend to use their own standard title blocks. These standard title blocks can be made available from a CAD library or as a template, and they are easily inserted into a CAD file. In many CAD programs, the title block text fields are already completed or are set up for easy modification. In parametric solid-modeling software, the title block text fields are associatively linked to the CAD database.

Drawings often contain more information than can be comfortably included on a single sheet. In this case a continuation title block can be used on all sheets after the first sheet. The continuation title block contains less information than the title block on the main sheet. Figure 12-12 shows an example of a continuation title block.

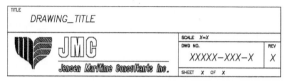

TITLE

DRAWING_TITLE

JMC
Jensen Maritime Consultants Inc.

SCALE	X=X		
DWG NO.			REV
XXXXX–XXX–X			X
SHEET	X OF X		

Figure 12-12 Continuation title block (Courtesy of Jensen Maritime Consultants, Inc.)

▊ BORDERS AND ZONES

All drawings include a rectangular border. Some drawings also include *zones*. Zones are regular ruled intervals along the edges of the drawing border that assist users in finding information on a complicated drawing. These zones are similar to the zones found on highway road maps, with numerals applied horizontally and letters vertically. Figure 12-13 shows a portion of a drawing where a detail view is labeled according to the zone in which it is located.

▊ REVISION BLOCKS

Changes are often made to engineering drawings to account for design changes, customer requests, errors, and so on. These changes, or *revisions*, are kept track of in a *revision block* typically located on the upper-right corner of the drawing. Figure 12-14 shows an example of a revision block. At a minimum, each modification recorded in the revision block includes the date, the name of the person responsible, and a brief description of the change. A revision number, normally placed in circle or triangle, is associated with each revision and placed both on the revision block and in the area on the drawing sheet where the revision has been made. In a zoned drawing, the zone of the affected area is listed in the revision block.

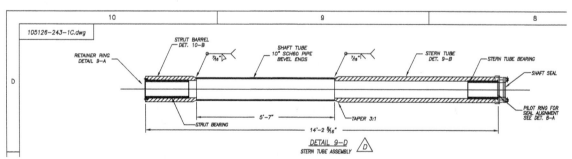

Figure 12-13 Drawing detail labeled by zone (Courtesy of Jensen Maritime Consultants, Inc.)

REVISIONS				
SYM	ZONE	DESCRIPTION	BY	DATE
P0	ALL	INITIAL ISSUE	YH	4/20 06
P1		1) OVERALL SHAFT LENGTH ADJUSTED PER UNDER WATER SHIPCHECK 4-26-06. 2) RUDDER AND EXISTING STRUT BEARING LOCATED. 3) NEW PROPELLER AND NOZZLE LOCATED TO CLEAR EXISTING RUDDER. 4) TRIMMING OFF 9-½" OF AFT EDGE OF SKEG PLATE SHOWN, TO SUIT THE LONGER CP PROPELLER HUB.	YH	4/29 06
P2		1) NUMBER AND LOCATION OF SHAFT BEARING REVISED. 2) ONE PILLOW BEARING REMOVED. 3) STERN TUBE AND BULKHEAD STUFFING BOXES REPLACED WITH SHAFT SEALS. 4) INDIVIDUAL LENGTHS OF TAIL SHAFT AND LINE SHAFT REVISED (OVERALL LENGTH REMAINS UNCHANGED) 5) SHAFT LINE MATERIALS CHANGED TO CARBON STEEL GR. 4 WITH BRONZE LINERS IN WAY OF STRUT AND STERN TUBE/SHAFT SEAL. 6) SHAFT DIAMETER AND STERN TUBE/STRUT DIAMETERS INCREASED. 7) RESERVATION NOTE #3 REMOVED.	YH	5/3 06
-		1) NEW ENGINE BLOCK RECEIVED FROM CUMMINS INSERTED IN DRAWING. 2) CRAFT MODEL S1 BCH 508 FL PILLOW BLOCK SHOWN. 3) SHAFT LINE ELEVATION (HEIGHT ABOVE BASE LINE) CORRECTED. 4) RESERVATION NOTE #2 REMOVED. GEN. NOTE #2 & #3 REVISED. 5) PRELIMINARY STATUS REMOVED.	YH	5/12 06
A		REVISED PROPELLER HUB FROM HUNDESTED INSERTED.	YH	5/19 06
B		1. LINE SHAFT BEARING RELOCATED PER HUNDESTED SUGGESTION. 2. PROPELLER NOZZLE TILTED 2 DEGREES.	YH	5/31 06
C		1. LINE SHAFT BEARING RELOCATED TO FRAME #39. 2. TAIL SHAFT LENGTH INCREASED, LINE SHAFT LENGTH DECREASED. 3. STERN TUBE AND STRUT DETAILS ADDED (SHT #2).	YH	6/12 06

Figure 12-14 Revision block (Courtesy of Jensen Maritime Consultants, Inc.)

DRAWING SCALE

CAD models, whether 2D and 3D, should always be created full-size, without concern for how large or small the actual object may be. In this way the CAD data are based on the true size of the object. Added dimensions will be true-size, query commands will report true-size data, and the CAD database will be suitable for export to downstream analysis and manufacturing applications.

Once the full-size model is complete, drawings are prepared and scaled to fit the designated drawing sheet size. Table 12-2 provides examples of some common scales used on engineering drawings.

The drawing scale is normally indicated on the title block. In the event that different views on the drawing sheet are drawn at different scales, the scale of the view is indicated on the view label.

Table 12-2 Common drawing scales

1:1	1/8" = 1'0"
1:2	1/4" = 1'0"
1:4	3/8" = 1'0"
1:8	1/2" = 1'0"
1:10	1" = 1'0"
1:20	1" = 10'
1:30	1" = 20'
1:40	1/4" = 1"
1:50	1/2" = 1"
1:100	1" = 1"

NOTES:

UNSPECIFIED TOLERANCES:

DECIMALS: .0000 ± .0005
.000 ± .005
.00 ± .01

FRACTIONS: ±1/32
ANGLES ±1°

SQUARENESS AND PARALLELISM
±.015 INCHES PER FOOT

CONCENTRICITY: T.I.R EQUALS
TOLERANCE ON DIAMETER

DIMENSIONING SYSTEM:
ANSI/ASME Y14.5M

Figure 12-15 General tolerance note (Courtesy of Cozzini, Inc.)

TOLERANCE NOTES

Many companies include a general tolerance note similar to the one shown in Figure 12-15 on their detail drawings. These tolerance notes refer to any dimensions that have not been specifically toleranced. Tolerance notes are typically found in the lower-right corner of the drawing.

STANDARD PARTS

Standard parts may either be purchased from outside or produced within the company. Typical standard parts include threaded fasteners, bushings, bearings, pins, keys, pumps, and valves. Though standard parts do not require a detail drawing, they are shown on the assembly drawing and included in the parts list. Most 3D parametric modeling programs have a built-in standard parts library, from which standard part models can be directly inserted into an assembly model. Figure 10-36 in Chapter 10 shows a screen capture of a CAD library interface within a parametric modeling program.

WORKING DRAWING CREATION USING PARAMETRIC MODELING SOFTWARE

In the remaining pages of this chapter, various techniques to develop working drawings are described. These parametric modeling techniques include the creation of part drawings, assembly models, sectioned assembly and exploded views, as well as adding a parts list and balloons to an exploded view.

The steps required to generate a detail drawing of a part using parametric modeling software are described.

Step 1
Insert a base view of the part model.

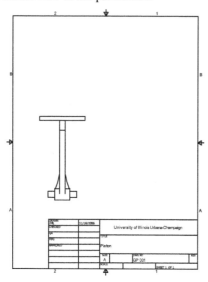

Step 2
Using the base view, project the other views.

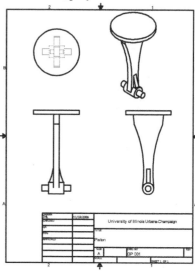

Step 3
Add centerlines.

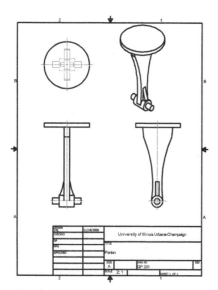

Step 4
Import parametric dimensions from part model. Note that these dimensions will probably need to be repositioned. In addition, at least some of these imported parametric dimensions will not be suitable for documentation purposes and will need to be replaced.

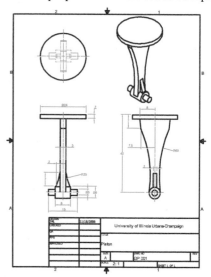

Figure 12-16 Extracting a detail drawing from a parametric model

The steps required to create a parametric assembly model are described, assuming that all of the part models have already been created.

Step 1

Import the base part into the assembly environment.

Step 2

Import the other component files into the assembly.

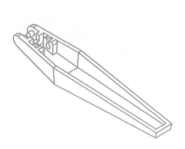

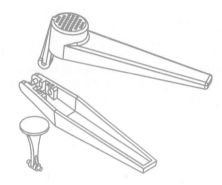

Step 3

Using assembly constraints like mate and insert, correctly position the components with respect to the base component and to one another.

Step 4

Properly constrained, moving parts will behave as they would in reality. The garlic press assembly model can be opened and closed, simulating the behavior of the actual device.

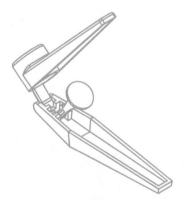

Figure 12-17 Creating an assembly model

WORKING DRAWING CREATION USING PARAMETRIC MODELING SOFTWARE

This section shows the steps needed to create a sectioned assembly drawing from a parametric assembly model.

Step 1

Import a base view of an assembly into the drawing environment.

Step 2

Using a sectioning tool, create the section view(s). Note that the angle of the section lining (ANSI 31) differs for the different parts through which the cutting plane passes. Note also that some conventions (such as not applying section lining to cut thin features) are not automatically adhered to.

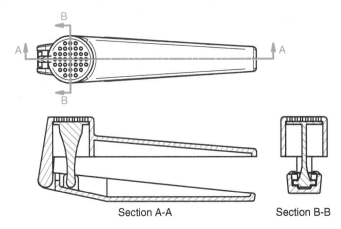

Step 3

Modify the hatch pattern to adhere to convention. Note also that the holes in the top of the cylindrical handle part have been suppressed for clarity at this stage.

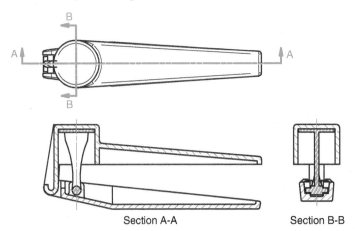

Figure 12-18 Extracting a sectioned assembly drawing from a parametric assembly model

This section shows the steps required to develop an exploded view of a parametric assembly model.

Step 1

Import an assembly model into the exploded-view creation environment.

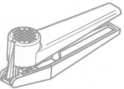

Step 2

Rotate open the cylindrical handle part.

Step 3

Translate the cylindrical handle part vertically.

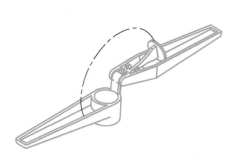
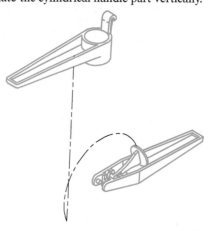

Step 4

Rotate the piston part to an upright position.

Step 5

Translate the piston part vertically.

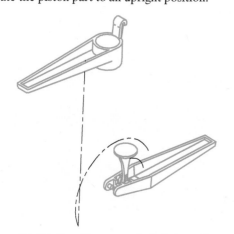
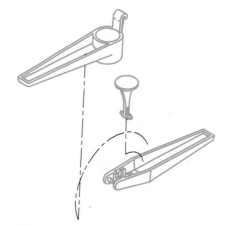

Figure 12-19 Creating an exploded view of a parametric assembly model

The process of generating an exploded-view drawing, parts list, and balloons is addressed, using parametric modeling software.

Step 1

Insert an exploded view into the drawing environment.

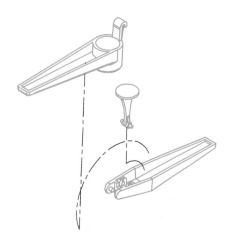

Step 2

Insert a parts list that is linked to the exploded view.

Parts List			
ITEM	QTY	PART NUMBER	DESCRIPTION
1	1	Handle	
2	1	Piston	
3	1	Cylinder_Handle	

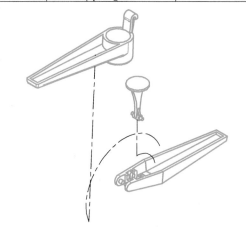

Step 3

Customize the parts list.

Bill of Materials					
ITEM	QTY	PART NUMBER	DESCRIPTION	MATERIAL	MASS
1	1	GP 001	Handle	Aluminum-6061	0.046 kg
2	1	GP 002	Piston	Aluminum-6061	0.006 kg
3	1	GP 003	Cylinder_Handle	Aluminum-6061	0.059 kg

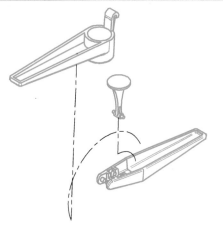

Step 4

Add balloons.

Bill of Materials					
ITEM	QTY	PART NUMBER	DESCRIPTION	MATERIAL	MASS
1	1	GP 001	Handle	Aluminum-6061	0.046 kg
2	1	GP 002	Piston	Aluminum-6061	0.006 kg
3	1	GP 003	Cylinder_Handle	Aluminum-6061	0.059 kg

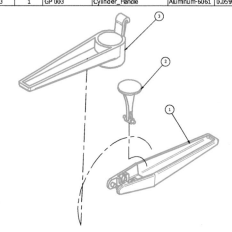

Figure 12-20 Creating an exploded-view drawing with parts list and balloons from a parametric assembly model

Fiskars Group is one of the oldest companies in the world. The company was founded in 1649 in the village Fiskars, Finland. Today the global headquarters are in Helsinki. Fiskars creates tools for use in and around the home, including household goods, garden tools, and craft tools.

The following figures show product documentation for a well-known Fiskars product, hand pruning shears. The documentation was created by the industrial design group at Fiskars. Figure 12-21 shows several concept sketches for the pruner. Figure 12-22 shows different orthographic views of the product, including removed section views of the handles. Figure 12-23 shows a rendered view of the product, and Figure 12-24 shows an exploded view and parts list.

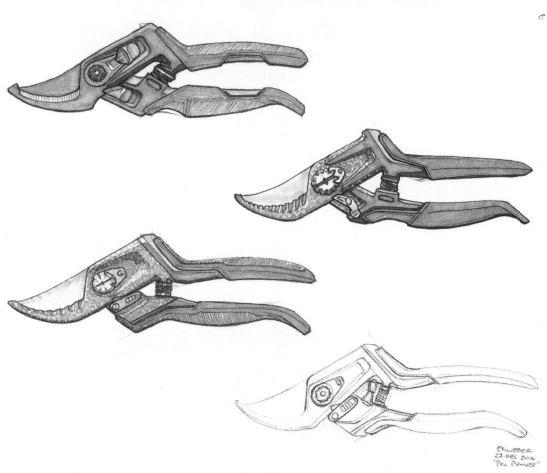

Figure 12-21 Fiskars Hand Pruner—concept sketches

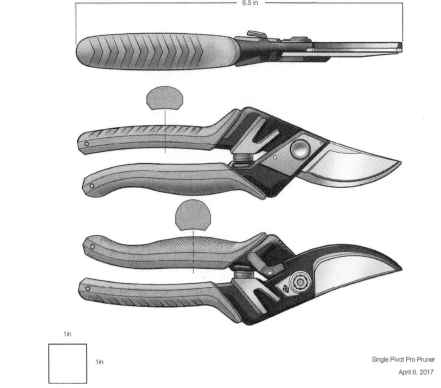

8.5 in

1in

1in

Figure 12-22 Fiskars Hand Pruner—orthographic views

Figure 12-23 Fiskars Hand Pruner—rendered view (Fiskars Group)

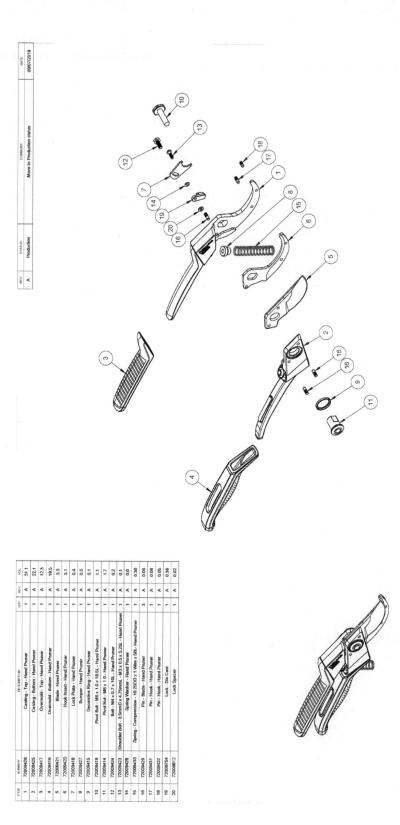

Figure 12-24 Fiskars Hand Pruner—exploded view

1. Given a dimensioned isometric view (Figures P10-1 through P10-24 in Chapter 10), create a dimensioned detail drawing of the part.

2. Given a dimensioned assembly view (Figures P12-1 through P12-5), create a complete set of working drawings of the assemblies.

    a. Create a solid model of each part.

b. Create dimensioned detail drawings for each of the individual parts, including auxiliary views and section views when appropriate.

c. Create an exploded pictorial assembly drawing, including balloons and parts list.

d. Where appropriate, create a sectioned assembly drawing.

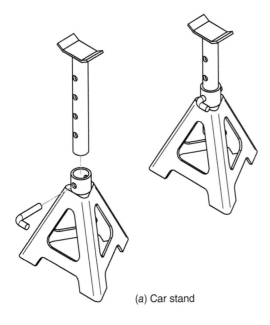

(a) Car stand

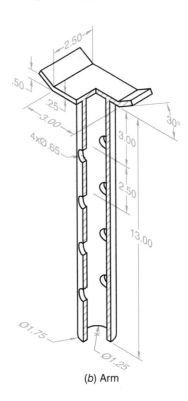

(b) Arm

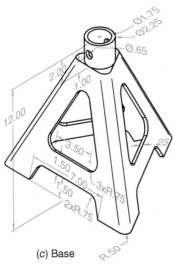

(c) Base

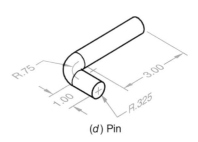

(d) Pin

Figure P12-1

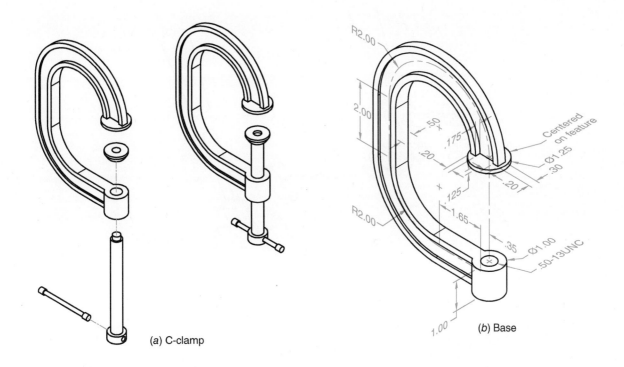

(a) C-clamp

(b) Base

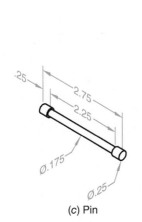

(c) Pin

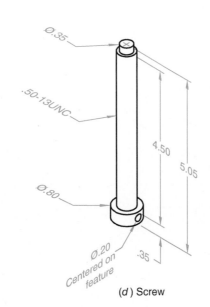

(d) Screw

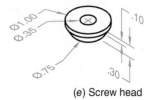

(e) Screw head

Figure P12-2

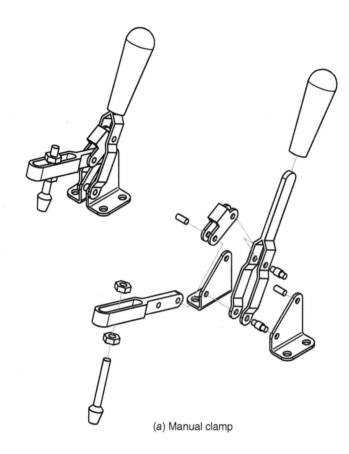

(a) Manual clamp

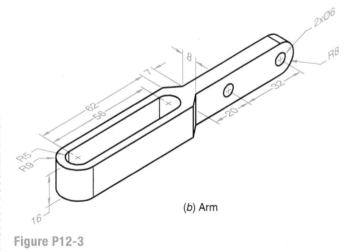

(b) Arm

Figure P12-3

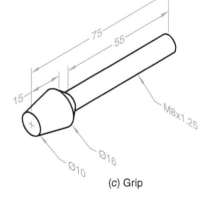

(c) Grip

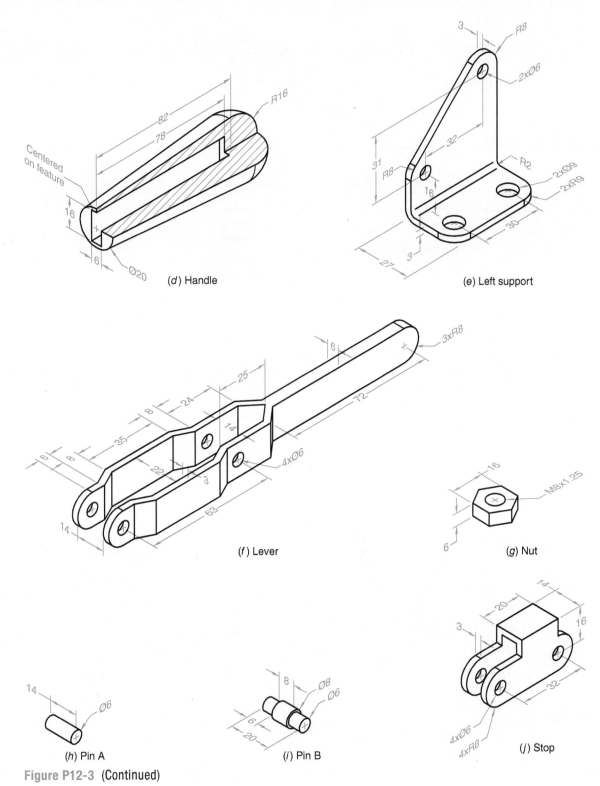

(d) Handle

(e) Left support

(f) Lever

(g) Nut

(h) Pin A

(i) Pin B

(j) Stop

Figure P12-3 (Continued)

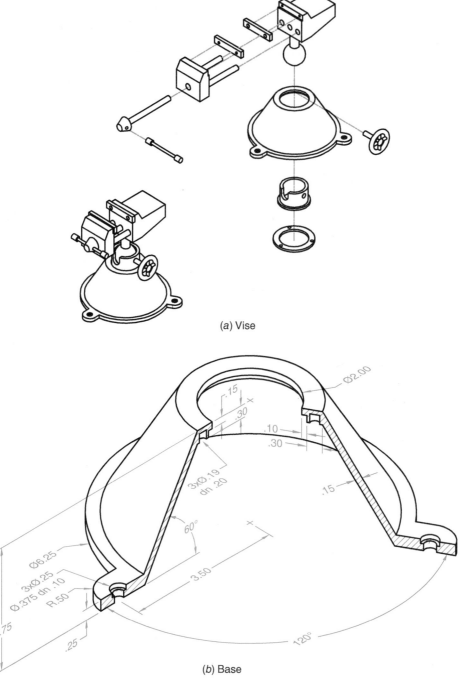

(a) Vise

(b) Base

Figure P12-4

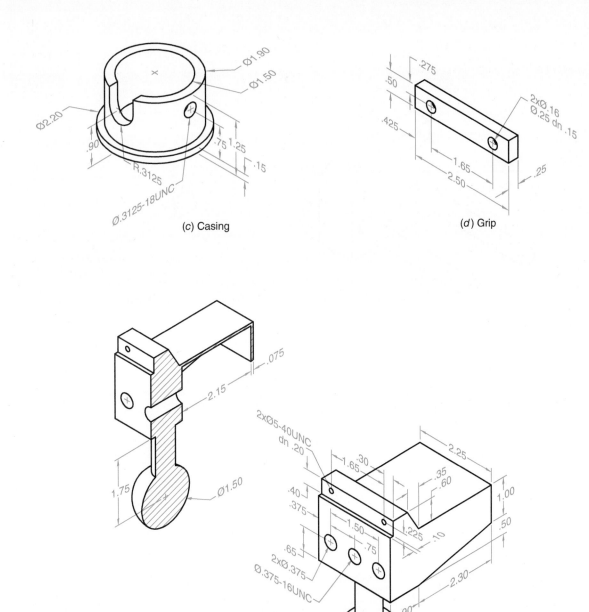

(c) Casing

(d) Grip

(e) Head

Figure P12-4 (Continued)

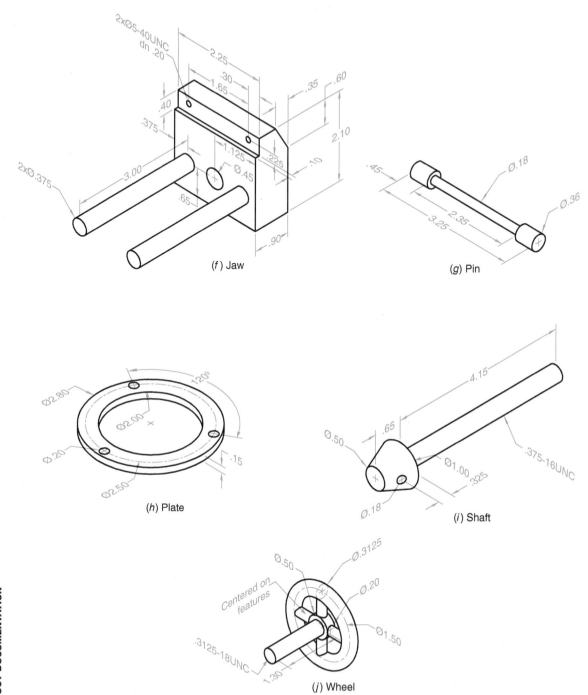

(f) Jaw

(g) Pin

(h) Plate

(i) Shaft

(j) Wheel

Figure P12-4 (Continued)

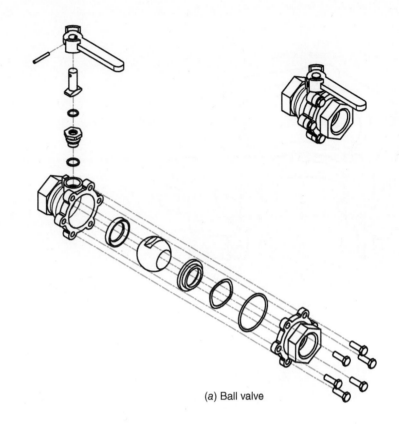

(a) Ball valve

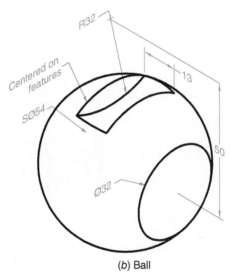

(b) Ball

Figure P12-5

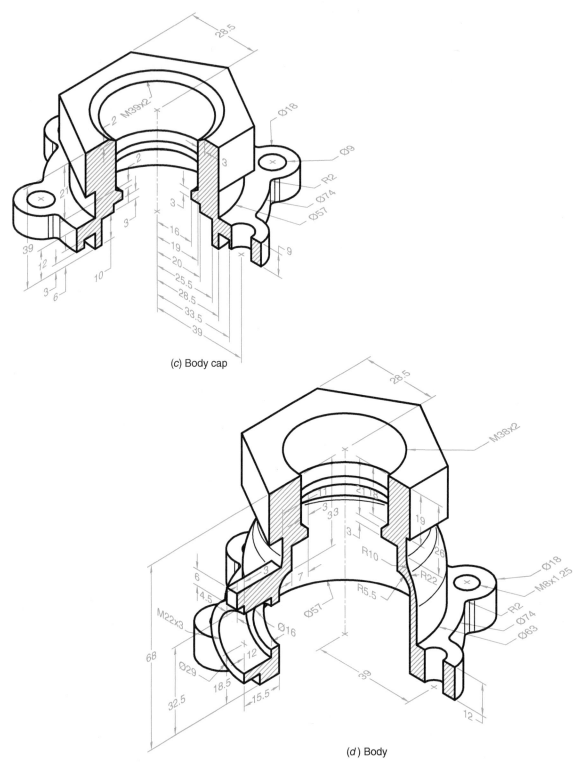

(c) Body cap

(d) Body

Figure P12-5 (Continued)

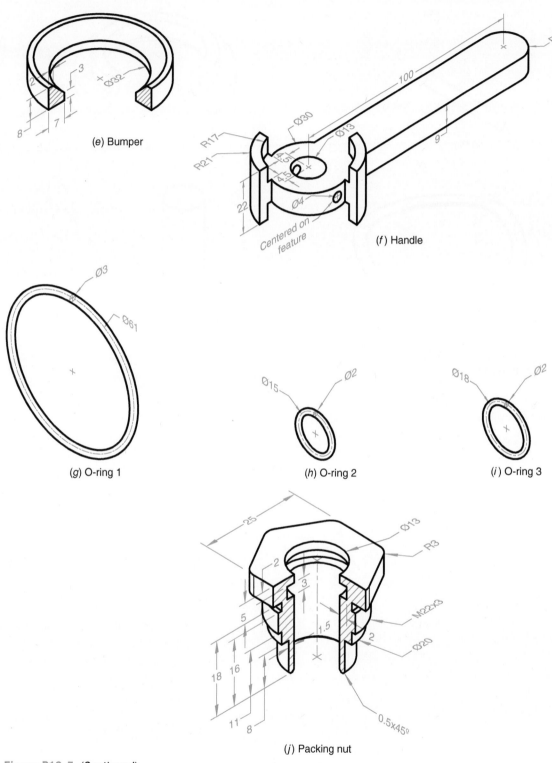

(e) Bumper

(f) Handle

(g) O-ring 1

(h) O-ring 2

(i) O-ring 3

(j) Packing nut

Figure P12-5 (Continued)

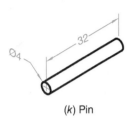

(k) Pin

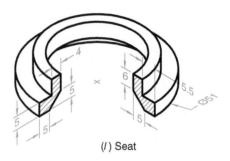

(l) Seat

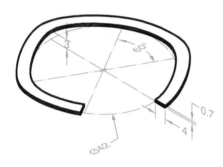

Uncompressed height is 7
compressed height is 3

(m) Spring

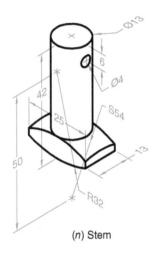

(n) Stem

Figure P12-5 (Continued)

CHAPTER

13 ADDITIVE MANUFACTURING

■ INTRODUCTION

Additive manufacturing (AM) refers to the fabrication of physical objects built layer by layer from CAD data. Additive manufacturing is a more descriptive term for what used to be called *rapid prototyping* (RP). Because "rapid prototyping" implies that the output is a prototype, and this is no longer necessarily the case (see paragraph on direct digital manufacturing below), the term "additive manufacturing" has been gaining ground in recent years. Additive manufacturing is popularly known as *3D printing*.

While many different additive manufacturing technologies have emerged since the 1980s, most all of them start with a digital CAD model.[1] The model geometry is then converted into cross-sectional slices, and the part is additively fabricated one physical cross section at a time.

In traditional subtractive manufacturing processes like machining, a computer numerical control (CNC) machine tool removes material from the workpiece to arrive at the final part. Additive manufacturing, as the name implies, is an additive process, with each part being built up layer by layer.

[1] An important exception where a CAD model is not required is 3D scanning and reality capture, discussed in Chapter 14. Here the STL file needed as input for an AM device is generated either from 3D scanning or using photogrammetric methods.

AM prototypes are used to evaluate the form, fit, and function of a design. A physical model, particularly one that can be quickly generated, enables all parties involved in the development process to visualize, discuss, and intelligently evaluate the form of a particular design. In this way, potential problems and misunderstandings are uncovered and resolved, thus avoiding costly mistakes that can go undetected until late in the product's development cycle.

3D prints are also used to evaluate whether components are being built to the tolerances required for assembly purposes. Finally, AM prototypes are used to verify that a design will function as intended. Common function verification tasks include:

- Demonstrating the practicality of an assembly. Many products are difficult—or even impossible—to assemble.

- Evaluating the kinematic performance of an assembly. Do moving parts perform as intended? Are there any unexpected interferences?

- Assessing aerodynamic performance. Here the geometric shape is of primary importance; a prototype made from a different material may be sufficient to perform this assessment.

In the event that such characteristics as the strength, fatigue, operational temperature limits, or corrosion resistance of a part are to be tested, the prototype must be made of the same material as the actual part. In this case, AM prototypes are sometimes used as patterns for other fabrication processes.

Direct digital manufacturing (DDM) refers to the use of additive manufacturing technologies for the production or manufacturing of end-use components. DDM is appropriate for parts with custom (unique, complex) geometries, small lot sizes, and where a fast turnaround is necessary. Examples include orthodontic treatment devices (e.g., clear braces) and hearing aid shells.

The most significant benefits of additive manufacturing are a compressed design cycle and improved product quality. Additive manufacturing dramatically reduces the time and expense required to take a new product from initial concept to final production, and it is also helpful in identifying design flaws.

Additive manufacturing speeds up the entire product development process. Building with an AM machine is typically performed in a single step. Traditional manufacturing processes require multiple, iterative stages to be carried out. For example, CNC machining requires careful planning and a sequential approach that may require the construction of fixtures before the part can be made. AM can eliminate or at least simplify many of these process stages.

In contrast to subtractive manufacturing techniques, AM technologies include the following unique characteristics[2]:

- **Design flexibility**—AM allows designers to create almost any shape; no tooling or fixturing is required, and minimal manufacturing constraints are placed upon the design.
- **Cost of geometric complexity**—when discussing AM, the phrase "complexity is free" is often used. Using AM, a simple shape (cube, cylinder) requires about the same time to fabricate as a complex geometric shape with the same enclosing volume. Furthermore, it may

not even be possible to produce the complex shape using subtractive methods.

- **Single-part assemblies**—using AM, it is possible to produce "single-part assemblies", that is, components that include working joints, both rotational and translational. This is accomplished by modeling the assembly in CAD with sufficient space between the parts so that, when printed, the gaps are filled with support material. Once the support material is removed, an assembly consisting of moving parts results (see Figure 13-1).
- **Time and cost efficiency** in production runs—because there is no startup tooling and low inventory costs, AM is well-suited for production involving low part quantities.

▋ AM TECHNOLOGIES

Although additive manufacturing has been around since the 1980s, in the past decade the industry has experienced an explosion in growth. As a consequence, only recently have attempts been made to categorize the different AM technologies. At the time of this writing, the following seven AM process categories are currently

Figure 13-1 Single-part assembly with moving parts

[2]W. Gao et al., *The Status, Challenges, and Future of Additive Manufacturing in Engineering*, 2015.

recognized by the International Standards Organization (ISO) and ASTM[3]:

1. ***Binder jetting***—a liquid bonding agent is selectively deposited to join powder materials.
2. ***Directed energy deposition***—focused thermal energy is used to fuse materials by melting as they are being deposited.
3. ***Material extrusion***—material is selectively dispensed through a nozzle or orifice.
4. ***Material jetting***—droplets of build material are selectively deposited.
5. ***Powder bed fusion***—thermal energy selectively fuses regions of a powder bed.
6. ***Sheet lamination***—sheets of material are bonded to form a part.
7. ***Vat photopolymerization***—liquid photopolymer in a vat is selectively cured by light-activated polymerization.

As the industry continues to develop and mature, these AM technology categories will continue to evolve. What follows is a brief description of four of these processes: vat photopolymerization, material extrusion, powder bed fusion, and material jetting.

Vat Photopolymerization

Stereolithography is a vat photopolymerization process used to produce parts from photopolymer materials in a liquid state using one or more lasers to selectively cure to a predetermined thickness and harden the material into shape layer upon layer. Stereolithography was developed by the 3D Systems Corporation in the 1980s and remains one of the most popular AM methods. In fact, the STL file format, originally developed by 3D Systems, is the industry standard for interfacing a virtual solid model with an AM machine. As seen in Figure 13-2, a stereolithography apparatus (SLA) builds

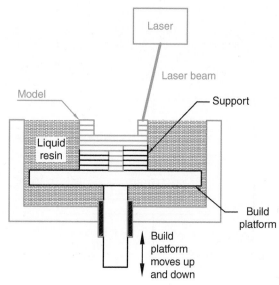

Figure 13-2 Stereolithography

plastic parts one layer at a time by tracing a concentrated beam of ultraviolet light onto the surface of a vat of liquid photopolymer resin. Inside the vat is a table that can move up and down. After each layer is completely traced, the worktable is lowered exactly one layer thickness into the vat of liquid, allowing fresh polymer to flow in and cover earlier work. Stereolithography requires a support structure when the part being built has undercuts—that is, when an upper cross section overhangs a lower cross section. A wide variety of photosensitive polymers are available for making parts, including clear, water-resistant, and flexible resins. Compared with other AM technologies, SLA has a high build speed. Parts made using stereolithography have a high resolution and good surface finish.

Material Extrusion

Fused deposition modeling (FDM) is a material extrusion process used to make thermoplastic parts through heated extrusion and deposition of materials layer by layer. The technology was developed by Stratasys, Inc. in the late 1980s.

[3]*ISO/ASTM 52900:2015 Standard Terminology for Additive Manufacturing—General Principles—Terminology.*

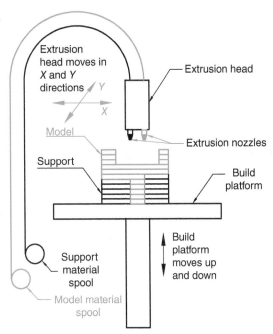

Figure 13-3 Fused deposition modeling (FDM)

Figure 13-4 Extruded parts with support material

A pre-heating chamber raises the material temperature to its melting point. Each layer is generated by extruding a thermoplastic material (e.g., ABS plastic) through a nozzle, as shown in Figure 13-3. As in stereolithography, parts are built on a platform or table that moves up and down. Once a complete layer is deposited, the table drops down and another layer begins. FDM prototypes also require a support structure, which is built automatically as the part builds. In FDM, two extrusion heads are often used, one for the modeling material and the other for a water-soluble support material. Figure 13-4 shows a recently printed part prior to removing the support material, while Figure 13-5 shows an agitation bath used to facilitate the removal of the support material.

FDM and other material extrusion processes are relatively inexpensive. Many FDM machines are office-friendly, and fairly fast when making small parts. FDM parts have good mechanical properties, to the point that functional parts can be produced. Recent patent expiration has led to a proliferation of low cost, consumer grade material extrusion products. This will be discussed in more detail in the section on low cost 3D printers.

Powder Bed Fusion

Powder bed fusion (PBF) processes utilize lasers and are known as laser sintering machines. *Laser sintering (LS)* is a powder bed fusion process used to produce objects from powdered materials using one or more lasers to selectively fuse or melt the particles at the surface, layer by layer, in an enclosed chamber. Some laser sintering machines work with polymers, while others work with metals, with fairly significant differences between the technologies employed by the two, but otherwise both adhere to the definition of LS provided above.

Selective laser sintering (SLS) was developed at the University of Texas at Austin to work with polymers and was later commercialized by the DTM Corporation in 1992. 3D Systems then purchased DTM in 2001. EOS GmbH—Electro Optical Systems, a German company, first manufactured an LS machine for producing plastic prototypes in 1994. With the expiration of several key patents related to plastic LS in 2014, several new companies have recently emerged, with new machines offering improvements over previous systems.

Laser sintering employs a powder that is selectively fused by a laser, as shown in Figure 13-6. After a counter-rotating roller spreads the powder, the laser is used to fuse a single layer

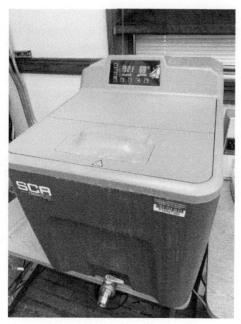

Figure 13-5 Agitation bath used to remove support material

of material. This process repeats until the part is complete. The process takes place inside an enclosed chamber filled with nitrogen gas in order to minimize oxidation and degradation of the powder material. Parts can be produced from a wide range of commercially available thermoplastic powder materials, including polyamide (i.e., nylon) and polystyrene. The mechanical

properties of plastic laser sintering parts produced using polyamide powders approach those of injection molded plastic parts.

In 1998 EOS released a direct metal laser sintering machine capable of processing metal powders. Today, in addition to EOS and 3D Systems, several European companies make laser-based systems for direct melting and sintering of metal powders.[4] Most any metal that can be welded is considered to be a good candidate for PBF processing, including various steels and titanium. Metal-based additive processes, including laser and electron beam, are among the fastest growing areas of AM. Parts produced using these technologies are increasingly common in aerospace and biomedical applications, owing to their excellent material properties and ability to handle geometric complexity.

As previously mentioned, PBF can, in contrast to many other AM processes, process a wide variety of materials, including polymers, metals,

Figure 13-6 Selective laser sintering

[4]In addition to powder-based fusion of metal powders using lasers, a system developed in Sweden (at Chalmers University of Technology, later commercialized by Arcon AB in 2001) employs electron beam melting to fuse metal powder particles.

ceramics, and composites. In the case of polymer-based PBF, support material is not necessary, because the loose powder provides sufficient support for cantilevered features. This can save time during part building, as well as allowing for advanced geometries (e.g., internal cooling channels) that would otherwise be difficult to post-process when supports are necessary. Supports, however, are required for most metal PBF processes. PBF processes are increasingly used for direct manufacturing of end-use products, and the material properties are comparable to many engineering-grade polymers, metals, and ceramics. The accuracy and surface finish of powder-based AM processes are typically inferior to liquid-based processes, and the total part construction time can take longer than other additive manufacturing processes. Polymer PBF processes do, however, have the ability to nest parts in three-dimensions, as no support structures are needed.

Material Jetting

Two of the seven AM technologies currently recognized by the ISO/ASTM, material jetting and binder jetting, are based on 2D printing technologies like inkjet printing. This section will focus on material jetting, where all of the part material is dispensed from a print head. In the case of binder jetting, a binder is printed onto a powder bed, with the powder forming the bulk of the completed part.

Material jetting is an AM process in which droplets of build material, currently either wax or acrylic photopolymer, are selectively deposited. The first commercially successful material jetting technologies used wax as the build material. Today, Solidscape, Inc., the originator of this technology, uses wax to make investment casting molds for jewelry and dentistry.

With respect to product design, however, it is material jetting using acrylic polymers that has had the greatest impact. In 2000, Objet Geometries of Israel introduced an additive machine that jets liquid using a print head with over 1,500 nozzles. The droplets are then exposed to ultraviolet light to promote polymerization.

In 2012, Objet was acquired by Stratasys. Today, Stratasys markets several material jetting machines. Figure 13-7 shows a Stratasys Polyjet machine. Each photopolymer layer is immediately cured by UV light as it is printed. Typical

Figure 13-7 Stratasys Polyjet machine

layer thickness is 0.0006 inches (0.015 mm). The print head contains more than 1,500 nozzles, thus allowing material to be deposited line-wise rather than point-wise. Support material is required and consists of a gel-like material.

Some Stratasys Polyjet machines are able to print in multiple materials, and even to vary material properties within a single part. Softer, rubber-like features can thus be combined with harder materials in a single print. See Figure 13-8.

AM machines based on printing technology have the advantages of low cost, high speed, ability to build with multiple materials, and color printing. Disadvantages include limited material selection (wax, acrylic photopolymers) and part accuracy.

▌ CLASSIFICATION OF AM TECHNOLOGIES

Figure 13-9 shows a two-dimensional method for classifying AM processes.[5] The horizontal dimension relates to the method by which the layers are deposited. These delivery methods include, from left to right, single point, two point, line (array), and layer. Earlier technologies (e.g., SLA, FDM, SLS) employed a single point

[5]This classification scheme was first introduced by Pham in 1998, and later expanded upon in Gibson, *Additive Manufacturing Technologies*, 2015.

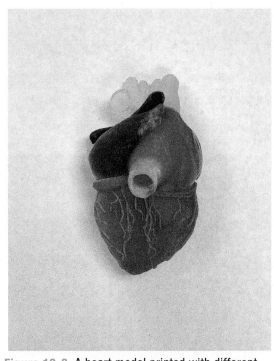

Figure 13-8 A heart model printed with different materials on a Stratasys Polyjet machine

source to draw with or upon the raw material, whereas later methods increased the number of sources in order to increase throughput. Two point sources, for example, allow for the simultaneous deposition of two different materials (e.g., model and support material). The Polyjet system,

Layer delivery methods →

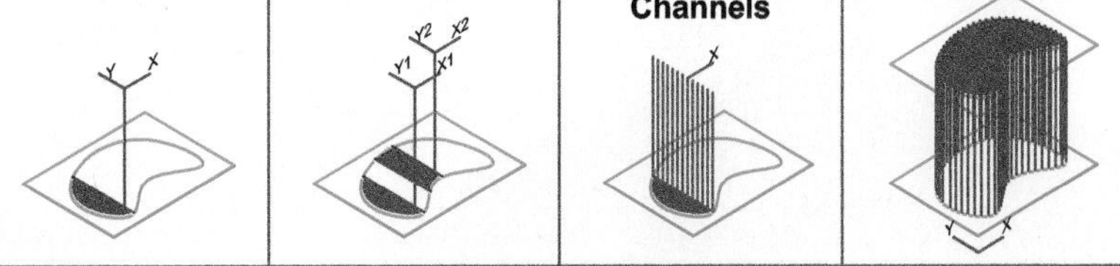

Raw material		1D Channel	2 x 1D Channels	Array of 1D Channels	2D Channel
		x, y	Y2 X2 Y1 X1	x, y	y, x
Liquid polymer		SLA (3D Systems 1980's)	Dual beam SLA (3D Systems)	Multijet (Objet 2000's/Stratasys)	DLP SLA (FormLabs)
Discrete particles		SLS (DTM/3D Systems 1990's), LS (EOS 1990's)	LS (EOS)	3D Printing (Z Corp 1990's/3D Systems)	
Molten material		FDM (Stratasys 1980's)	FDM (Stratasys model, support)		
Solid sheets		LOM			

Figure 13-9 Classification of AM processes

originally developed by Objet (now Stratasys), uses inkjet print heads to deliver a line, or array of raw material. Layer processing, where an entire layer is cured at the same time, is exemplified by digital light processing (DLP) SLA systems like Formlabs Form 3 printer.

The vertical dimension describes the raw material employed, that is, liquid polymer, discrete particles, molten material, and solid sheets. Vat photopolymerization (e.g., 3D Systems SLA) uses a liquid polymer. Discrete particles are employed in powder bed fusion and binder jetting technologies. Material extrusion uses a molten material, and sheet lamination uses solid sheets of material.

■ 3D PRINTER FILE FORMATS

The de facto standard format used to pass model geometry to a 3D printer is the STereoLithography, or STL, file. All CAD programs can save directly to the STL file format. In an STL file, the external boundary of an object is represented by a mesh of triangles, called a tessellated or faceted object. Figure 13-10 shows an STL file of a joystick. Each triangle, or face,

has an inside surface and an outside surface. The vertices of each triangle are ordered so that, in traversing the vertices in a counterclockwise (CCW) direction, the surface normal defined by this procedure identifies the outside face, as seen in Figure 13-11.

STL files can be saved in either binary (compact) or ASCII (text editable) form. Figure 13-12 shows a CAD model of a tetrahedron, together with the associated STL ASCII file.

The simplicity of the STL file format, as a list of triangles, makes file conversion to and from STL particularly easy. Many standard surface triangulation algorithms have been developed in computer graphics and are used to assist in developing file translation routines. In addition, the accuracy to which the STL file represents the actual geometry can be controlled by specifying the minimum size of these triangles within the CAD software. The tradeoff for this increase in accuracy is increased file size. Any manifold solid can be represented in STL, and simple slicing algorithms are available to convert the watertight triangular mesh into cross-sectional slices.

The disadvantages of the STL format include data redundancy, errors due to approximation, and information loss. In an STL file each facet normal is explicitly stored, despite the fact that this information can be determined from the vertex order. In addition, vertices are stored multiple times, one for each facet. Because they can only be represented as planar triangles, STL files do a poor job in representing curved surfaces. Finally, because the STL file has no provision for specifying units, color, material, and other feature information, this information is lost when converting to STL.

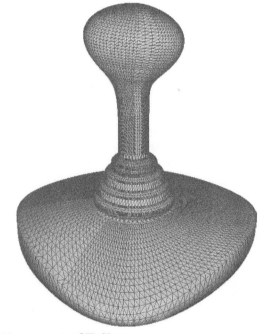

Figure 13-10 STL file

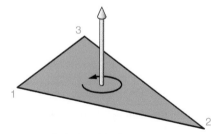

Figure 13-11 Relationship between vertex order and surface normal direction on an STL triangle

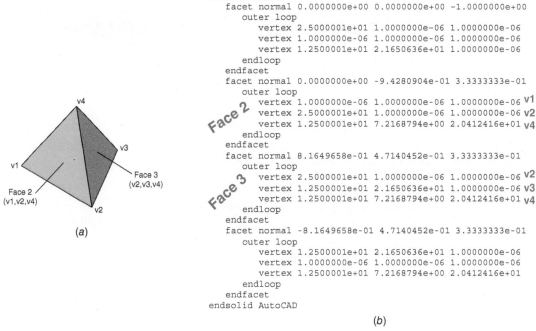

```
solid AutoCAD
    facet normal 0.0000000e+00  0.0000000e+00 -1.0000000e+00
        outer loop
            vertex 2.5000001e+01 1.0000000e-06 1.0000000e-06
            vertex 1.0000000e-06 1.0000000e-06 1.0000000e-06
            vertex 1.2500001e+01 2.1650636e+01 1.0000000e-06
        endloop
    endfacet
    facet normal 0.0000000e+00 -9.4280904e-01 3.3333333e-01
        outer loop
            vertex 1.0000000e-06 1.0000000e-06 1.0000000e-06 v1
            vertex 2.5000001e+01 1.0000000e-06 1.0000000e-06 v2
            vertex 1.2500001e+01 7.2168794e+00 2.0412416e+01 v4
        endloop
    endfacet
    facet normal 8.1649658e-01 4.7140452e-01 3.3333333e-01
        outer loop
            vertex 2.5000001e+01 1.0000000e-06 1.0000000e-06 v2
            vertex 1.2500001e+01 2.1650636e+01 1.0000000e-06 v3
            vertex 1.2500001e+01 7.2168794e+00 2.0412416e+01 v4
        endloop
    endfacet
    facet normal -8.1649658e-01 4.7140452e-01 3.3333333e-01
        outer loop
            vertex 1.2500001e+01 2.1650636e+01 1.0000000e-06
            vertex 1.0000000e-06 1.0000000e-06 1.0000000e-06
            vertex 1.2500001e+01 7.2168794e+00 2.0412416e+01
        endloop
    endfacet
endsolid AutoCAD
```

(b)

Figure 13-12 (a) CAD model of a tetrahedron and (b) its associated ASCII STL file

Currently, the preferred 3D printer file format for multicolor printing is OBJ. Like STL, the OBJ format has been in existence for decades, is open-source, stores geometry as tessellated triangles,[6] and supports both binary and ASCII encodings. The OBJ file format can, additionally, encode information about the color, materials, and texture of the part, as well as geometry. The OBJ format stores color and texture information in a companion file format called Material Template Library (MTL). However, because OBJ files come in pairs (OBJ and MTL), the MTL file frequently gets separated from the OBJ file, thus resulting in the loss of color and texture information. Since OBJ is more complicated than STL, it is more difficult to repair, and there is not a similar level of support.

To address the shortcomings of the STL format—large file size, slow, error-prone, no scale information, no color, texture, and material

information—in 2011 the American Society of Testing and Materials (ASTM) released the Additive Manufacturing File Format (AMF), dubbing it "STL 2.0." AMF is an XML-based format, meaning that it is human-readable. AMF supports curved (not just straight-sided) triangles, meaning that it is more accurate than STL, and with a smaller file size. AMF also supports units, as well as color, texture, and materials. The biggest shortcoming of AMF is limited adoption; the 3D printing industry has been slow to embrace this new file format.

In 2015, Microsoft formed a consortium consisting of nearly all the major CAD (Autodesk, Dassault Systems, Siemens) and AM (Stratasys, 3D Systems, HP, GE, Materialize, Shapeways) companies in order to develop a new open-source 3D printer file format, called the 3D Manufacturing Format (3MF). 3MF is loosely inspired by AMF. It is XML-based (human readability), supports colors, materials, and textures, and ensures that files are 100% manifold, that is, manufacturable. Like the AMF file format, the 3MF file format is still new, and consequently

[6]Unlike STL, the OBJ format can alternatively support freeform schemes for curves and surfaces, potentially making the file size smaller and more accurate than STL.

3D PRINTER FILE FORMATS

365

has not yet been widely adopted. At the time of this writing, it remains to be seen which new 3D printer file format, AMF or 3MF, will eventually replace STL.

▌STL REPAIR TOOLS

STL files may contain errors that cause the print to fail, or to print incorrectly. A good triangle mesh that can be used for 3D printing needs to be *valid, closed, oriented,* and without *self-intersections.*[7] For a mesh to be valid, each triangle in the mesh must share two of its vertices with each adjoining triangle. That is, a vertex cannot intersect the side of an edge, as shown in Figure 13-13*a* on the left. A surface used for 3D printing must be closed, or watertight. Figure 13-14*a* on the left shows a closed surface with no boundary edges, while Figure 13-14*b* on the right shows an open surface with multiple boundary edges.

As stated previously, the orientation of a triangle is determined by the order of its vertices (see Figure 13-15). Starting with any vertex and proceeding counterclockwise, the right-hand rule

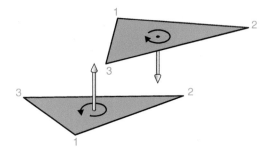

Figure 13-15 Orientation of a triangle determined by order of its vertices

defines the surface normal, as well as the outside of the face. In a closed solid mesh to be printed, all faces should be oriented so that the surface normal points toward the outside of the solid.

Self-intersections occur when two bodies cut through one another, as seen in Figure 13-16. These self-intersections lead to failures during slicing, where the 3D data (i.e., the STL file) must be converted into 2.5-dimensional[8]

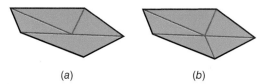

| (a) | (b) |

Figure 13-13 Comparison of (*a*) invalid triangulation on the left and (*b*) valid triangulation on the right

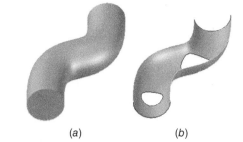

| (a) | (b) |

Figure 13-14 Comparison of (*a*) closed on the left and (*b*) open on the right surfaces

[7]Autodesk Knowledge Network.

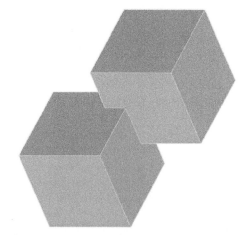

Figure 13-16 Self-intersections

[8]In CNC machining, 2.5 dimensions (2.5D) means that a machine tool can only be programmed to cut in two directions, x and y, for a given feature, or sequence. The depth (z) of the cut is fixed for that feature. In true three dimensions (3D), the movement of the machine tool can be controlled along x, y, and z during the cut, meaning that smooth 3D contouring can be accomplished. With 2.5D, stair steps will result along the z axis direction, rather than a smooth contour. 3D printers are 2.5D.

slice data. Self-intersections in the original data result in self-intersection in the slices. These self-intersections may cause constructional failures or instabilities. To avoid these self-intersections, make sure that all solid bodies are combined.

In order to detect and then fix these errors (a nonmanifold mesh, holes or gaps in the mesh, flipped face orientation, self-intersecting meshes), as well as to perform many other useful operations (mesh simplification, scaling, adjusting wall thickness, boundary edge smoothing), several STL repair software tools are available. Two of the most prominent are Autodesk Meshmixer, which is free, and Autodesk Netfabb.

CHARACTERISTICS OF AM SYSTEMS

Important characteristics common to most AM systems include part orientation, support structure, and hatch style.

Part orientation

Before converting the STL file into layered cross sections, the user decides[9] how to orient the model. Part orientation can affect build time, part resolution, and surface finish. Higher resolutions of curved surfaces can be obtained by choosing the orientation so that the object's most significant curvature appears in the individual layers, or slices.

As an example, consider a solid cylinder. If the cylinder is oriented so that the axis of the cylinder is normal to the build platform, as seen in Figure 13-17a, then all of the cross sections will be circles, and the curved surface of the resulting prototype will be very smooth. If, however, the cylinder is oriented so that the axis is parallel to the build platform (see Figure 13-17b), the resulting cross sections will be rectangular, and the curved surface of the cylinder will have a stair step appearance.

[9]In reality, most 3D printer interface applications include an option to allow the software to optimally orient the part prior to slicing.

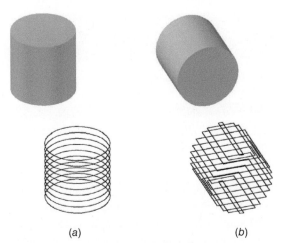

(a) (b)

Figure 13-17 Cylindrical cross-sections: (a) circular versus (b) rectangular

Support Structure

Many AM technologies require additional support structures in order to successfully build parts. These support structures are, generally speaking, automatically generated by the software. Support structures are required (1) to assist in the easy removal of the part from the base platform (see Figure 13-18) and (2) when islands or cantilevered sections exist. An island is a portion of a layer that

Figure 13-18 Base, anchor support structure, and parts

Figure 13-19 Support structure

is not connected to any other portion of the same layer. Examples of cantilevered features include overhangs and arches. In both instances, upper cross sections have a larger area than lower cross sections. Figure 13-19 shows, on the left, a printed part prior to removing the support structure; on the right the same part is shown after the support structure has been removed.

Hatch Style

In most AM technologies, the user has at least some limited control over the type of internal hatching employed on the interior of solids. For example, a sparse interior can be specified, where solid boundaries are created first and then a honeycomb interior. Sparse models use less material, reduce fabrication time, and are lighter in weight than prototypes made with a solid interior. Figure 13-20 shows the internal structure of a block printed with a sparse fill.

Figure 13-20 Sparse fill

▌ AM COMMERCIAL CATEGORIES

There are currently three recognizable AM commercial categories: (1) industrial, (2) professional, and (3) consumer. Originally, all AM machines were in the industrial category, that is, machines costing more than $100k USD. In fact, some metal additive machines cost well over $500k. 3D printers ranging in cost from roughly $10k to $60k fall into the professional category. This category emerged around the year 2000. In recent years a third category, consumer, has appeared. The cost of consumer category 3D printers is under $5k.

Industrial Category AM

Until around the year 2000, almost all rapid prototyping machines cost more than S100k. Perhaps the only exception were the 3D printing[10] machines from Z Corp, based on a technology developed at and licensed from MIT. Significant developments in the industrial AM sector in this century include more emphasis on direct manufacturing, the emergence of metal additive devices, and market consolidation through acquisition. As seen in Table 13-1, almost all of the main AM technologies are owned by either 3D Systems or Stratasys. The only exception is the German company EOS, whose direct metal laser sintering process is used to make metal parts directly from metal powders. 3D Systems developed stereolithography in 1980s (see Figure 13-21) and acquired SLS technology through the purchase of DTM in 2000. 3D Systems has also developed their own material jetting machines, and in 2012 they acquired binder jetting technology through the purchase of Z Corp. 3D Systems is also the dominant player in the reverse engineering/3D scanning industry, both in hardware and software. Stratasys developed FDM (see Figure 13-22) in 1980s, and in 2012 acquired material jetting technology with the purchase of Objet.

Figures 13-21 and 13-22 show, respectively, industry category stereolithography and FDM machines.

[10]The MIT/Z Corp binder jetting machines were called 3D printers. Later other low-cost machines also came to be referred to as 3D printers.

Table 13-1 **Major AM technologies and companies**

AM Technology Category	Original Technology	Company
Vat photopolymerization (VP)	Stereolithography (SLA)	3D Systems
Material extrusion (ME)	Fused deposition modeling (FDM)	Stratasys
Powder bed fusion (PBF)	Laser sintering (LS)	3D Systems, EOS
Material jetting	lnkjet printing	Stratasys (Objet)

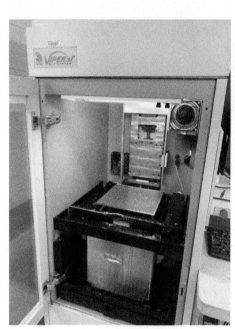

Figure 13-21   3D Systems industry category Viper si2 SLA printer

Professional Category 3D Printers

After the year 2000, a second professional category of commercial AM began to emerge. Industrial AM machines are too expensive for most midsize design offices and product development firms to justify. In addition, industrial AM machines are not intended for operation in an office environment. In response to these drawbacks, a number of AM service bureaus emerged to provide customers with physical prototypes within a short turnaround time.

Another alternative, particularly attractive with design offices and in educational institutions, are professional class 3D printers. These machines are significantly less expensive and can be used in an office environment. Professional category machines offer a favorable price-to-performance ratio compared with the industrial AM market. The cost of professional AM machines roughly ranges from $10k to $60k. In addition, professional category 3D printers offer much of the functionality of industrial AM technology.

Although industrial machines on the whole produce more accurate parts than their professional class counterparts, the differences are not dramatic. Industry machines have slightly larger build volumes than 3D printers. Furthermore, the software provided with industrial category machines gives the user more control over the output than professional class 3D printer software does. Perhaps the most significant advantage of the more expensive

Figure 13-22 Stratasys industry category Fortus 360mc FDM printer

devices is the ability to use different materials. By and large, professional class AM machines can only work with a single material.

Although they are more complicated than a conventional printer, professional category 3D printers do not require extensive training in order to operate. By not requiring a special operating environment, these 3D printers are coming to be viewed as standard design office equipment, comparable to a plotter or large photocopier. Some are actually faster than their more expensive predecessors, where a physical prototype can typically be ready in a day or less.

Like the industrial category of AM devices, the professional category is dominated by two U.S. companies, 3D Systems and Stratasys. Examples include, from 3D Systems, stereolithography and binder jetting, and from Stratasys, FDM and material jetting.

Consumer (Home and Hobby, Desktop) Category 3D Printers

In the 2013 U.S. State of the Union address, President Obama said that ". . 3D printing has the potential to revolutionize the way we make almost everything." As the year unfolded more and more, media attention was showered on this "new," disruptive technology, with stories about 3D printed guns, candy, and human body parts, until seemingly everyone knew about 3D printing. So what happened to cause this rapid ascent of 3D printing into public awareness?

More than anything else, the expiration of key patents protecting various 3D printing technologies gave rise to this explosion of interest. Many of the original AM technologies, developed in the 1980s, were protected by patents that prevented others from using the technology. In the

United States and Europe, patents have a duration of 20 years from the time of their initial filing. Chuck Hull, co-founder of 3D Systems, filed his first patent for stereolithography in 1984, which expired in 2004. The impact of this expiration was not immediately felt, however, because laser-cured photopolymer systems like stereolithography are relatively expensive and difficult to replicate. More important was the expiration of the original 1989 patent for fused deposition modeling held by Scott Crump, co-founder of Stratasys, in 2009. Material extrusion is much easier to replicate at a low cost than laser-cured photopolymer systems.

Almost immediately, a huge number of variants of this technology appeared due to the lapse of this and other FDM patents. The ensuing competition has driven the cost of these systems down to the point where individual owners can now afford to buy a 3D printer for home use.

Another important factor driving the expansion of 3D printing has been the emergence of the maker movement. 3D printing, more than any other technology, is closely identified with makers and maker spaces. One example of this close connection between the two is the RepRap project. RepRap is an open source, DIY initiative started in England in 2005 with the aim of developing a low-cost 3D printer capable of making most of its own components.

Today, the low-cost 3D printer market includes dozens of competitors, the vast majority of which use material extrusion processes. Some examples are shown in Figure 13-23. The low-cost, consumer class of 3D printers is the fastest growing segment of the AM market. As additional AM patents continue to lapse,[11] the market for low-cost AM should continue to grow.

[11]Some key droplet deposition and powder sintering by laser patents expired in late 2014 and early 2015.

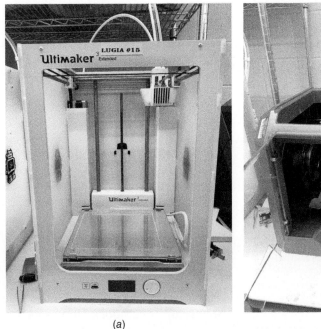

(a)   (b)

Figure 13-23 Low-cost 3D printers: (a) Ultimaker 3; (b) Dremel Digilab

▌ DESIGN FOR ADDITIVE MANUFACTURING

Design for Conventional Manufacturing Processes

Traditional manufacturing processes like CNC machining (for metals) and injection molding (for plastics) (see Figures 13-24 and 13-25 respectively) typically require multiple operations, extensive tooling and fixturing, considerable setup time, and process planning. The more complex the part, the more stages, tooling, and planning are needed to produce the part. As a result, parts made using conventional manufacturing processes need to be produced in large batch sizes in order to justify the setup costs. A large batch size means that product customization is difficult, if not impossible, to achieve.

Additionally, conventional manufacturing processes like CNC machining impose significant constraints upon design. In order to effectively design products, design engineers need to be familiar with the limitations of the manufacturing process, as well as with efficient assembly techniques, supplier capabilities, dimensional tolerances, material behavior, and so on. Design for manufacture and assembly, or simply design for manufacture (DFM), is the practice of designing parts, components, and products so that manufacturing difficulties are minimized, and production costs are reduced.

All CNC machining processes, for example, are subject to the following limitations: (1) almost all parts must be made in stages, often requiring multiple passes for material removal, and (2) all machining is performed from a single approach direction. In the case of CNC milling, for example, this means that the raw stock must be held in a particular orientation (e.g., top side up) while the various features accessible in that orientation are machined. Upon completion of this operation the stock must be flipped (e.g., bottom side up) so that the second operation, machining of the bottom features, can be carried out. A further complication is that in each operation the workpiece must somehow be rigidly held in place, while still allowing access for machining.

For a simple part (see Figure 13-26) that can be machined using a single setup operation, CNC machining is very fast and cost-effective. But for parts with even a moderately complex geometry, like the one shown in Figure 13-27, production using CNC milling would require two setup operations, top and bottom. In general, when using conventional manufacturing processes, the cost-effectiveness of part

Figure 13-24 Two CNC machines, a lathe in left foreground, and a mill in right background

Figure 13-25 Low-cost injection molding machine used for education

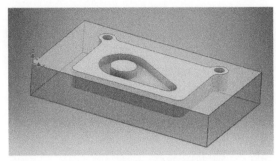

Figure 13-26 Part requiring a single setup for CNC milling

fabrication decreases as the complexity of the part geometry increases.

Design for AM

Unlike conventional manufacturing, AM can in most cases be done in a single step, with little if any planning, tooling, or setup time. With AM, conventional manufacturing process constraints can be ignored. AM is most cost-effective with small batch sizes, and even a batch size of one is feasible. AM enables both the use of design customization (see Figure 13-28) and geometric complexity without incurring time or cost penalties (see Figure 13-29).

Another important benefit of design for AM is the capacity to fabricate operational mechanisms, that is, linkages with rotational and translational joints, in a single build. This can be accomplished by printing an assembly with built in clearances between the joints. During the build, the void spaces are filled with dissolvable support material. Once the support material is dissolved, the joints articulate as seen in Figure 13-1 on page 358.

Other unique build capabilities of AM include multimaterial printing and embedded foreign

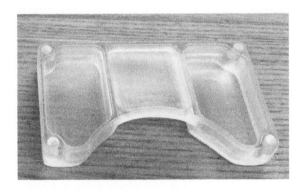

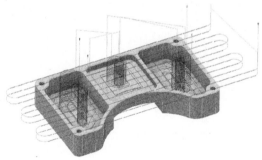

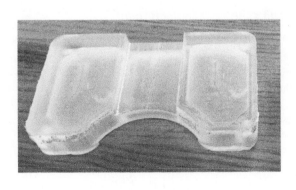

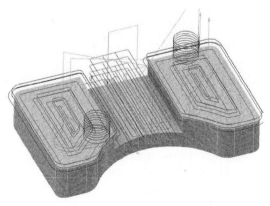

Figure 13-27 Part requiring two setup operations for CNC milling

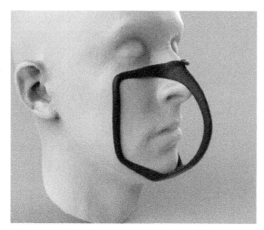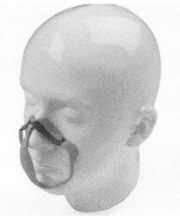

Figure 13-28 Face mask frames customized for a specific individual (special thanks to Mike Halloran)

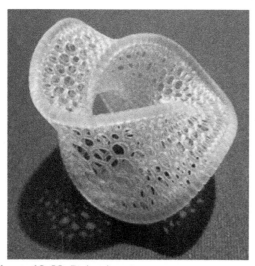

Figure 13-29 Ember logo

components. Using material jetting technology, multimaterial parts can be fabricated where a soft, rubber-like material is combined with a hard plastic in the same build (or a solid region in combination with a lattice structure). In AM, it is also possible to design a void within a part, pause the build at the void and embed a component (e.g., sensor, circuit, magnet), and then fully encapsulate the foreign component once the build resumes.

Table 13-2 compares various design considerations when considering both conventional subtractive manufacturing and additive manufacturing.

Table 13-2 **Design considerations for traditional and additive manufacturing**

Design considerations	Conventional manufacturing	Additive manufacturing
Operations, stages, steps	Multiple	One
Setup time, process planning	Significant	Minimal
Tooling, fixtures	Required	Unnecessary
Impose constraints upon design	Yes	Minimal
Preferred lot (batch) size	Large	Small
Cost of geometric complexity	Expensive	Free
Cost of design customization	Expensive	Free
Single-part assemblies, functional mechanisms	No	Yes
Multimaterial printing	No	Yes
Embedded components	No	Yes

QUESTIONS

SHORT ANSWER

1. What are four characteristics that differentiate subtractive manufacturing techniques from AM technologies?

2. Name three additive manufacturing technologies.

TRUE OR FALSE

3. Fused deposition modeling (FDM) builds parts one layer at a time by tracing a laser beam on the surface of a vat of liquid photopolymer resin.

4. CNC machining is typically used with metals.

5. Injection molding is typically used with metals.

6. The object shown in the figure below can be oriented for additive manufacture so that no support material is required.

Figure P13-1 Object to be printed

7. Stereolithography does not require a CAD file (i.e., solid or watertight surface model) in order to make a physical prototype of a part.

8. Selective laser sintering (SLS) does not require a support structure.

9. Additive manufacturing is sometimes called 3D printing.

MULTIPLE CHOICE

10. Which of the following is not true of STL `SS` files?
    a. A list of triangles
    b. Conversion to and from STL is easy
    c. Does a good job on curves surfaces
    d. Data redundancy is a disadvantage
    e. Accuracy controlled the number of triangles

11. The expiration of key patents has resulted in the significant growth of which AM market category?
    a. Industrial
    b. Professional
    c. Consumer (home and hobby)
    d. All of the above
    e. None of the above

12. Additive manufacturing holds significant `SS` advantages over conventional (e.g., machining, injection molding) manufacturing. Which of the following is not an advantage of AM when compared with traditional manufacturing?
    a. Minimal tooling and setup costs
    b. Ability to handle large batch sizes
    c. Minimal design constraints
    d. Ability to handle part complexity
    e. Ability to handle part customization

13. Choose the best answer. Layered additive manufacturing technologies are classified according to which of the following:
    a. Raw material used
    b. Ways in which the layers are constructed
    c. Raw material used and way in which the layers are constructed
    d. Thickness of layer
    e. Type of layer used

CHAPTER

14

3D SCANNING

∎ REVERSE ENGINEERING

Reverse engineering is a systematic approach used to analyze the design of an existing device or system, often with the aim of duplicating or enhancing the device or system. Reverse engineering[1] techniques are commonly used in industry, either to benchmark a product or as a first step in a product's redesign.

World-class companies employ reverse engineering in order to gain competitive advantage. In particular, reverse engineering is used (1) to understand a product's design, (2) to promote product improvement, and (3) for competitive benchmarking. Reverse engineering techniques offer a systematic approach to fully understanding and representing a product's design—either one's own or that of a competitor. It also provides a means to explore new avenues for the continuous improvement, evolution, and redesign of a product or product family. Finally, reverse engineering is a benchmarking technique that makes possible a comprehensive understanding of competitor products. This knowledge can be leveraged to identify *best-in-class* approaches to common problems, which are then used to develop better products.

The term *reverse engineering* also refers to the use of 3D scanning technologies used to create a digital model from a physical object. Once the digital model exists, additive manufacturing can be used to reverse the process by generating a physical part from the 3D model. The topic of this chapter is 3D scanning. Additive manufacturing is discussed in the previous chapter.

∎ 3D SCANNING

Introduction

3D scanning is the process of either measuring or creating a digital representation of a physical object. This digital representation of the physical object may be referred to as a ***3D scan,*** a ***mesh***, or scan data A ***3D scanner*** is the device used to take 3D measurements or create 3D scans of a physical object.

This section addresses the way in which CAD models are generated from scan data.[2] From the perspective of 3D scanning, reverse engineering is

---

[1]When applied to software, the term *reverse engineering* refers to reversing a program's machine code back to the original source code that it was written in, using program language statements. Software reverse engineering is used when the original source code is not available.

[2]For more information about 3D scanning, see *Reverse Engineering: an Industrial Perspective* (*Springer Series on Advanced Manufacturing*), a collection of essays edited by Vinesh Raja and Kiran J. Fernandes, Springer-Verlag, London, 2008.

the process of obtaining a geometric model from data points acquired by scanning or digitizing existing parts. This physical-to-digital pipeline is used for a variety of purposes, including:

- To create a digital model from a physical object
- For part inspection, by comparing the fabricated part to its CAD description
- For product improvement
- To create 3D data from an individual, a model,[3] or a sculpture. These scan data are then used to create, scale, or reproduce artwork, or for animation in games and movies
- For construction documentation of architecture, engineering, and construction (AEC) projects
- To generate data to support the creation of dental implants, prostheses, and orthoses

3D scanning is used in the making of films like *Avatar* and the computer vision technology used, for example, on the *Mars Rover*, as well as in magnetic resonance imaging (MRI) and CT (computerized tomography) scans. On these pages, however, the discussion will focus on design and reverse engineering applications of 3D scanning.

The primary goals of 3D scanning are:

- Reverse engineering of a component or object by extracting dimensions in order to produce a digital model
- Quality control of a component by measurement for analysis and/or dimensional inspection
- Reality capture (i.e., reverse engineering) of building or site by extracting dimensions in order to produce a digital model

3D Scanner Pipeline

The reverse engineering scan process is represented by the scanner pipeline shown in Figure 14-1. In the first phase, data collection, a 3D scanner is used to gather ***point cloud*** data. The output of this hardware scanning phase is one or more point cloud data sets.[4]

Both the second and the third output phase are conducted in a reverse engineering software

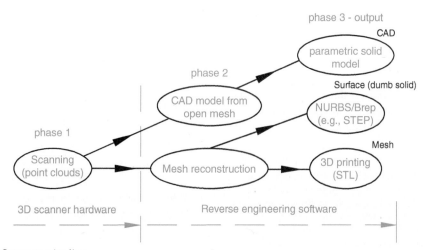

Figure 14-1 Scanner pipeline

---

[3]Scanning a clay model of a car body, for example, is a common workflow employed in the automotive industry.

[4]In reality, the scanner software is often able to produce STL files (i.e., a triangular mesh), so the output of this stage may come in STL file format.

package, for example, Geomagic Design X[5] or VXelements.[6] The second phase includes two paths. In mesh reconstruction the scan data are cleaned, merged, repaired, and optimized. If the resulting mesh is closed, i.e., watertight, then this mesh can be exported as an STL file and used to produce a 3D print. STL files are discussed in Chapter 13. In addition, the mesh reconstruction output can be used to generate a NURBS surface that can be opened in any CAD application. If the mesh output is closed and error free, then the surface is converted to a featureless solid. This path is described further in the section NURBS Surface Model from Scan Data.

The 3D scan data obtained in phase 1 can also be used in Geomagic Design X to create a feature-based CAD model. In this case the mesh captured by the 3D scanner does not necessarily need to be cleaned, merged, repaired, optimized, or closed. Instead the mesh is segmented into different geometric regions based on the curvature and features of the scan data. This in turn allows for the identification of construction features that can be used to build the CAD model. 2D sketch curves derived from the mesh are then used to develop a history-based parametric solid model. This stage is described further in the section Parametric CAD Model from Scan Data, and Figure 14-25 on page 388.

Mesh Terminology

A mesh is a polygonal model consisting of vertices, edges, faces (usually triangles), and boundaries. See Figure 14-2. Resolution refers to the

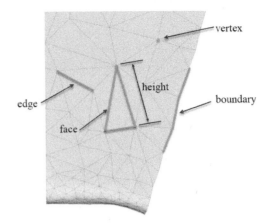

Figure 14-2 Mesh terminology

level of detail that the sensor can acquire. A high resolution mesh creates a large file size and can be difficult to use in scan data processing software. A low resolution mesh is easy to use in software but may not adequately represent the feature information of an object. The resolution specified when scanning corresponds to the height of a typical triangle, or face, as seen in Figure 14-2. Scanner specifications include *mesh resolution*, that is the highest resolution (i.e., smallest triangles) at which the scanner can capture data. If a scanning session is saved, then the mesh resolution of the output file may still be changed.

Figure 14-3 illustrates the difference between *resolution* and *accuracy*, where resolution refers to the distance between measurements and accuracy is the distance (error) between the measurements and the actual part.

3D Scanning Technologies

In the following sections, several different scanning technologies, both short and long range, will be discussed. These are:

- Contact-based (short range)
- Laser triangulation (short range)
- Structured light (short range)
- Photogrammetry (long and short range)
- Laser pulse, also called time of flight, or Lidar (long range)

[5]Geomagic Design X is owned by 3D Systems. Geomagic Design X was originally developed as Rapidform XOR by INUS Technology in Seoul, South Korea. Rapidform was a groundbreaking software package, in that it combines 3D scan data processing and parametric solid modeling. Geomagic, another scan data processing company operating in North Carolina, purchased Rapidform in 2012. 3D Systems acquired Geomagic in 2013, and began marketing Rapidform as Geomagic Design X.

[6]VXelements is a scan data processing software package used with Creaform 3D scanners. Creaform is a technology company based in Quebec, Canada. VXmodel is a scan-to-CAD software module, also developed by Creaform, that works with VXelements.

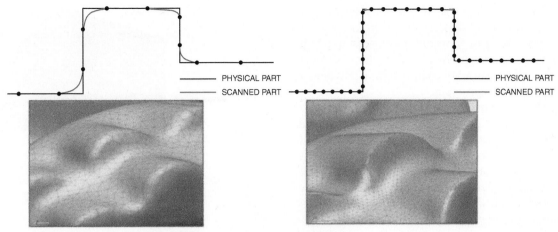

Figure 14-3 Mesh resolution versus accuracy (courtesy of Creaform)

CONTACT-BASED SCANNERS

Contact scanners, sometimes called digitizers, employ contact probes that automatically (or manually) follow the contours of a physical surface. Contact scanners are typically stationary and are used to scan objects that are small enough to be moved to the scanner. Contact scanning is widely used for performing quality inspection of parts. Contact scanners are based on ***coordinate measuring machine (CMM)*** technology. A portable CMM is shown in Figure 14-4.

Figure 14-4 Portable coordinate measuring machine (Reproduced by permission of Revware, Inc. © 2013 Revware, Inc. All rights reserved)

A CMM consists of a probe supported on three mutually perpendicular axes, where each axis has a built-in reference standard. The probe allows accurate measurements along each axis relative to the standard. A digitizer generates 3D coordinate points as the probe moves across the surface, with typical tolerances ranging from 0.01 to 0.02 mm. Because the probe may deflect to register a point, soft tactile materials (e.g., rubber) cannot be easily or accurately digitized.

The advantages of contact scanners include high accuracy, insensitivity to color or transparency, and the ability to measure deep slots and pockets. Disadvantages include slow data collection, the fact that soft objects may have distorted probe readings, and difficulty in working with organic, freeform shapes.

Modern contact scanners used for automatic part inspection use a combination of optical scanning equipment and robotic arms. This hybrid approach (i.e., contact and noncontact optical) reduces operator variability and human errors all while saving a significant amount of time and providing more data.

NONCONTACT SCANNERS

Noncontact scanners use lasers, optics, and charged-coupled device (CCD) sensors to capture point data. No physical contact is necessary between the scanner and the object being scanned. The most significant disadvantage

3D SCANNING

379

of these computer vision systems is lower accuracy; typical tolerances range from 0.025 to 0.2 mm. Because noncontact devices use light, they encounter problems with shiny surfaces, although a temporary coating of fine powder can help solve this problem.

The advantages of noncontact scanners include fast digitizing, no physical contact between the scanner and the object, reasonable accuracy for common applications, the ability to detect colors, and the ability to scan highly detailed objects. Disadvantages include lower accuracy, and some limitations for colored, transparent, or reflective surfaces.

LASER TRIANGULATION

Most laser scanners use geometric ***triangulation*** to determine the 3D coordinates of an object. In this method, shown in Figure 14-5, a charge-coupled device (CCD) camera detects the laser light reflected off the object and, using trigonometric triangulation, calculates the distance from the object to the scanner. Triangulation scanners are short range, used for scanning mechanical parts and artifacts over short distances.

Illumination techniques used with laser triangulation scanners include a single laser point and sheet of light. ***Sheet-of-light*** lasers project a plane of light on the object. As the plane intersects the object, a line is formed on the object. Scanning speeds for sheet-of-light scanners are considerably faster than those for point lasers. Figure 14-6 illustrates sheet-of-light scanning.

A problem common to computer vision systems is ***occlusion***, where part of an object blocks,

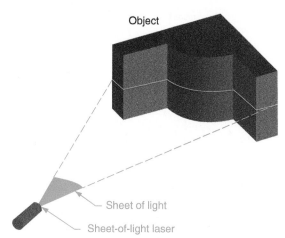

Object

Sheet of light

Sheet-of-light laser

Figure 14-6 Sheet of light

or occludes, the view of another part of the object. This can be as simple as the front occluding the back of an object, or it can be a more complicated situation like the one illustrated in Figure 14-7, where an object feature occludes either the laser or the camera, preventing them from seeing the same surface. The occlusion problem can be addressed by taking multiple scans and then merging, or registering, them.

Some sheet-of-light laser triangulation systems are handheld, as seen in Figure 14-8, with the camera and laser mounted on a wand. In order for the handheld scanner to collect data, the position of the scanner must be known. The scanner position can be determined with an external tracking system or by using reference features on the surface being scanned. Adhesive reflective tabs like those shown in Figure 14-9 often serve as reference features.

Figure 14-10 shows the handheld laser scanning process in action. As the scanner passes over the individual being scanned, the point cloud is captured in real time on the computer screen. Note the adhesive tabs used to track the location of the scanner. Because hair is difficult to scan, the individual is wearing a bathing cap.

STRUCTURED LIGHT

Structured light scanners do not employ laser light, but they do use triangulation. Structured light scanners project a series of linear patterns

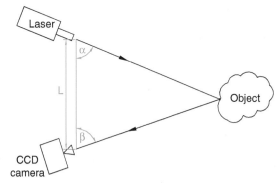

Laser

α

L

β

Object

CCD camera

Figure 14-5 Laser triangulation

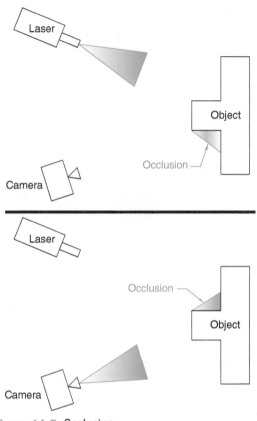

Figure 14-9 Handheld scanner with reflective tabs

Figure 14-10 Handheld laser scanning process in action (special thanks to Mike Halloran)

Figure 14-7 Occlusion

Figure 14-8 Handheld laser scanner

onto an object. By examining the edges of each line in the pattern, the distance from the scanner to the object's surface can be calculated. See Figure 14-11. The structured light can be either white or blue and is generated by a projector. Like laser triangulation scanners, structured light scanners can either be handheld (see Figure 14-12) or stationary (see Figure 14-13), and they are short range, used for scanning relatively small objects over short distances. Figure 14-14 shows the structured light process in action.

Structured light scanners have some advantages over laser systems; for example, the data acquisition is very fast, color texture information is available, and they do not use a laser. For these

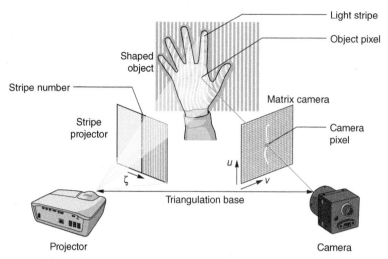

Figure 14-11 Structured light scanning technology

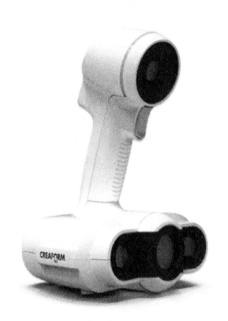

Figure 14-12 Handheld structured light scanner

reasons, structured light systems are favored for digitizing images of human beings.

REALITY CAPTURE

Reality capture is a relatively new term used to describe the use of either photo-based photogrammetry (aerial, terrestrial) or laser-based Lidar[7] to

[7]Lidar is a surveying method used to measure distance to a target by illuminating the target with pulsed laser light and measuring the reflected pulses with a sensor.

Figure 14-13 Desktop structured light scanner

produce a digital model of an object, building, or site. These technologies are gradually replacing traditional surveying methods, which are time consuming and used for making 3D measurements, not digital models. Using either photogrammetry or laser scanning, millions of surface points can be captured, measured, and mapped in a single day.

Reality capture hardware (e.g., camera-mounted drone, Lidar) is used to gather detailed data for the project, be it an object, structure, or infrastructure. The data is then registered (i.e., merged, and aligned with a coordinate system), cleaned, and analyzed in reality capture software like Autodesk's ReCap Pro. The digital model is then imported into CAD or BIM, typically for design and renovation projects.

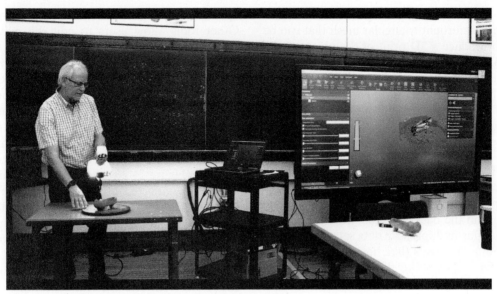

Figure 14-14 Handheld structured light scanning in operation

PHOTOGRAMMETRY

Photogrammetry is the art and science of extracting 3D information from photographs. The practice requires taking overlapping photographs of an object, structure, or site, and then converting them into a digital model. Photogrammetry is used by surveyors, architects, engineers, and contractors to create topographic maps, 3D meshes and point clouds, or drawings based on the real world.

There are two main types of photogrammetry: aerial and close-range. The use of camera-mounted drones to produce aerial photography is increasingly common. Drones can safely capture hard-to-access or inaccessible areas where traditional surveying would be dangerous or impractical. Close-range photogrammetry refers to the capture of images using a handheld camera or with a camera mounted to a tripod. The aim of this method is to make a 3D model of a smaller object.

TIME OF FLIGHT

Time-of-flight scanners (also known as laser pulse, or Lidar) collect 3D data from an object by measuring the time that a light pulse takes to travel to the target and back again. Because the speed of light is precisely known, this back-and-forth time can be used to calculate the exact distance between the scanner and the object. Since each measurement only collects a single data point, a tripod-mounted time-of-flight scanner must cast its laser beam in a wide range: the scanner head commonly rotates 360 degrees horizontally while a mirror tilts vertically. The laser beam is used to measure the distance to the first object on its path. These scanners are capable of capturing more than 100,000 points per second. Time-of-flight scanners are best suited for digitizing large, distant objects such as buildings (both interior and exterior), bridges, and dams, and so they are typically used in civil engineering and architectural settings.

Time-of-flight scanners include both laser pulse and phase shift lasers. Phase shift laser scanners are a subcategory of laser pulse scanners. In addition to pulsing the laser, the phase shift systems also modulate the power of the laser beam. Phase shift lasers offer a better overall performance.

Reverse Engineering Software

Once an object has been scanned (Phase 1 of the scanner pipeline process shown in Figure 14-1 on page 377), the resulting point cloud is imported into a reverse engineering software package.

Potential reverse engineering software outputs include a watertight triangular mesh that can be used to produce a 3D print of the object, and a NURBS model of the object. The former provides a first-order approximation of the object, while the latter generates a higher-order description of the object's geometry that can be used directly in CAD.

MESH RECONSTRUCTION (OR POINT PROCESSING)

Phase 1, 3D scanning, has been discussed. Phase 2 includes mesh reconstruction, which involves (1) importing the point cloud data, (2) reducing the noise in the data collected, (3) aligning and then merging multiple scan data sets, if necessary, (4) eliminating errors (like holes) in the merged scan, and (5) *sampling* the data to define the optimum number of points and their relative density. Sampling is a compromise between data size and object coverage. A small data set is easy to work with, but a large data set is more accurate. Point-processing algorithms are used to set the sampling requirements and then to generate a suitable polygon mesh.

Figure 14-15 shows two point cloud data sets of a wooden clog after being imported into Geomagic Design X. Figure 14-16 shows the same point clouds after some noise has been deleted. Figure 14-17 shows two scan data sets in the process of being merged.

At this point the polygon mesh will probably still have a number of defects. Polygonal mesh construction tools are available to fill holes (see Figure 14-18), to smooth data, including ragged boundary edges, and to perform basic polygon operations such as offset, trim, shell, and thicken. Other tools allow users to edit the polygonal mesh structure and to work with sharp edges, since scanning devices have trouble capturing sharp features.

Figure 14-19(*a*) shows the actual wooden clog, while Figure 14-19(*b*) shows 3D print of the clog.

Figure 14-20 shows another example of the mesh construction process. Note that Figure 14-9 shows the scanning process in action for this example. Figure 14-20*a* shows the raw scan data after being imported into Geomagic Design X. In Figure 14-20*b* some of the noise has been deleted from the scan, and the Split tool with the cap ends option active has been applied in order to close off the mesh. In Figure 14-20*c*, most of the holes have been filled, and the edges of the bathing cap above the eyes and ears have been smoothed. Figure 14-20*d* shows the completed mesh, after the ripples from the bathing cap have been defeatured and smoothed. Figure 14-21 shows a 3D print of the mesh shown in Figure 14-20*d*.

Figure 14-22 shows another workflow that can be used to create a life mask of a person's face. In this case, a Venetian Carnival mask has been scanned (Figure 14-22*a*), rather than a person's face, but the process is the same. Figure 14-22*b* shows the mesh after mesh reconstruction (remove noise, fill holes, trim boundary edge, sampling). Figure 14-22*c* shows a copy of the mesh in Figure 14-22*b*, viewed from the side so that it is clear that the mesh is open. Figure 14-22*d* shows the same mesh after thickening. Note that the mesh is now closed, so a 3D print can be made.

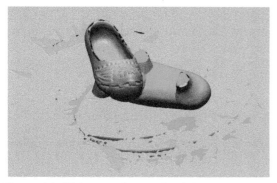

Figure 14-15 Imported point cloud data

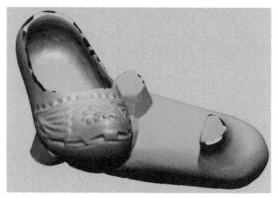

Figure 14-16 Point cloud after noise removal

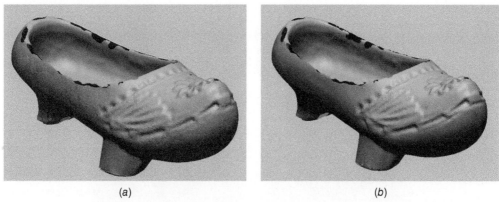

(a)   (b)

Figure 14-17 Two scans (a) aligned and (b) merged into a single scan

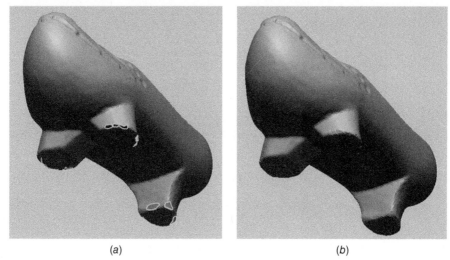

(a)   (b)

Figure 14-18 (a) Highlighted holes to be filled, (b) holes filled

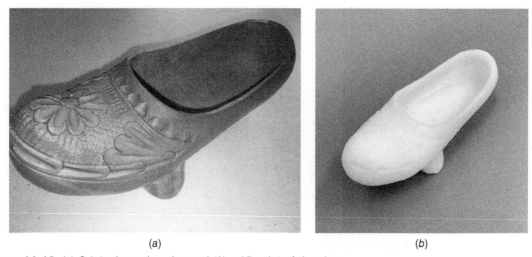

(a)   (b)

Figure 14-19 (a) Original wooden clog and (b) a 3D print of the clog

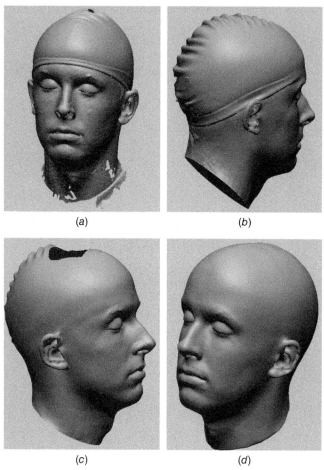

Figure 14-20 (*a*) Raw scan data, (*b*) mesh split with end capped, (*c*) mesh holes filled and smoothed above ears and eyes, and (*d*) mesh defeatured, smoothed, and optimized (courtesy of Mike Halloran)

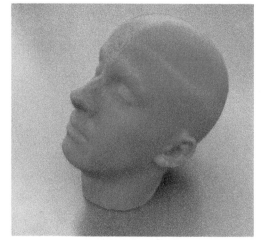

Figure 14-21 3D print of head scan (special thanks to Mike Halloran)

Figure 14-23*a* shows a photo of the actual mask, while Figure 14-23*b* shows a 3D print of the mask made from the thickened mesh.

One possible output of the mesh reconstruction process is a closed mesh STL file. This file can be directly employed for 3D printing, as seen in Figures 14-19*b*, Figure 14-21, and Figure 14-23*b*. Further, this mesh may be used as the basis for constructing a NURBS boundary representation (B-rep) surface, and even to create a parametric solid model, as discussed in the following sections.

NURBS SURFACE MODEL FROM SCAN DATA

In addition to the triangular mesh output—a first-order polygonal approximation of the object geometry—that is derived from the mesh

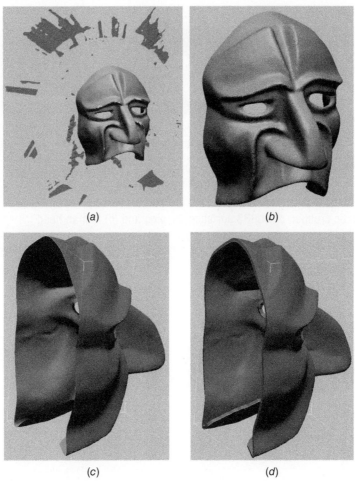

(a)   (b)

(c)   (d)

Figure 14-22 Venetian Carnival mask: (*a*) raw scan data, (*b*) cleaned mesh, (*c*) copy of cleaned mesh viewed from back side, and (*d*) thickened mesh

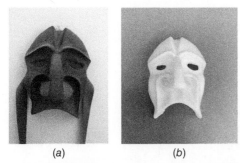

(a)   (b)

Figure 14-23 Venetian Carnival mask: (*a*) photo of actual mask; (*b*) 3D print of mask

reconstruction phase (as shown at the bottom right of the scanner pipeline, Figure 14.1 on page 377), a higher-order NURBS surface description can also be derived. Powerful surface-fitting algorithms are available in reverse engineering software that are able to generate a network of NURBS curves and surfaces. These curves and surface patches can be either manually defined by the operator or automatically defined. The latter is a one-step process, suitable for most geometries. Figure 14-24 shows, respectively, automatically generated NURBS surface models for both the wooden clog and the human head. Regardless of the approach employed, the objective is to prepare a patch structure that will support quadrilateral NURBS surfaces, all beginning from a triangular mesh.

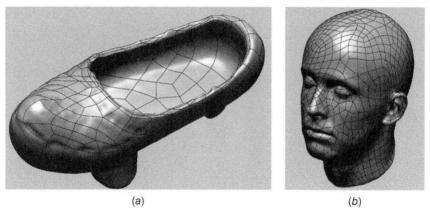

Figure 14-24 Automatic creation of NURBS surfaces: (a) clog and (b) human head

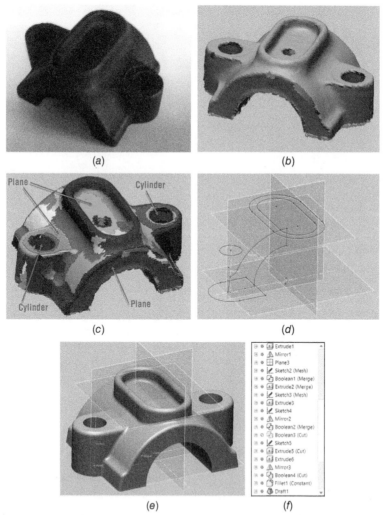

Figure 14-25 Bearing retainer: (*a*) bearing retainer photo, (*b*) raw scan data, (*c*) region groups after segmentation, (*d*) 2D CAD sketches, (*e*) parametric solid model, (*f*) partial feature tree

PARAMETRIC CAD MODEL FROM SCAN DATA

As shown in the scanner pipeline figure (Figure 14-1 on page 377), it is also possible to derive feature-based CAD models from raw scan data. Reverse engineering software applications like Geomagic Design X and even VXmodel from Creaform have the ability to do this. Figure 14-25 shows this workflow using Geomagic. Figure 14-25*a* shows a photograph of a retainer bearing. Figure 14-25*b* shows raw scan data after being imported into Geomagic Design X. Note that it is not necessary for the scan data to be cleaned, closed, or enhanced. Figure 14-25*c* shows the mesh regions after segmentation. Each colored region indicates a different geometry (i.e., planar, cylindrical, revolved, freeform), and can be used to capture design intent. For example, a planar region can be used to derive a datum plane and a cylindrical region to derive the construction axis for a hole. Figure 14-25*d* shows the 2D sketches used to create parametric features. These CAD sketches are derived from mesh sketches extracted from the mesh. Figure 14-25*e* shows the parametric solid model within Geomagic. Figure 14-25*f* shows a portion of the feature tree within Geomagic for this part. This model can be live transferred to a variety of CAD applications, including SolidWorks and Inventor.

▌ QUESTIONS

SHORT ANSWER

1. To gain competitive advantage, companies use reverse engineering for what three things?

TRUE OR FALSE

2. A noncontact scanner is sometimes called a digitizer.

MULTIPLE CHOICE

3. Contact scanners have a number of advantages. Which of the following is not an advantage of a contact scanner?
   a. High accuracy
   b. Rapid data collection
   c. Low cost
   d. Ability to measure deep slots and pockets

4. Which of the following is an illumination technique associated with laser triangulation scanners? `SS`
   a. Sheet of light
   b. Flood light
   c. LED
   d. Charge-couple device (CCD)

5. Which of the following is an example of a contact scanning technology?
   a. Laser triangulation
   b. Coordinate measuring machine (CMM)
   c. Time of flight (TOF)
   d. Structured light

6. Sophisticated scan data processing software (like Geomagic) can be used to produce which of the following outputs? `SS`
   a. 3D printed parts
   b. NURBS surface models
   c. Parametric, feature-based models
   d. All of the above
   e. None of the above

7. Of the following scanner categories, which one captures independent area (image) scans?
   a. Time of flight
   b. Structured light
   c. Sheet of light laser
   d. All of the above

8. Of the following scanner categories, which is most appropriate for scanning smaller objects? `SS`
   a. Time of flight
   b. Structured light
   c. Sheet of light laser
   d. Both a and b
   e. Both a and c
   f. Both b and c

9. Of the following scanner categories, which is most appropriate for scanning a building?
   a. Time of flight
   b. Structured light
   c. Sheet of light laser
   d. None of the above

CHAPTER

15 SIMULATION

■ UPFRONT ANALYSIS

In the past decade, simulation software has improved dramatically. Known also as computer-aided engineering (CAE) software, simulation software includes finite element analysis (FEA), dynamics simulation, computational fluid dynamics (CFD), and manufacturing analysis, including plastic injection molding simulation. Improvements to simulation software include (1) CAD integration, (2) a modern user interface, (3) automated functions, and (4) a robust underlying technology. Simulation software is now embedded within all of the mid-range parametric modeling packages, allowing designers to use these powerful analysis tools in a familiar environment. Even standalone analysis programs now leverage, and are associatively linked with, the CAD environment. The modern simulation software interface guides the user, anticipating the next steps and providing feedback on the status of the analysis. Critical and/or labor-intensive tasks are now automated—for example, automatic meshing in FEA. Well-documented benchmark problems allow comparison of FEA results with known results based on first principles and have consistently verified the reliability of the software. Many simulation software packages have been under development for almost 40 years now, allowing their developers to continuously improve upon and demonstrate the reliability of their solvers.

As a direct consequence of these improvements, manufacturers now use simulation early in the design process to develop improved products in a shortened time frame, and to minimize the time spent in the verification and testing phase of the product development process. Costs can be reduced by identifying questions and concerns early in the design process. By using simulation early in the design phase, rather than later in the product development process, it is possible to eliminate additional rounds of prototyping and testing. Best-in-class manufacturers:

1. Use more simulation, while making fewer prototypes than their competitors
2. Systematically use simulation tools at regular intervals in order to support critical design decisions
3. Provide access to simulation capabilities for their engineers directly through CAD applications, and through independent pre-processors only when necessary

■ FINITE ELEMENT ANALYSIS

Although some problems have an exact solution, other problems are too complicated, and no precise solution exists. For these complex problems, numerical methods have been developed to find approximate solutions. These problems, frequently iterative in nature, are well suited to solution by computer. *Finite element analysis (FEA)* is a numerical technique used to solve physical problems that cannot be solved analytically. FEA was originally developed in the 1940s to solve structural analysis problems in civil and aeronautical engineering, but today other applications include heat transfer, fluid flow, and electromagnetism. The discussion of FEA in this text will be limited to structural, or *stress analysis*, problems.

The basis for FEA is Hooke's Law, an approximation stating that the extension of a spring is directly proportional to the load applied to it. Many materials (e.g., aluminum, steel) obey this law as long as the elastic limit of the material is not exceeded. Mathematically, Hooke's Law is expressed as follows:

$F = kx$, where
F = force
k = a proportional constant
x = amount of stretching (deformation)

This equation is the foundation for understanding linear finite element stress analysis.

Hooke's Law can also be expressed in terms of stress and strain; that is, below a material's elastic limit, strain is directly proportional to stress. *Stress* is the measure of the average amount of force per unit area within a material that develops as a result of an externally applied force. *Strain* is the relative deformation produced by stress.

$$\sigma = \frac{F}{A}, \quad \varepsilon = \frac{\Delta l}{l}, \quad \sigma = E\varepsilon$$

Rearranging yields

$$F = \left(\frac{EA}{l}\right)\Delta l, \quad \text{similar to } F = kx$$

where

σ = stress	Δl = change in length
F = force	l = initial length
A = area	E = Young's Modulus
ε = strain	(Modulus of Elasticity)

Imagine that a fully constrained component, either a part or an assembly, is subjected to a force (or a combination of force, pressure, and moment). If the force is large enough, the part may break. That is, the internal stress in the component will have exceeded the allowable stress, or *yield stress*, of the component material.

The forces, pressures, and moments acting on the part are called *loads*. The constraints used to remove the six degrees of freedom that hold the part in place are called supports, or simply constraints. The loads and supports together are called *boundary conditions*.

Unless the component geometry, material, and boundary conditions are all extremely simple, it will not be possible to directly determine the level of stress in the component. It will, however, be possible to determine approximate levels of deflection, strain, and stress using finite element analysis.

The component, whether it is made of steel, plastic, or some other material, can generally be thought of as a continuous elastic structure. FEA works by breaking down this geometric continuum into an extremely large, but finite, number of *elements* and then analyzing the behavior of each element.

Elements are connected together at *nodes*, and nodes have degrees of freedom (DOF). A simple one-dimensional rod element will have nodes with only one DOF, whereas a complex hexahedral element may have nodes with as many as six DOFs.

The behavior (deflection, strain, stress) of individual elements is predicted by mathematical equations, and the summation of the individual behaviors predicts the behavior of the actual component. The process of breaking geometry down into elements and nodes is called *meshing*. FEA is traditionally regarded as having three phases: (1) preprocessing, (2) solution, and (3) postprocessing. Preprocessing includes the following stages:

- Geometry—where the part geometry is imported from CAD or, sometimes, created from scratch.
- Modeling—there are several possible modeling analysis approaches, including beam, plane stress, plane strain, axisymmetric, shell, and solid. Only solid modeling analysis is considered here.
- Meshing—either manual or automatic.

- Material properties—the material of the part being analyzed must be specified.
- Boundary conditions—both supports and loads.

In the solution phase, the FEA engine is used to simultaneously solve linear (and even nonlinear) matrix equations to determine the unknown DOFs—that is, the nodal displacements. Note that solution run time is proportional to the number of elements and nodes.

Traditionally, postprocessing consisted of the derivation of strains and stresses from the basic quantities—that is, the nodal displacements. In modern FEA software packages, however, strains, stresses, and even safety factors are all calculated and available once the solution is complete. Therefore, today the term *postprocessing* refers more to the visualization of the resulting data, as well as to the generation of reports summarizing the findings.

Modeling and Meshing

Finite elements come in a variety of dimensional shapes, including 1D beams, 2D plates and curved shells, and 3D tetrahedrons and hexahedrons. For accurate results, the choice of element dimension should be governed by the geometry of the component being analyzed. Use 1D beam elements for beam or truss-like structures such as the car chassis shown in Figure 15-1. For thin-walled structures such as the bracket shown in Figure 15-2, use 2D shell elements. For a bulky,

Figure 15-2 Use 2D shell elements for thin-walled structures

low-aspect-ratio, "potato-shaped" object like the one shown in Figure 15-3, use 3D solid elements.

Another element parameter is order—linear, quadratic, and cubic. A linear element can have only straight edges, whereas higher-order elements can have curved edges that better capture the underlying geometry.

Finally, the nodes of an element can have varying degrees of freedom, from one to six. Degrees of freedom represent the ability of the element to transmit or react to a load. The number of degrees of freedom in a model determines the number of equations required to define the model. In fact, the number of nodal DOFs is the best indication of the size of the model.

Solid elements can be either hexahedral (bricks) or tetrahedral. Both can be either linear or quadratic. Tetrahedral elements are normally used in automatic meshing, discussed below. Brick elements are more accurate than tetrahedral elements, but automatic meshing with bricks only works for certain geometries (i.e., extrusions, revolutions). Quadratic tetrahedral meshes can provide acceptable results.

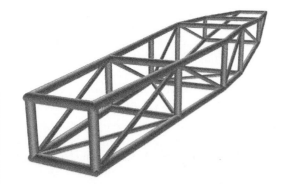

Figure 15-1 Use 1D beam elements for beam or truss-like structures

Figure 15-3 Use 3D solid elements for bulky, low-aspect-ratio objects

Good mesh quality is essential to an accurate mathematical representation of the physical model. The quality of a mesh can often be gauged by its visual appearance. Characteristics of a high-quality mesh include uniform, undistorted element shapes and smooth, gradual transitions in mesh density. Ideal element shapes include equilateral triangles, squares, and bricks. In any case, well-shaped elements should have a low aspect ratio. Mesh elements have trouble handling significant geometric curvature, as well as in capturing rapidly varying strains. In these regions, the mesh needs to be tighter in order to capture the change.

It is also important that the mesh be sufficiently dense to accurately capture the behavior of the continuous elastic structure that it attempts to simulate. **Convergence**, although not covered here, is the process of successfully refining a mesh in order to produce optimal FEA results.

The denser the mesh, the longer the solution time, so sound judgment gained through experience is essential. Selecting the appropriate modeling analysis approach is important, as is the use of symmetry to limit the model's size.

Most shell and solid meshing used for upfront simulation is now done automatically. The initial mesh is based on an initial default element size. Tools are available to refine the mesh locally—for example, by specifying mesh size on surfaces, edges, and vertices. Figure 15-4 shows some meshes.

Boundary Conditions

Boundary conditions are either loads or constraints (supports) that act on the model. Boundary conditions represent everything in the operating environment that is not explicitly modeled. For difficult problems, boundary conditions pose the most complicated aspect of a finite element analysis. Considerable experience is needed to correctly model the loads and constraints acting on a component.

If a model does not behave as it should, then it is probably either over- or understiffened.[1] An

---

[1] *Stiffness* is a measure of the resistance to deformation in an elastic body. Stiffness depends on both the type of material and the geometry of the body. To accurately capture the behavior of a body in a finite element analysis, it is essential to model the stiffness accurately.

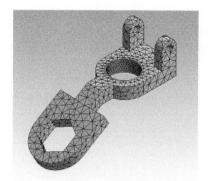

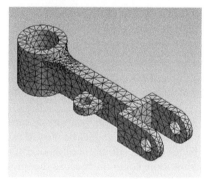

Figure 15-4 Example tetrahedral meshes

overstiffened model's ability to react to loading as it should is prevented by either excessive or redundant constraints, while an underconstrained model may not have enough constraints to prevent rigid body motion. Understiffening can also occur when a load is used to represent an assembly component that is not modeled; the load imparts no stiffness, whereas the actual structure may.

Loads are inputs to the system. They include forces, moments, pressures, temperatures, and accelerations. In defining a load, it may be necessary to define its magnitude, orientation, distribution, or time dependence.

Constraints remove spatial degrees of freedom. FEA models must be fully constrained in all six DOFs in order remove the possibility of rigid body motion; otherwise, the analysis fails. Constraints also react to the applied loads. Constraint types include fixed, cylindrical, pinned, and frictionless.

Results

In solving an FEA simulation, fundamental quantities like nodal displacements and temperatures are calculated. Other quantities (e.g., strains, stresses, heat flux, and safety factors) are then derived from the basic quantities. The fundamental quantities are consequently more accurate, while the derived quantities are less reliable.

In a stress analysis, nodal displacement is the fundamental quantity, and for this reason, a displacement (i.e., deformation) contour plot should be the first result to be analyzed. It is good practice to animate the displacement contour plot to verify that the component behaves as expected, given the applied boundary conditions. If it does not, check the boundary conditions. View the displacement contour plot. Do the magnitudes of the deformations seem reasonable? If not, review the load magnitudes, material properties, and units.

Next review the stress contour plot. Is the contour quality acceptable? Poor contour quality—that is, hard discontinuities and jagged distributions—may indicate that there are problems with the mesh. Recall that a good mesh is characterized by well-shaped elements with a low aspect ratio, as well as by smooth and gradual transitions between different mesh densities. It may be necessary to refine the mesh by increasing the number of elements. Finally, review any specific results, such as stress magnitudes and safety factors.

There are many kinds of stress, including normal and shear, axial, bending and torsion, and tensile and compressive. Equivalent stress, or ***von Mises stress***, is often used in design work because it allows any arbitrary 3D stress state to be represented as a single positive stress value. Equivalent stress is part

Contour plot

The results of an FEA simulation are viewed using a contour plot, seen in Figure 15-5. These plots use color to represent the different magnitudes of the quantity under consideration, typically deformation, strain, or stress. The plots should be viewed using contour bands (distinct differentiation of color), rather than smooth contours, because the smooth contours suggest that the underlying mesh is of high quality, which may not be the case. Regarding color, although red contours do, for example, signify areas of high stress (or deformation, strain, or the like), and blue contours signify regions of low stress, the actual stress level depends on the legend, which serves to relate color to numerical values. Even if the overall stress level is low, there may still be red regions on the contour plot. Also, the legend can be adjusted, thus changing the color patterns.

Contour plots can also be animated. The animation shows the component response as the boundary conditions are applied. Note that by default, contour plot animations are scaled so that the deformation movement is obvious to the viewer. This scaling can be adjusted and even set to the actual movement.

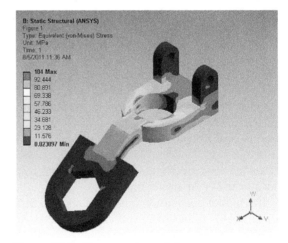

Figure 15-5 Stress contour plot

of the maximum equivalent stress failure theory used to predict yielding in a ductile material.

Failure theories exist that are used to predict the failure of both ductile and brittle materials. These failure theories are often available for use in commercial FEA software. The most commonly available failure theories are:

- Maximum equivalent stress
- Maximum shear stress
- Mohr–Coulomb stress
- Maximum tensile stress

The maximum equivalent stress and maximum shear stress failure theories are used primarily for ductile (e.g., steel, aluminum, brass) materials, and the Mohr–Coulomb stress and maximum tensile stress failure theories are used for brittle materials. All of them calculate a specific stress and compare it to an allowable stress. All of them are typically expressed in the form of a **safety factor**, where the safety factor is the allowable, or yield stress divided by the calculated stress. As long as the safety factor is greater than 1, the criterion is met.

FEA workflow

The strength of a handle used on a ball valve is evaluated. The ball valve and handle are shown in Figure 15-6. Note that because of stops on the handle and the valve housing, the handle is free to rotate only 90 degrees. Figure 15-7 shows a detail with the handle rotated so that the stops are in contact. The handle, made of steel, will be analyzed with the stops in contact and a person pushing against the end of the handle.

Figure 15-6 Ball valve and handle

Figure 15-7 Ball valve and handle detail

The handle, seen in Figure 15-8, is imported into ANSYS Workbench, an important CAD-embedded finite element analysis program. Figure 15-9 shows the part mesh.

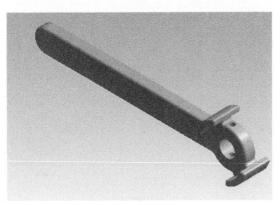

Figure 15-8 Ball valve handle

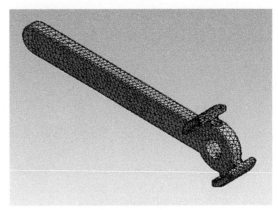

Figure 15-9 Meshing of ball valve handle

Next the boundary conditions are assigned, as shown in Figure 15-10. First, a cylindrical support is added to the surface of the hole labeled A. Cylindrical supports prevent a cylindrical face from moving or deforming in a combination of radial, axial, and tangential directions. In this case the tangential direction is left free so that the handle is able to move, rotate, and deform tangentially, while being otherwise constrained. To fix this last remaining degree of freedom, a displacement support is applied to the face of the stop, labeled B in Figure 15-10. Surface B is constrained so that it cannot move (displacement = 0) in a direction normal to the surface. The handle is now fully constrained in all six DOFs.

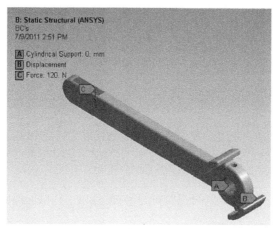

Figure 15-10 Handle boundary conditions

To simulate the loading, a force of 120 Newtons (120 N) is applied to a surface patch at the end of the handle, in order to simulate someone pushing against the handle while the handle stop is in contact with the valve housing stop. Loads and supports are typically applied to surface areas. If the available surface area on the model is too large, a smaller surface patch can be created in CAD software using a split tool. This was done to create the surface patch C, shown in Figure 15-10.

In ANSYS Workbench it is necessary to specify the results to be solved for. In this case, the total deformation and the equivalent (von Mises) stress have been solved for, as seen in Figures 15-11 and 15-12, respectively. The maximum deflection is 0.15 mm, and the maximum equivalent stress is 112.5 MPa.

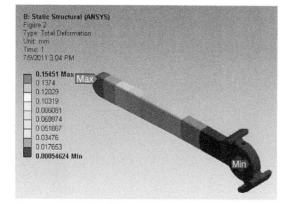

Figure 15-11 Total deformation

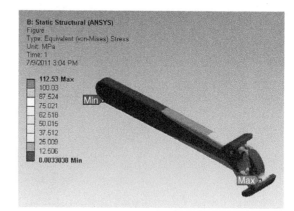

Figure 15-12 von Mises stress

In addition, a safety factor tool based on the maximum equivalent stress failure criteria has been included, as seen in Figure 15-13. The safety factor is 2.2.

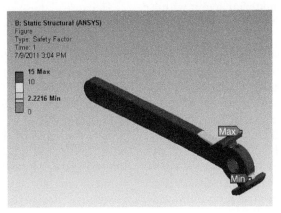

Figure 15-13 Safety factor for handle

In order to save some material, the original CAD file is modified to include symmetrical pockets to both sides of the handle, as seen in Figure 15-14, and the analysis is run again. Figure 15-15 shows that the safety factor is now 1.8, which is deemed acceptable, and the change is approved.

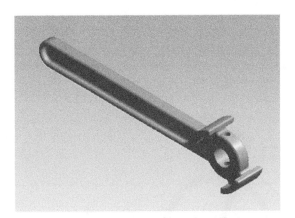

Figure 15-14 Ball valve handle with pockets

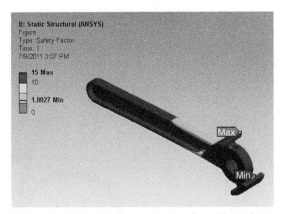

Figure 15-15 Safety factor for handle with pockets

■ GENERATIVE DESIGN

Generative design is an iterative design exploration process. Starting with the definition of high-level goals and constraints, generative design leverages the power of computationally driven artificial intelligence to automatically explore a wide design space in order to identify the best design options. Originally used in architecture, generative design is currently being enthusiastically investigated and employed by companies like ANSYS, Autodesk, General Motors, and Carbon. This trend may well represent a sea change in the way products are designed going forward.

Generative design is related to ***topology optimization*** but is a much newer concept. Topology optimization is a mathematical method used to optimize material layout within a given design space. For a given set of loads and constraints, the goal of topology optimization is to maximize system performance (by reducing mass, avoiding resonance, minimizing thermal stress or deformation). Topology optimization uses finite element methods to evaluate design performance. It is a foundational technology; one that has been used since 1990s. Topology optimization is used primarily by stress analysts.

Like topology optimization, generative design is most commonly used for structural optimization, allowing for the design of light weight parts with sufficient strength, stiffness, and fatigue resistance characteristics. While topology optimization provides a single solution to a specific problem, generative design leverages artificial intelligence and machine learning to explore a much wider range of possible solutions,

comparing the results of thousands of simulations to close in on a design that delivers the most favorable combination of attributes.

Because these algorithms automatically adjust part geometry between simulations, no manual refinement is necessary. This means that design engineers can productively use generative design, not just analysis specialists. Still, engineers must validate the generated output through analysis and testing, and to ensure that the design is manufacturable.

By now, most of the major CAD companies have integrated generative design tools into their software. Because these generative algorithms are so computationally demanding, generative design calculations are well suited for cloud-based computation. Generative design and additive manufacturing work well together, since AM machines deal well with the complex, organic shapes that typically emerge from generative algorithms. The following section provides an example of a typical generative design workflow.

Generative Design Workflow

The goal of this example is to create a bike bracket similar to the one depicted in Figure 15-16.

Figure 15-16 Example of a wall-mount bike bracket that is currently on the market

The first step is to identify and create the critical structures of the component, the preserved geometry and the obstacle geometry. The minimum preserved geometry of the bike bracket are the hooks where the bike would hang and the mounting holes for the screws as seen in Figure 15-17. Additionally, since the bracket is intended to be affixed to a wall, the wall needs to be included as obstacle geometry to prevent interferences with the environment (Figure 15-18).

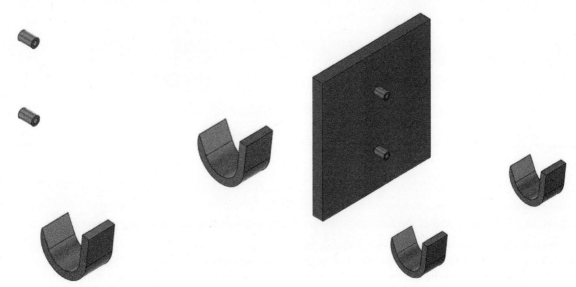

Figure 15-17 Isometric view of bike bracket preserved features: hooks and mounting screw holes

Figure 15-18 Isometric view of bike bracket preserved geometry with a wall added as obstacle geometry

Once the initial shapes of the bracket have been created, the expected loads must be applied (Figure 15-19) to emulate what kind of stress the component will be under.

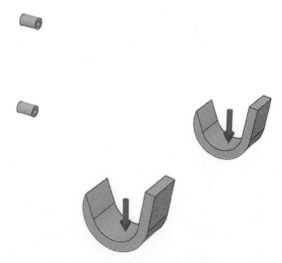

Figure 15-19 Pictorial view of preserved geometry with an applied force of 20 lbs. to simulate a bike

Now the structural constraints must be added to provide information on how the design interfaces with its surroundings. In this case, a fixed constraint has been applied to the ends of the mounting screw holes as this is where the bracket would be fixed to the wall (Figure 15-20).

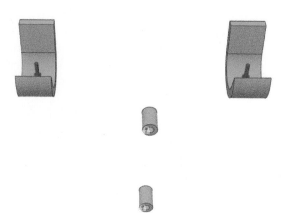

Figure 15-20 Pictorial view of the fixed constraints applied to the ends of the mounting screw holes

Next, the design criteria must be defined. Fusion 360 can be set to minimize the mass or maximize the stiffness of the structure. Additionally, the safety factor can be adjusted according to the use of the product. Finally, the possible materials must be chosen along with the method(s) of manufacturing: die casting, 2-axis cutting, milling, additive, or simply unrestricted. With the data input, the software is now able to generate a set of outcomes to explore, as seen in Figure 15-21.

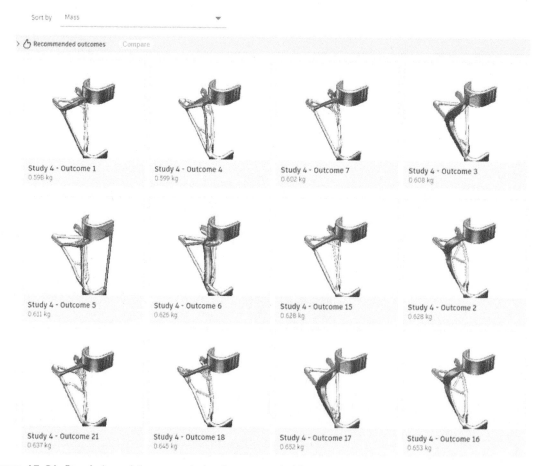

Figure 15-21 Panel view of the generated outcomes sorted by mass

These outcomes can now be sorted by criteria such as mass, minimum factor of safety, volume, or even the estimated piece part cost of production. The model of any outcome can be exported for further design (Figure 15-22), or the study may have produced results that call for iteration. At which point, the designer must return to the first steps and make tweaks to the design criteria: create additional obstacle geometry, modify the materials used, or add in load cases that were not initially accounted for.

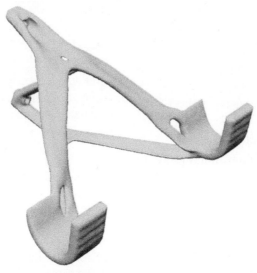

Figure 15-22 Model view of one of the generated outcomes

■ DYNAMICS SIMULATION SOFTWARE

Dynamics simulation software, also called motion analysis software, is used to model the time-varying behavior of an assembly in motion under various load conditions. Physics-based dynamics simulation software is used to investigate the dynamics of moving parts, to determine how loads and forces are distributed throughout a mechanical system, and to improve and optimize the performance of commercial products. Motion analysis software is used to analyze the large rotations and other highly nonlinear motion typical of mechanical systems that would be difficult to solve using finite element analysis. Dynamics simulation software is based on Newton's second law of motion, $F = ma$. The dynamics of complex mechanical systems can typically be described by ordinary or partial differential equations. Numerical methods are employed to solve these equations. The numerical simulation is conducted by dividing time into segments and then calculating the dynamic equilibrium of the mechanism at each time step.

The automotive, aerospace, and heavy machinery industries are all important users of dynamics simulation software. In product design it is important to have a solid understanding of how different moving components interact with one another in an assembly. Dynamics simulation is used in the design of six-DOF vehicles, steam turbines, robotic arms, and so on. Simulation is also used in 3D gaming systems, where a *physics engine* is employed to accelerate the action.

Software packages used for dynamics simulation include MSC Adams, a high-end multibody dynamics and motion analysis software, and Working Model, an engineering simulation software product useful for basic physics simulations. Other simulation packages are included in MCAD programs such as SolidWorks and Autodesk Inventor. Both of the dynamics simulation environments available in these parametric solid modelers were obtained through the acquisition of other engineering simulation software companies.

Some motion analysis software capabilities include:

- Convert assembly constraints to motion joints
- Access a large library of motion joints
- Define external forces and moments
- Create motion simulations based on position, velocity, acceleration, and torque as functions of time in joints
- Visualize 3D motion using the trace of the motion of a part
- Export results to Microsoft® Excel
- Export load conditions at any motion state to an FEA program
- Calculate the force required to keep a dynamics simulation in static equilibrium
- Use friction, damping, stiffness, and elasticity as functions of time when defining joints

In a parametric assembly modeling environment, **constraints** (e.g., mate) are used to control component motion. When first placed in an assembly, components have six degrees of freedom. Constraints are then added to restrict these degrees of freedom. In a dynamics simulation environment, by contrast, **joints** are used to control component motion. Here components initially have zero degrees of freedom. Mechanical joints are then built to create these degrees of freedom. To bridge these differences, a common first task when running a motion analysis is to convert assembly constraints into mechanical joints. Assembly constraints can be converted either automatically or manually into joints, or the joints can be created from scratch. Each of these approaches has its advantages and disadvantages. Figure 15-23 shows a mechanical joint to assembly constraint conversion table.

The Kinematic Models for Design Digital Library (KMODDL),[2] maintained by Cornell University, is "a collection of mechanical models and related resources for teaching the principles

[2]The KMODDL URL is http://kmoddl.library.cornell.edu.

Joint	Name of Joint	Bound DOF	Assembly Constraint Equivalent	Open DOF
	Revolution	5	Insert (circular edges)	1 - Rotation
	Prismatic	5	Combining 2 Mates (plane, plane) nonparallel	1 - Translation
	Cylindrical	4	Mate (line, line)	1 - Translation 1 - Rotation
	Spherical	3	Mate (point, point)	3 - Rotation
	Planar	3	Mate (plane, plane)	2 - Translation 1 - Rotation
	Point Line	2	Mate (line, point)	1 - Translation 3 - Rotation
	Line Plane	2	Mate (plane, line)	2 - Translation 2 - Rotation
	Point Plane	1	Mate (plane, point)	2 - Translation 3 - Rotation
	Spatial	0	NO constraints	3 - Translation 3 - Rotation
	Weld	6	Combining 3 constraints or 2 Inserts	NONE

Figure 15-23 Mechanical joint to assembly constraint conversion table (Courtesy of Fernando Class–Morales)

of kinematics—the geometry of pure motion." One of the collection's most important models is the Peaucellier–Lipkin linkage, which represents a culminating solution to the nineteenth-century mathematical problem of how to convert circular motion into linear motion. Figure 15-24 shows a screen capture of this linkage in a motion analysis

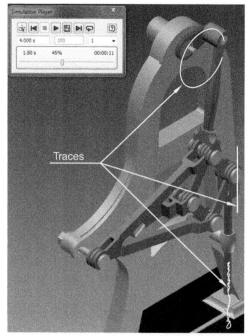

Traces

Figure 15-24 Peaucellier–Lipkin linkage with traces (Courtesy of Zanxi An)

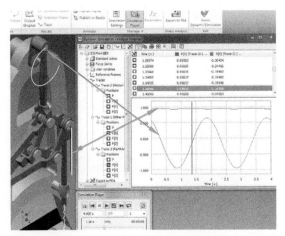

Figure 15-25 Peaucellier–Lipkin linkage with graphical output (Courtesy of Zanxi An)

application. In this figure, *traces* are used to help visualize the conversion from circular to linear motion. Figure 15-25 shows another important capability of dynamics simulation software: the ability to create graphical outputs of different dynamic quantities such as positions, velocities, accelerations, forces, and torques.

A typical simulation workflow includes the following steps:

1. Create mechanical joints between components that have relative motion between them. There are two stages:
   a. Stage 1—create standard joints. Standard joints are joints that can be created from assembly constraints, like the ones shown in Figure 15-23. As we noted earlier, there are three possibilities:
      i. Automatically convert assembly constraints to standard joints.
      ii. Manually convert assembly constraints to standard joints.
      iii. Create standard joints from scratch.
   b. Stage 2—create nonstandard joints. Nonstandard joints do not use assembly constraints. Nonstandard joints include rolling (e.g., gears), sliding, spring, and contact.
2. Create environmental conditions to simulate reality. The possibilities include:
   a. Define a joint's starting position.
   b. Apply friction to joints.

   c. Apply external loads (e.g., force, torque).
   d. Impose motion on predefined joints. Imposing motion on a joint is used to simulate the presence of a motor.
3. Analyze the results.

Results include reaction forces, velocities, and accelerations. A dynamics simulation reveals how parts respond structurally to dynamic loads at any point within the range of motion of the assembly. The next section provides a demonstration showing how dynamics simulation software is used.

Dynamics Simulation Software Demonstration

In this demonstration, the locking pliers assembly shown in Figure 15-26 will be analyzed in a dynamics simulation program to determine the force required to open and close the pliers. Note that these forces are not large, because locking pliers are designed to be quickly opened and closed.

The constrained assembly is brought into the dynamics simulation software package. The assembly constraints are automatically converted into mechanical joints, and those parts that do not move relative to one another are joined into welded groups.

A spring joint is added to the model, connecting the upper jaw to the fixed handle. The actual spring is used to obtain the length and stiffness needed for the spring joint inputs.

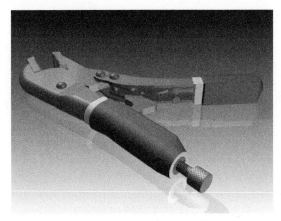

Figure 15-26 Rendered image of locking pliers

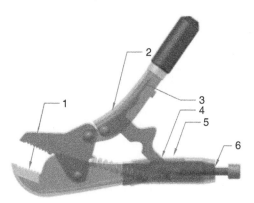

Figure 15-27 3D contacts on locking pliers
(Courtesy of Yinye Yang)

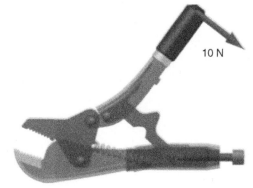

Figure 15-29 10 N force fails to close locking pliers
(Courtesy of Yinye Yang)

Next six 3D contact joints are manually added. The contact joints are used between parts that come into contact with one another, as shown in Figure 15-27.

For the external forces, gravity is first added, as shown in Figure 15-28 pointing down. Next the force used to open and close the locking pliers is added. This force can also be seen in Figure 15-28. As specified, the force to close the pliers remains perpendicular to the handle surface as the handle moves.

Friction is also added to each of the four revolute joints. A friction coefficient of 0.5 is used.

The simulation is now ready to run, using a time of 0.2 second, with 200 steps. The simulation is run a number of times, varying the force at each run. As seen in Figure 15-29, the pliers do not close when a force of 10 N is applied. When

the force is increased to 15 N, the pliers close, as shown in Figure 15-30.

Next the locking pliers are closed, and the direction of the force is reversed in order to determine the force needed to open the pliers. As seen in Figure 15-31, a force of 3 N is needed to open the pliers.

Figure 15-30 15 N force closes locking pliers
(Courtesy of Yinye Yang)

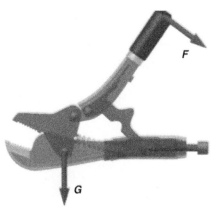

Figure 15-28 External forces on locking pliers
(Courtesy of Yinye Yang)

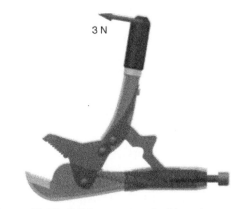

Figure 15-31 3 N force opens locking pliers
(Courtesy of Yinye Yang)

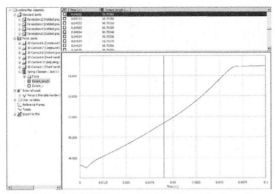

Figure 15-32 Graph of spring extension as locking pliers open (Courtesy of Yinye Yang)

These results are reasonable, because the actual force required to open the locking pliers and that required to close them are both quite small, and in fact the actual force required to open the pliers is clearly less than the force needed close them.

Finally, Figure 15-32 shows a graph of the spring length as the locking pliers go from closed to open.

■ QUESTIONS

TRUE OR FALSE

1. Simulation software is also known as computer-aided engineering (CAE) software.
2. Finite element analysis was first developed in the 1960s.
3. If the safety factor is positive, then actual stress is less than the allowable stress.
4. The von Mises stress is always positive.
5. Finite element analysis cannot be used for assemblies.
6. Dynamics simulation software is also called motion analysis software.
7. Joints in a dynamic simulation are identical to constraints in an assembly.

MULTIPLE CHOICE

`SS` 8. Which of the following are not FEA phases?
   a. Preprocessing
   b. Analysis
   c. Solution
   d. Postprocessing

9. Which of the following are solid elements used in finite element analysis?
   a. Tetrahedrons
   b. Hexahedrons
   c. Octahedrons
   d. Polyhedrons
   e. Both a and b
   f. Both c and d

10. What is the first quantity to be calcu- `SS` lated when solving an FEA stress analysis problem?
   a. Displacements
   b. Strains
   c. Stresses
   d. Safety factor

11. For joints in a dynamic analysis simulation, which of the following can be specified?
   a. A joint's starting position
   b. Friction
   c. External loads (e.g., force, torque)
   d. Imposed motion
   e. All of the above

12. Automatic meshing algorithms for solids `SS` typically employ what kind of elements?
   a. Tetrahedrons
   b. Bricks
   c. Wedges
   d. Pyramids
   e. All of the above
   f. None of the above

13. What is the best way to have confidence in an FEA solution?
   a. Use tetrahedral elements
   b. Verify that the percent difference in stress levels between convergence runs is getting smaller
   c. Simplify the boundary conditions
   d. Ensure that there are no singularities in the model

A

ANSI PREFERRED ENGLISH LIMITS AND FITS

ANSI RUNNING AND SLIDING FITS (RC)–ENGLISH UNITS

American National Standard Running and Sliding Fits (ANSI B4.1–1967, R1979)

Tolerance limits given in body of table are added to or subtracted from basic size (as indicated by + or – sign) to obtain maximum and minimum sizes of mating parts.

Values shown below are in thousandths of an inch.

Nominal Size Range, Inches Over To	Class RC1 Clearance*	Class RC1 Hole H5	Class RC1 Shaft g4	Class RC2 Clearance*	Class RC2 Hole H6	Class RC2 Shaft g5	Class RC3 Clearance*	Class RC3 Hole H7	Class RC3 Shaft f6	Class RC4 Clearance*	Class RC4 Hole H8	Class RC4 Shaft f7
0–0.12	0.1 / 0.45	+0.2 / 0	-0.1 / -0.25	0.1 / 0.55	+0.25 / 0	-0.1 / -0.3	0.3 / 0.95	+0.4 / 0	-0.3 / -0.55	0.3 / 1.3	+0.6 / 0	-0.3 / -0.7
0.12–0.24	0.15 / 0.5	+0.2 / 0	-0.15 / -0.3	0.15 / 0.65	+0.3 / 0	-0.15 / -0.35	0.4 / 1.12	+0.5 / 0	-0.4 / -0.7	0.4 / 1.6	+0.7 / 0	-0.4 / -0.9
0.24–0.40	0.2 / 0.6	+0.25 / 0	-0.2 / -0.35	0.2 / 0.85	+0.4 / 0	-0.2 / -0.45	0.5 / 1.5	+0.6 / 0	-0.5 / -0.9	0.5 / 2.0	+0.9 / 0	-0.5 / -1.1
0.40–0.71	0.25 / 0.75	+0.3 / 0	-0.25 / -0.45	0.25 / 0.95	+0.4 / 0	-0.25 / -0.55	0.6 / 1.7	+0.7 / 0	-0.6 / -1.0	0.6 / 2.3	+1.0 / 0	-0.6 / -1.3
0.71–1.19	0.3 / 0.95	+0.4 / 0	-0.3 / -0.55	0.3 / 1.2	+0.5 / 0	-0.3 / -0.7	0.8 / 2.1	+0.8 / 0	-0.8 / -1.3	0.8 / 2.8	+1.2 / 0	-0.8 / -1.6
1.19–1.97	0.4 / 1.1	+0.4 / 0	-0.4 / -0.7	0.4 / 1.4	+0.6 / 0	-0.4 / -0.8	1.0 / 2.6	+1.0 / 0	-1.0 / -1.6	1.0 / 3.6	+1.6 / 0	-1.0 / -2.0
1.97–3.15	0.4 / 1.2	+0.5 / 0	-0.4 / -0.7	0.4 / 1.6	+0.7 / 0	-0.4 / -0.9	1.2 / 3.1	+1.2 / 0	-1.2 / -1.9	1.2 / 4.2	+1.8 / 0	-1.2 / -2.4
3.15–4.73	0.5 / 1.5	+0.6 / 0	-0.5 / -0.9	0.5 / 2.0	+0.9 / 0	-0.5 / -1.1	1.4 / 3.7	+1.4 / 0	-1.4 / -2.3	1.4 / 5.0	+2.2 / 0	-1.4 / -2.8
4.73–7.09	0.6 / 1.8	+0.7 / 0	-0.6 / -1.1	0.6 / 2.3	+1.0 / 0	-0.6 / -1.3	1.6 / 4.2	+1.6 / 0	-1.6 / -2.6	1.6 / 5.7	+2.5 / 0	-1.6 / -3.2
7.09–9.85	0.6 / 2.0	+0.8 / 0	-0.6 / -1.2	0.6 / 2.6	+1.2 / 0	-0.6 / -1.4	2.0 / 5.0	+1.8 / 0	-2.0 / -3.2	2.0 / 6.6	+2.8 / 0	-2.0 / -3.8
9.85–12.41	0.8 / 2.3	+0.9 / 0	-0.8 / -1.4	0.8 / 2.9	+1.2 / 0	-0.8 / -1.7	2.5 / 5.7	+2.0 / 0	-2.5 / -3.7	2.5 / 7.5	+3.0 / 0	-2.5 / -4.5
12.41–15.75	1.0 / 2.7	+1.0 / 0	-1.0 / -1.7	1.0 / 3.4	+1.4 / 0	-1.0 / -2.0	3.0 / 6.6	+2.2 / 0	-3.0 / -4.4	3.0 / 8.7	+3.5 / 0	-3.0 / -5.2
15.75–19.69	1.2 / 3.0	+1.0 / 0	-1.2 / -2.0	1.2 / 3.8	+1.6 / 0	-1.2 / -2.2	4.0 / 8.1	+2.5 / 0	-4.0 / -5.6	4.0 / 10.5	+4.0 / 0	-4.0 / -6.5

ANSI RUNNING AND SLIDING FITS (RC)—ENGLISH UNITS

Values shown below are in thousandths of an inch.

Nominal Size Range, Inches Over To	RC5 Clearance*	RC5 Hole H8	RC5 Shaft e7	RC6 Clearance*	RC6 Hole H9	RC6 Shaft e8	RC7 Clearance*	RC7 Hole H9	RC7 Shaft d8	RC8 Clearance*	RC8 Hole H10	RC8 Shaft c9	RC9 Clearance*	RC9 Hole H11	RC9 Shaft
0–0.12	0.6	+0.6	−0.6	0.6	+1.0	−0.6	1.0	+1.0	−1.0	2.5	+1.6	−2.5	4.0	+2.5	−4.0
	1.6	0	−1.0	2.2	0	−1.2	2.6	0	−1.6	5.1	0	−3.5	8.1	0	−5.6
0.12–0.24	0.8	+0.7	−0.8	0.8	+1.2	−0.8	1.2	+1.2	−1.2	2.8	+1.8	−2.8	4.5	+3.0	−4.5
	2.0	0	−1.3	2.7	0	−1.5	3.1	0	−1.9	5.8	0	−4.0	9.0	0	−6.0
0.24–0.40	1.0	+0.9	−1.0	1.0	+1.4	−1.0	1.6	+1.4	−1.6	3.0	+2.2	−3.0	5.0	+3.5	−5.0
	2.5	0	−1.6	3.3	0	−1.9	3.9	0	−2.5	6.6	0	−4.4	10.7	0	−7.2
0.40–0.71	1.2	+1.0	−1.2	1.2	+1.6	−1.2	2.0	+1.6	−2.0	3.5	+2.8	−3.5	6.0	+4.0	−6.0
	2.9	0	−1.9	3.8	0	−2.2	4.6	0	−3.0	7.9	0	−5.1	12.8	0	−8.8
0.71–1.19	1.6	+1.2	−1.6	1.6	+2.0	−1.6	2.5	+2.0	−2.5	4.5	+3.5	−4.5	7.0	+5.0	−7.0
	3.6	0	−2.4	4.8	0	−2.8	5.7	0	−3.7	10.0	0	−6.5	15.5	0	−10.5
1.19–1.97	2.0	+1.6	−2.0	2.0	+2.5	−2.0	3.0	+2.5	−3.0	5.0	+4.0	−5.0	8.0	+6.0	−8.0
	4.6	0	−3.0	6.1	0	−3.6	7.1	0	−4.6	11.5	0	−7.5	18.0	0	−12.0
1.97–3.15	2.5	+1.8	−2.5	2.5	+3.0	−2.5	4.0	+3.0	−4.0	6.0	+4.5	−6.0	9.0	+7.0	−9.0
	5.5	0	−3.7	7.3	0	−4.3	8.8	0	−5.8	13.5	0	−9.0	20.5	0	−13.5
3.15–4.73	3.0	+2.2	−3.0	3.0	+3.5	−3.0	5.0	+3.5	−5.0	7.0	+5.0	−7.0	10.0	+9.0	−10.0
	6.6	0	−4.4	8.7	0	−5.2	10.7	0	−7.2	15.5	0	−10.5	24.0	0	−15.0
4.73–7.09	3.5	+2.5	−3.5	3.5	+4.0	−3.5	6.0	+4.0	−6.0	8.0	+6.0	−8.0	12.0	+10.0	−12.0
	7.6	0	−5.1	10.0	0	−6.0	12.5	0	−8.5	18.0	0	−12.0	28.0	0	−18.0
7.09–9.85	4.0	+2.8	−4.0	4.0	+4.5	−4.0	7.0	+4.5	−7.0	10.0	+7.0	−10.0	15.0	+12.0	−15.0
	8.6	0	−5.8	11.3	0	−6.8	14.3	0	−9.8	21.5	0	−14.5	34.0	0	−22.0
9.85–12.41	5.0		−5.0	5.0	+5.0	−5.0	8.0	+5.0	−8.0	12.0	+8.0	−12.0	18.0	+12.0	−18.0
	10.0		−7.0	13.0	0	−8.0	16.0	0	−11.0	25.0	0	−17.0	38.0	0	−26.0
12.41–15.75	6.0	+3.5	−6.0	6.0	+6.0	−6.0	10.0	+6.0	−10.0	14.0	+9.0	−14.0	22.0	+14.0	−22.0
	11.7	0	−8.2	15.5	0	−9.5	19.5	0	−13.5	29.0	0	−20.0	45.0	0	−31.0
15.75–19.69	8.0	+4.0	−8.0	8.0	+6.0	−8.0	12.0	+6.0	−12.0	16.0	+10.0	−16.0	25.0	+16.0	−25.0
	14.5	0	−10.5	18.0	0	−12.0	22.0	0	−16.0	32.0	0	−22.0	51.0	0	−35.0

Note: In the RC5 block for 9.85–12.41 the Hole H8 values are +3.0 / 0.

Data in boldface type are in accordance with American–British–Canadian (ABC) Agreements. Symbols H5, g4, etc. are hole and shaft designations in ABC system. Limits for sizes above 19.69 inches are also given in the ANSI Standard.

*Pairs of values shown represent minimum and maximum amounts of clearance resulting from application of standard tolerance limits.

Source: Courtesy of The American Society of Mechanical Engineers.

APPENDIX A ANSI PREFERRED ENGLISH LIMITS AND FITS

ANSI CLEARANCE LOCATION FITS (LC)–ENGLISH UNITS

American National Standard Clearance Locational Fits (ANSI B4.1–1967, R1979)

Tolerance limits given in body of table are added or subtracted to basic size (as indicated by + or – sign) to obtain maximum and minimum sizes of mating parts.

Values shown below are in thousandths of an inch.

Nominal Size Range, Inches Over To	Class LC1			Class LC2			Class LC3			Class LC4			Class LC5		
	Clearance*	Hole H6	Shaft h5	Clearance*	Hole H7	Shaft h6	Clearance*	Hole H8	Shaft h7	Clearance*	Hole H10	Shaft h9	Clearance*	Hole H7	Shaft g6
0–0.12	0 / 0.45	+0.25 / 0	0 / −0.2	0 / 0.65	+0.4 / 0	0 / −0.25	0 / 1	+0.6 / 0	0 / −0.4	0 / 2.6	+1.6 / 0	0 / −1.0	0.1 / 0.75	+0.4 / 0	−0.1 / −0.35
0.12–0.24	0 / 0.5	+0.3 / 0	0 / −0.2	0 / 0.8	+0.5 / 0	0 / −0.3	0 / 1.2	+0.7 / 0	0 / −0.5	0 / 3.0	+1.8 / 0	0 / −1.2	0.15 / 0.95	+0.5 / 0	−0.15 / −0.45
0.24–0.40	0 / 0.65	+0.4 / 0	0 / −0.25	0 / 1.0	+0.6 / 0	0 / −0.4	0 / 1.5	+0.9 / 0	0 / −0.6	0 / 3.6	+2.2 / 0	0 / −1.4	0.2 / 1.2	+0.6 / 0	−0.2 / −0.6
0.40–0.71	0 / 0.7	+0.4 / 0	0 / −0.3	0 / 1.1	+0.7 / 0	0 / −0.4	0 / 1.7	+1.0 / 0	0 / −0.7	0 / 4.4	+2.8 / 0	0 / −1.6	0.25 / 1.35	+0.7 / 0	−0.25 / −0.65
0.71–1.19	0 / 0.9	+0.5 / 0	0 / −0.4	0 / 1.3	+0.8 / 0	0 / −0.5	0 / 2	+1.2 / 0	0 / −0.8	0 / 5.5	+3.5 / 0	0 / −2.0	0.3 / 1.6	+0.8 / 0	−0.3 / −0.8
1.19–1.97	0 / 1.0	+0.6 / 0	0 / −0.4	0 / 1.6	+1.0 / 0	0 / −0.6	0 / 2.6	+1.6 / 0	0 / −1	0 / 6.5	+4.0 / 0	0 / −2.5	0.4 / 2.0	+1.0 / 0	−0.4 / −1.0
1.97–3.15	0 / 1.2	+0.7 / 0	0 / −0.5	0 / 1.9	+1.2 / 0	0 / −0.7	0 / 3	+1.8 / 0	0 / −1.2	0 / 7.5	+4.5 / 0	0 / −3	0.4 / 2.3	+1.2 / 0	−0.4 / −1.1
3.15–4.73	0 / 1.5	+0.9 / 0	0 / −0.6	0 / 2.3	+1.4 / 0	0 / −0.9	0 / 3.6	+2.2 / 0	0 / −1.4	0 / 8.5	+5.0 / 0	0 / −3.5	0.5 / 2.8	+1.4 / 0	−0.5 / −1.4
4.73–7.09	0 / 1.7	+1.0 / 0	0 / −0.7	0 / 2.6	+1.6 / 0	0 / −1.0	0 / 4.1	+2.5 / 0	0 / −1.6	0 / 10.0	+6.0 / 0	0 / −4	0.6 / −3.2	+1.6 / 0	−0.6 / −1.6
7.09–9.85	0 / 2.0	+1.2 / 0	0 / −0.8	0 / 3.0	+1.8 / 0	0 / −1.2	0 / 4.6	+2.8 / 0	0 / −1.8	0 / 11.5	+7.0 / 0	0 / −4.5	0.6 / 3.6	+1.8 / 0	−0.6 / −1.8
9.85–12.41	0 / 2.1	+1.2 / 0	0 / −0.9	0 / 3.2	+2.0 / 0	0 / −1.2	0 / 5	+3.0 / 0	0 / −2.0	0 / 13.0	+8.0 / 0	0 / −5	0.7 / 3.9	+2.0 / 0	−0.7 / −1.9
12.41–15.75	0 / 2.4	+1.4 / 0	0 / −1.0	0 / 3.6	+2.2 / 0	0 / −1.4	0 / 5.7	+3.5 / 0	0 / −2.2	0 / 15.0	+9.0 / 0	0 / −6	0.7 / 4.3	+2.2 / 0	−0.7 / −2.1
15.75–19.69	0 / 2.6	+1.6 / 0	0 / −1.0	0 / 4.1	+2.5 / 0	0 / −1.6	0 / 6.5	+4 / 0	0 / −2.5	0 / 16.0	+10.0 / 0	0 / −6	0.8 / 4.9	+2.5 / 0	−0.8 / −2.4

ANSI CLEARANCE LOCATION FITS (LC)—ENGLISH UNITS

Values shown below are in thousandths of an inch.

Nominal Size Range, Inches Over To	Class LC6 Clearance*	Class LC6 Std Tol Hole H9	Class LC6 Std Tol Shaft f8	Class LC7 Clearance*	Class LC7 Std Tol Hole H10	Class LC7 Std Tol Shaft e9	Class LC8 Clearance*	Class LC8 Std Tol Hole H10	Class LC8 Std Tol Shaft d9	Class LC9 Clearance*	Class LC9 Std Tol Hole H11	Class LC9 Std Tol Shaft c10	Class LC10 Clearance*	Class LC10 Std Tol Hole H12	Class LC10 Std Tol Shaft	Class LC11 Clearance*	Class LC11 Std Tol Hole H13	Class LC11 Std Tol Shaft
0–0.12	0.3 / 1.9	+1.0 / 0	−0.3 / −0.9	0.6 / 3.2	+1.6 / 0	−0.6 / −1.6	1.0 / 2.0	+1.6 / 0	−1.0 / −2.0	2.5 / 6.6	+2.5 / 0	−2.5 / −4.1	4 / 12	+4 / 0	−4 / −8	5 / 17	+6 / 0	−5 / −11
0.12–0.24	0.4 / 2.3	+1.2 / 0	−0.4 / −1.1	0.8 / 3.8	+1.8 / 0	−0.8 / −2.0	1.2 / 4.2	+1.8 / 0	−1.2 / −2.4	2.8 / 7.6	+3.0 / 0	−2.8 / −4.6	4.5 / 14.5	+5 / 0	−4.5 / −9.5	6 / 20	+7 / 0	−6 / −13
0.24–0.40	0.5 / 1.8	+1.4 / 0	−0.5 / −1.4	1.0 / 4.6	+2.3 / 0	−1.0 / −2.4	1.6 / 5.2	+2.2 / 0	−1.6 / −3.0	3.0 / 8.7	+3.5 / 0	−3.0 / −5.2	5 / 17	+6 / 0	−5 / −11	7 / 25	+9 / 0	−7 / −16
0.40–0.71	0.6 / 3.2	+1.6 / 0	−0.6 / −1.6	1.3 / 5.6	+2.8 / 0	−1.2 / −2.8	2.0 / 6.4	+2.8 / 0	−2.0 / −3.6	3.5 / 10.3	+4.0 / 0	−3.5 / −6.3	6 / 20	+7 / 0	−6 / −13	8 / 28	+10 / 0	−8 / −18
0.71–1.19	0.8 / 4.0	+2.0 / 0	−0.8 / −2.0	1.6 / 7.1	+3.5 / 0	−1.6 / −3.6	3.5 / 8.0	+3.5 / 0	−2.5 / −4.5	4.5 / 13.0	+5.0 / 0	−4.5 / −8.0	7 / 23	+8 / 0	−7 / −15	10 / 34	+12 / 0	−10 / −22
1.19–1.97	1.0 / 5.1	+2.5 / 0	−1.0 / −2.6	1.0 / 8.5	+4.0 / 0	−2.0 / −4.5	3.6 / 9.5	+4.0 / 0	−3.0 / −5.5	5.0 / 15.0	+6 / 0	−5.0 / −9.0	8 / 28	+10 / 0	−8 / −18	12 / 44	+16 / 0	−12 / −28
1.97–3.15	1.2 / 6.0	+3.0 / 0	−1.0 / −3.0	2.5 / 10.0	+4.5 / 0	−2.5 / −5.5	4.0 / 11.5	+4.5 / 0	−4.0 / −7.0	6.0 / 17.5	+7 / 0	−6.0 / −10.5	10 / 34	+12 / 0	−10 / −22	14 / 50	+18 / 0	−14 / −32
3.15–4.73	1.4 / 7.1	+3.5 / 0	−1.4 / −3.6	3.0 / 11.5	+5.0 / 0	−3.0 / −6.5	5.0 / 13.5	+5.0 / 0	−5.0 / −8.5	7 / 21	+9 / 0	−7 / −12	11 / 39	+14 / 0	−11 / −25	16 / 60	+22 / 0	−16 / −38
4.73–7.09	1.6 / 8.1	+4.0 / 0	−1.6 / −4.1	3.5 / 13.5	+6.0 / 0	−3.5 / −7.5	6 / 16	+6 / 0	−6 / −10	8 / 24	+10 / 0	−8 / −14	12 / 44	+16 / 0	−12 / −28	18 / 68	+25 / 0	−18 / −43
7.09–9.85	2.0 / 9.3	+4.5 / 0	−2.0 / −4.8	4.0 / 15.5	+7.0 / 0	−4.0 / −8.5	7 / 18.5	+7 / 0	−7 / −11.5	10 / 29	+12 / 0	−10 / −17	16 / 52	+18 / 0	−16 / −34	22 / 78	+28 / 0	−22 / −50
9.85–12.41	2.2 / 10.2	+5.0 / 0	−2.2 / −5.2	4.5 / 17.5	+8.0 / 0	−4.5 / −9.5	7 / 20	+8 / 0	−7 / −12	12 / 32	+12 / 0	−12 / −20	20 / 60	+20 / 0	−20 / −40	28 / 88	+30 / 0	−28 / −58
12.41–15.75	2.5 / 12.0	+6.0 / 0	−2.5 / −6.0	5.0 / 20.0	+9.0 / 0	−5 / −11	8 / 23	+9 / 0	−8 / −14	14 / 37	+14 / 0	−14 / −23	22 / 66	+22 / 0	−22 / −44	30 / 100	+35 / 0	−30 / −65
15.75–19.69	2.8 / 12.8	+6.0 / 0	−2.8 / −6.8	5.0 / 31.0	+10.0 / 0	−5 / −11	9 / 25	+10 / 0	−9 / −15	16 / 42	+16 / 0	−16 / −26	25 / 75	+25 / 0	−25 / −50	35 / 115	+40 / 0	−35 / −75

Data in boldface type are in accordance with American–British–Canadian (ABC) agreements. Symbols H6, H7, f 6, etc. are hole and shaft designations in ABC system. Limits for sizes above 19.69 inches are not covered by ABC agreements but are given in the ANSI Standard.

*Pairs of values shown represent minimum and maximum amounts of interference resulting from application of standard tolerance limits.

Source: Courtesy of The American Society of Mechanical Engineers.

APPENDIX A   ANSI PREFERRED ENGLISH LIMITS AND FITS

ANSI TRANSITION LOCATION FITS (LT)–ENGLISH UNITS

ANSI Standard Transition Location Fits (ANSI B4.1–1967, R1979)

Values shown below are in thousandths of an inch.

Nominal Size Range, Inches Over To	Class LT1 Fit*	Class LT1 Hole H7	Class LT1 Shaft js6	Class LT2 Fit*	Class LT2 Hole H8	Class LT2 Shaft js7	Class LT3 Fit*	Class LT3 Hole H7	Class LT3 Shaft k6	Class LT4 Fit*	Class LT4 Hole H7	Class LT4 Shaft n6	Class LT5 Fit*	Class LT5 Hole H7	Class LT5 Shaft n6	Class LT6 Fit*	Class LT6 Hole H7	Class LT6 Shaft n7
0–0.12	−0.12 / +0.52	+0.4 / 0	+0.12 / −0.12	−0.2 / +0.8	+0.6 / 0	+0.2 / −0.2							−0.5 / +0.15	+0.4 / 0	+0.5 / +0.25	−0.65 / +0.15	+0.4 / 0	+0.65 / +0.25
0.12–0.24	−0.15 / +0.65	+0.5 / 0	+0.15 / −0.15	−0.25 / +0.95	+0.7 / 0	+0.25 / −0.25							−0.6 / +0.2	+0.5 / 0	+0.6 / +0.3	−0.8 / +0.2	+0.5 / 0	+0.8 / +0.3
0.24–0.40	−0.2 / +0.8	+0.6 / 0	+0.2 / −0.2	−0.3 / +1.2	+0.9 / 0	+0.3 / −0.3	−0.5 / +0.5	+0.6 / 0	+0.5 / +0.1	−0.7 / +0.8	+0.9 / 0	+0.7 / +0.1	−0.8 / +0.2	+0.6 / 0	+0.8 / +0.4	−1.0 / +0.2	+0.6 / 0	+1.0 / +0.4
0.40–0.71	−0.2 / +0.9	+0.7 / 0	+0.2 / −0.2	−0.35 / +1.35	+1.0 / 0	+0.35 / −0.35	−0.5 / +0.6	+0.7 / 0	+0.5 / +0.1	−0.8 / +0.9	+1.0 / 0	+0.8 / +0.1	−0.9 / +0.2	+0.7 / 0	+0.9 / +0.5	−1.2 / +0.2	+0.7 / 0	+1.2 / +0.5
0.71–1.19	−0.25 / +1.05	+0.8 / 0	+0.25 / −0.25	−0.4 / +1.6	+1.2 / 0	+0.4 / −0.4	−0.6 / +0.7	+0.8 / 0	+0.6 / +0.1	−0.9 / +1.1	+1.2 / 0	+0.9 / +0.1	−1.1 / +0.2	+0.8 / 0	+1.1 / +0.6	−1.4 / +0.2	+0.8 / 0	+1.4 / +0.6
1.19–1.97	−0.3 / +1.3	+1.0 / 0	+0.3 / −0.3	−0.5 / +2.1	+1.6 / 0	+0.5 / −0.5	−0.7 / +0.9	+1.0 / 0	+0.7 / +0.1	−1.1 / +1.5	+1.6 / 0	+1.1 / +0.1	−1.3 / +0.3	+1.0 / 0	+1.3 / +0.7	−1.7 / +0.3	+1.0 / 0	+1.7 / +0.7
1.97–3.15	−0.3 / +1.5	+1.2 / 0	+0.3 / −0.3	−0.6 / +2.4	+1.8 / 0	+0.6 / −0.6	−0.8 / +1.1	+1.2 / 0	+0.8 / +0.1	−1.3 / +1.7	+1.8 / 0	+1.3 / +0.1	−1.5 / +0.4	+1.2 / 0	+1.5 / +0.8	−2.0 / +0.4	+1.2 / 0	+2.0 / +0.8
3.15–4.73	−0.4 / +1.8	+1.4 / 0	+0.4 / −0.4	−0.7 / +2.9	+2.2 / 0	+0.7 / −0.7	−1.0 / +1.3	+1.4 / 0	+1.0 / +0.1	−1.5 / +2.1	+2.2 / 0	+1.5 / +0.1	−1.9 / +0.4	+1.4 / 0	+1.9 / +1.0	−2.4 / +0.4	+1.4 / 0	+2.4 / +1.0
4.73–7.09	−0.5 / +2.1	+1.6 / 0	+0.5 / −0.5	−0.8 / +3.3	+2.5 / 0	+0.8 / −0.8	−1.1 / +1.5	+1.6 / 0	+1.1 / +0.1	−1.7 / +2.4	+2.5 / 0	+1.7 / +0.1	−2.2 / +0.4	+1.6 / 0	+2.2 / +1.2	−2.8 / +0.4	+1.6 / 0	+2.8 / +1.2
7.09–9.85	−0.6 / +2.4	+1.8 / 0	+0.6 / −0.6	−0.9 / +3.7	+2.8 / 0	+0.9 / −0.9	−1.4 / +1.6	+1.8 / 0	+1.4 / +0.2	−2.0 / +2.6	+2.8 / 0	+2.0 / +0.2	−2.6 / +0.4	+1.8 / 0	+2.6 / +1.4	−3.2 / +0.4	+1.8 / 0	+3.2 / +1.4
9.85–12.41	−0.6 / +2.6	+2.0 / 0	+0.6 / −0.6	−1.0 / +4.0	+3.0 / 0	+1.0 / −1.0	−1.4 / +1.8	+2.0 / 0	+1.4 / +0.2	−2.2 / +2.8	+3.0 / 0	+2.2 / +0.2	−2.6 / +0.6	+2.0 / 0	+2.6 / +1.4	−3.4 / +0.6	+2.0 / 0	+3.4 / +1.4
12.41–15.75	−0.7 / +2.9	+2.2 / 0	+0.7 / −0.7	−1.0 / +4.5	+3.5 / 0	+1.0 / −1.0	−1.6 / +2.0	+2.2 / 0	+1.6 / +0.2	−2.4 / +3.3	+3.5 / 0	+2.4 / +0.2	−3.0 / +0.6	+2.2 / 0	+3.0 / +1.6	−3.8 / +0.6	+2.2 / 0	+3.8 / +1.6
15.75–19.69	−0.8 / +3.3	+2.5 / 0	+0.8 / −0.8	−1.2 / +5.2	+4.0 / 0	+1.2 / −1.2	−1.8 / +2.3	+2.5 / 0	+1.8 / +0.2	−2.7 / +3.8	+4.0 / 0	+2.7 / +0.2	−3.4 / +0.7	+2.5 / 0	+3.4 / +1.8	−4.3 / +0.7	+2.5 / 0	+4.3 / +1.8

Data in boldface type are in accordance with American–British–Canadian (ABC) Agreements. Symbols H7, js6, etc. are hole and shaft designations in ABC system.

*Pairs of values shown represent maximum amount of interference (−) and maximum amount of clearance (+) resulting from application of standard tolerance limits.

Source: Courtesy of The American Society of Mechanical Engineers.

ANSI INTERFERENCE LOCATIONAL FITS (LT)—ENGLISH UNITS

Nominal Size Range, Inches Over To	Class LN1 Standard Limits			Class LN2 Standard Limits			Class LN3 Standard Limits		
	Limits of Interference*	Hole H 6	Shaft n 5	Limits of Interference*	Hole H 7	Shaft p 6	Limits of Interference*	Hole H 7	Shaft r 6
	Values shown below are in thousandths of an inch.								
0–0.12	0 0.45	+0.25 0	+0.45 +0.25	0 0.65	+0.4 0	+0.65 +0.4	0.1 0.75	+0.4 0	+0.75 +0.5
0.12–0.24	0 0.5	+0.3 0	+0.5 +0.3	0 0.8	+0.5 0	+0.8 +0.5	0.1 0.9	+0.5 0	+0.9 +0.6
0.24–0.40	0 0.65	+0.4 0	+0.65 +0.4	0 1.0	+0.6 0	+1.0 +0.6	0.2 1.2	+0.6 0	+1.2 +0.8
0.40–0.71	0 0.8	+0.4 0	+0.8 +0.4	0 1.1	+0.7 0	+1.1 +0.7	0.3 1.4	+0.7 0	+1.4 +1.0
0.71–1.19	0 1.0	+0.5 0	+1.0 +0.5	0 1.3	+0.8 0	+1.3 +0.8	0.4 1.7	+0.8 0	+1.7 +1.2
1.19–1.97	0 1.1	+0.6 0	+1.1 +0.6	0 1.6	+1.0 0	+1.6 +1.0	0.4 2.0	+1.0 0	+2.0 +1.4
1.97–3.15	0.1 1.3	+0.7 0	+1.3 +0.8	0.2 2.1	+1.2 0	+2.1 +1.4	0.4 2.3	+1.2 0	+2.3 +1.6
3.15–4.73	0.1 1.6	+0.9 0	+1.6 +1.0	0.2 2.5	+1.4 0	+2.5 +1.6	0.6 2.9	+1.4 0	+2.9 +2.0
4.73–7.09	0.2 1.9	+1.0 0	+1.9 +1.2	0.2 2.8	+1.6 0	+2.8 +1.8	0.9 3.5	+1.6 0	+3.5 +2.5
7.09–9.85	0.1 2.2	+1.2 0	+2.2 +1.4	0.2 3.2	+1.8 0	+3.2 +2.0	1.2 4.2	+1.8 0	+4.2 +3.0
9.85–12.41	0.2 2.3	+1.2 0	+2.3 +1.4	0.2 3.4	+2.0 0	+3.4 +2.2	1.5 4.7	+2.0 0	+4.7 +3.5
12.41–15.75	0.2 2.6	+1.4 0	+2.6 +1.6	0.3 3.9	+2.2 0	+3.9 +2.5	2.3 5.9	+2.2 0	+5.9 +4.5
15.75–19.69	0.2 1.8	+1.6 0	+2.8 +1.8	0.3 4.4	+2.5 0	+4.4 +2.8	2.5 6.6	+2.5 0	+6.6 +5.0

All data in this table are in accordance with American–British–Canadian (ABC) agreements.

Limits for sizes above 19.69 inches are not covered by ABC agreements but are given in the ANSI Standard. Symbols H7, p 6, etc. are hole and shaft designations in ABC system.

*Pairs of values shown represent minimum and maximum amounts of interference resulting from application of standard tolerance limits.

*Source: Courtesy of The American Society of Mechanical Engineers.

APPENDIX A ANSI PREFERRED ENGLISH LIMITS AND FITS

ANSI FORCE AND SHRINK FITS (FN)–ENGLISH UNITS

ANSI Standard Force and Shrink Fits (ANSI B4.1–1967, R1979)

Values shown below are in thousandths of an inch.

Nominal Size Range, Inches Over To	Class FN1 Interference*	Class FN1 Hole H6	Class FN1 Shaft	Class FN2 Interference*	Class FN2 Hole H7	Class FN2 Shaft s6	Class FN3 Interference*	Class FN3 Hole H7	Class FN3 Shaft t6	Class FN4 Interference*	Class FN4 Hole H7	Class FN4 Shaft u6	Class FN5 Interference*	Class FN5 Hole H8	Class FN5 Shaft x7
0–0.12	0.05 0.5	+0.25 0	+0.5 +0.3	0.2 0.85	+0.4 0	+0.85 +0.6				0.3 0.95	+0.4 0	+0.95 +0.7	0.3 1.3	+0.6 0	+1.3 +0.9
0.12–0.24	0.1 0.6	+0.3 0	+0.6 +0.4	0.2 1.0	+0.5 0	+1.0 +0.7				0.4 1.2	+0.5 0	+1.2 +0.9	0.5 1.7	+0.7 0	+1.7 +1.2
0.24–0.40	0.1 0.75	+0.4 0	+0.75 +0.5	0.4 1.4	+0.6 0	+1.4 +1.0				0.6 1.6	+0.6 0	+1.6 +1.2	0.5 2.0	+0.9 0	+2.0 +1.4
0.40–0.56	0.1 0.8	+0.4 0	+0.8 +0.5	0.5 1.6	+0.7 0	+1.6 +1.2				0.7 1.8	+0.7 0	+1.8 +1.4	0.6 2.3	+1.0 0	+2.3 +1.6
0.56–0.71	0.2 0.9	+0.4 0	+0.9 +0.6	0.5 1.6	+0.7 0	+1.6 +1.2				0.7 1.8	+0.7 0	+1.8 +1.4	0.8 2.5	+1.0 0	+2.5 +1.8
0.71–0.95	0.2 1.1	+0.5 0	+1.1 +0.7	0.6 1.9	+0.8 0	+1.9 +1.4				0.8 2.1	+0.8 0	+2.1 +1.6	1.0 3.0	+1.2 0	+3.0 +2.2
0.95–1.19	0.3 1.2	+0.5 0	+1.2 +0.8	0.6 1.9	+0.8 0	+1.9 +1.4	0.8 2.1	+0.8 0	+2.1 +1.6	1.0 2.3	+0.8 0	+2.3 +1.8	1.3 3.3	+1.2 0	+3.3 +2.5
1.19–1.58	0.3 1.3	+0.6 0	+1.3 +0.9	0.8 2.4	+1.0 0	+2.4 +1.8	1.0 2.6	+1.0 0	+2.6 +2.0	1.5 3.1	+1.0 0	+3.1 +2.5	1.4 4.0	+1.6 0	+4.0 +3.0
1.58–1.97	0.4 1.4	+0.6 0	+1.4 +1.0	0.8 2.4	+1.0 0	+2.4 +1.8	1.2 2.8	+1.0 0	+2.8 +2.2	1.8 3.4	+1.0 0	+3.4 +2.8	2.4 5.0	+1.6 0	+5.0 +4.0
1.97–2.56	0.6 1.8	+0.7 0	+1.8 +1.3	0.8 2.7	+1.2 0	+2.7 +2.0	1.3 3.2	+1.2 0	+3.2 +2.5	2.3 4.2	+1.2 0	+4.2 +3.5	3.2 6.2	+1.8 0	+6.2 +5.0
2.56–3.15	0.7 1.9	+0.7 0	+1.9 +1.4	1.0 2.9	+1.2 0	+2.9 +2.2	1.8 3.7	+1.2 0	+3.7 +3.0	2.8 4.7	+1.2 0	+4.7 +4.0	4.2 7.2	+1.8 0	+7.2 +6.0
3.15–3.94	0.9 2.4	+0.9 0	+2.4 +1.8	1.4 3.7	+1.4 0	+3.7 +2.8	2.1 4.4	+1.4 0	+4.4 +3.5	3.6 5.9	+1.4 0	+5.9 +5.0	4.8 8.4	+2.2 0	+8.4 +7.0
3.94–4.73	1.1 2.6	+0.9 0	+2.6 +2.0	1.6 3.9	+1.4 0	+3.9 +3.0	2.6 4.9	+1.4 0	+4.9 +4.0	4.6 6.9	+1.4 0	+6.9 +6.0	5.8 9.4	+2.2 0	+9.4 +8.0

Values shown below are in thousandths of an inch.

Nominal Size Range, Inches Over To	Class FN (FN1) Interference*	Hole H6	Shaft	Class FN (FN2) Interference*	Hole H7	Shaft s6	Class FN (FN3) Interference*	Hole H7	Shaft t6	Class FN (FN4) Interference*	Hole H7	Shaft u6	Class FN (FN5) Interference*	Hole H8	Shaft x7
4.73–5.52	1.2 / 2.9	+1.0 / 0	+2.9 / +2.2	1.9 / 4.5	+1.6 / 0	+4.5 / +3.5	3.4 / 6.0	+1.6 / 0	+6.0 / +5.0	5.4 / 8.0	+1.6 / 0	+8.0 / +7.0	7.5 / 11.6	+2.5 / 0	+11.6 / +10.0
5.52–6.30	1.5 / 3.2	+1.0 / 0	+3.2 / +2.5	2.4 / 5.0	+1.6 / 0	+5.0 / +4.0	3.4 / 6.0	+1.6 / 0	+6.0 / +5.0	5.4 / 8.0	+1.6 / 0	+8.0 / +7.0	9.5 / 13.6	+2.5 / 0	+13.6 / +12.0
6.30–7.09	1.8 / 3.5	+1.0 / 0	+3.5 / +2.8	2.9 / 5.5	+1.6 / 0	+5.5 / +4.5	4.4 / 7.0	+1.6 / 0	+7.0 / +6.0	6.4 / 9.0	+1.6 / 0	+9.0 / +8.0	9.5 / 13.6	+2.5 / 0	+13.6 / +12.0
7.09–7.88	1.8 / 3.8	+1.2 / 0	+3.8 / +3.0	3.2 / 6.2	+1.8 / 0	+6.2 / +5.0	5.2 / 8.2	+1.8 / 0	+8.2 / +7.0	7.2 / 10.2	+1.8 / 0	+10.2 / +9.0	11.2 / 15.8	+2.8 / 0	+15.8 / +14.0
7.88–8.86	2.3 / 4.3	+1.2 / 0	+4.3 / +3.5	3.2 / 6.2	+1.8 / 0	+6.2 / +5.0	5.2 / 8.2	+1.8 / 0	+8.2 / +7.0	8.2 / 11.2	+1.8 / 0	+11.2 / +10.0	13.2 / 17.8	+2.8 / 0	+17.8 / +16.0
8.86–9.85	2.3 / 4.3	+1.2 / 0	+4.3 / +3.5	4.2 / 7.2	+1.8 / 0	+7.2 / +6.0	6.2 / 9.2	+1.8 / 0	+9.2 / +8.0	10.2 / 13.2	+1.8 / 0	+13.2 / +12.0	13.2 / 17.8	+2.8 / 0	+17.8 / +16.0
9.85–11.03	2.8 / 4.9	+1.2 / 0	+4.9 / +4.0	4.0 / 7.2	+2.0 / 0	+7.2 / +6.0	7.0 / 10.2	+2.0 / 0	+10.2 / +9.0	10.0 / 13.2	+2.0 / 0	+13.2 / +12.0	15.0 / 20.0	+3.0 / 0	+20.0 / +18.0
11.03–12.41	2.8 / 4.9	+1.2 / 0	+4.9 / +4.0	5.0 / 8.2	+2.0 / 0	+8.2 / +7.2	7.0 / 10.2	+2.0 / 0	+10.2 / +9.0	12.0 / 15.2	+2.0 / 0	+15.2 / +14.0	17.0 / 22.0	+3.0 / 0	+22.0 / +20.0
12.41–13.98	3.1 / 5.5	+1.4 / 0	+5.5 / +4.5	5.8 / 9.4	+2.2 / 0	+9.4 / +8.0	7.8 / 11.4	+2.2 / 0	+11.4 / +10.0	13.8 / 17.4	+2.2 / 0	+17.4 / +16.0	18.5 / 24.2	+3.5 / 0	+24.2 / +22.0
13.98–15.75	3.6 / 6.1	+1.4 / 0	+6.1 / +5.0	5.8 / 9.4	+2.2 / 0	+9.4 / +8.0	9.8 / 13.4	+2.2 / 0	+13.4 / +12.0	15.8 / 19.4	+2.2 / 0	+19.4 / +18.0	21.5 / 27.2	+3.5 / 0	+27.2 / +25.0
15.75–17.72	4.4 / 7.0	+1.6 / 0	+7.0 / +6.0	6.5 / 10.6	+2.5 / 0	+10.6 / +9.0	9.5 / 13.6	+2.5 / 0	+13.6 / +12.0	17.5 / 21.6	+2.5 / 0	+21.6 / +20.0	24.0 / 30.5	+4.0 / 0	+30.5 / +28.0
17.72–19.69	4.4 / 7.0	+1.6 / 0	+7.0 / +6.0	7.5 / 11.6	+2.5 / 0	+11.6 / +10.0	11.5 / 15.6	+2.5 / 0	+15.6 / +14.0	19.5 / 23.6	+2.5 / 0	+23.6 / +22.0	26.0 / 32.5	+4.0 / 0	+32.5 / +30.0

Data in boldface type are in accordance with American–British–Canadian (ABC) agreements. Symbols H6, H7, s6, etc. are hole and shaft designations in ABC system. Limits for sizes above 19–69 inches are not covered by ABC agreements but are given in the ANSI standard.

*Pairs of values shown represent minimum and maximum amounts of interference resulting from application of standard tolerance limits.

Source: Courtesy of The American Society of Mechanical Engineers.

ANSI FORCE AND SHRINK FITS (FN)–ENGLISH UNITS

B ANSI PREFERRED METRIC LIMITS AND FITS

■ ANSI PREFERRED HOLE BASIS METRIC CLEARANCE FITS—METRIC UNITS

American National Standard Preferred Hole Basis Metric Clearance Fits (ANSI B4.2–1978, R1984)

Basic Size		Loose-Running			Free-Running			Close-Running			Sliding			Locational Clearance		
		Hole H11	Shaft c11	Fit†	Hole H9	Shaft d9	Fit†	Hole H8	Shaft f7	Fit†	Hole H7	Shaft g6	Fit†	Hole H7	Shaft h6	Fit†
1	Max	1.060	0.940	0.180	1.025	0.980	0.070	1.014	0.994	0.030	1.010	0.998	0.018	1.010	1.000	0.016
	Min	1.000	0.880	0.060	1.000	0.955	0.020	1.000	0.984	0.006	1.000	0.992	0.002	1.000	0.994	0.000
1.2	Max	1.260	1.140	0.180	1.225	1.180	0.070	1.214	1.194	0.030	1.210	1.198	0.018	1.210	1.200	0.016
	Min	1.200	1.080	0.060	1.200	1.155	0.020	1.200	1.184	0.006	1.200	1.192	0.002	1.200	1.194	0.000
1.6	Max	1.660	1.540	0.180	1.625	1.580	0.070	1.614	1.594	0.030	1.610	1.598	0.018	1.610	1.600	0.016
	Min	1.600	1.480	0.060	1.600	1.555	0.020	1.600	1.584	0.006	1.600	1.592	0.002	1.600	1.594	0.000
2	Max	2.060	1.940	0.180	2.025	1.980	0.070	2.014	1.994	0.030	2.010	1.998	0.018	2.010	2.000	0.016
	Min	2.000	1.880	0.060	2.000	1.955	0.020	2.000	1.984	0.006	2.000	1.992	0.002	2.000	1.994	0.000
2.5	Max	2.560	2.440	0.180	2.525	2.480	0.070	2.514	2.494	0.030	2.510	2.498	0.018	2.510	2.500	0.016
	Min	2.500	2.380	0.060	2.500	2.455	0.020	2.500	2.484	0.006	2.500	2.492	0.002	2.500	2.494	0.000
3	Max	3.060	2.940	0.180	3.025	2.980	0.070	3.014	2.994	0.030	3.010	2.998	0.018	3.010	3.000	0.016
	Min	3.000	2.880	0.060	3.000	2.955	0.020	3.000	2.984	0.006	3.000	2.992	0.002	3.000	2.994	0.000
4	Max	4.075	3.930	0.220	4.030	3.970	0.090	4.018	3.990	0.040	4.012	3.996	0.024	4.012	4.000	0.020
	Min	4.000	3.855	0.070	4.000	3.940	0.030	4.000	3.978	0.010	4.000	3.988	0.004	4.000	3.992	0.000
5	Max	5.075	4.930	0.220	5.030	4.970	0.090	5.018	4.990	0.040	5.012	4.996	0.024	5.012	5.000	0.020
	Min	5.000	4.855	0.070	5.000	4.940	0.030	5.000	4.978	0.010	5.000	4.988	0.004	5.000	4.992	0.000
6	Max	6.075	5.930	0.220	6.030	5.970	0.090	6.018	5.990	0.040	6.012	5.996	0.024	6.012	6.000	0.020
	Min	6.000	5.855	0.070	6.000	5.940	0.030	6.000	5.978	0.010	6.000	5.988	0.004	6.000	5.992	0.000
8	Max	8.090	7.920	0.260	8.036	7.960	0.112	8.022	7.987	0.050	8.015	7.995	0.029	8.015	8.000	0.024
	Min	8.000	7.830	0.080	8.000	7.924	0.040	8.000	7.972	0.013	8.000	7.986	0.005	8.000	7.991	0.000
10	Max	10.090	9.920	0.260	10.036	9.960	0.112	10.022	9.987	0.050	10.015	9.995	0.029	10.015	10.000	0.024
	Min	10.000	9.830	0.080	10.000	9.924	0.040	10.000	9.972	0.013	10.000	9.986	0.005	10.000	9.991	0.000
12	Max	12.110	11.905	0.315	12.043	11.956	0.136	12.027	11.984	0.061	12.018	11.994	0.035	12.018	12.000	0.029
	Min	12.000	11.795	0.095	12.000	11.907	0.050	12.000	11.966	0.016	12.000	11.983	0.006	12.000	11.989	0.000
16	Max	16.110	15.905	0.315	16.043	15.950	0.136	16.027	15.984	0.061	16.018	15.994	0.035	16.018	16.000	0.029
	Min	16.000	15.795	0.095	16.000	15.907	0.050	16.000	15.966	0.016	16.000	15.983	0.006	16.000	15.989	0.000
20	Max	20.130	19.890	0.370	20.052	19.935	0.169	20.033	19.980	0.074	20.021	19.993	0.042	20.021	20.000	0.034
	Min	20.000	19.760	0.110	20.000	19.883	0.065	20.000	19.959	0.020	20.000	19.980	0.007	20.000	19.987	0.000

Basic Size		Loose-Running			Free-Running			Close-Running			Sliding			Locational Clearance		
		Hole H11	Shaft c11	Fit[†]	Hole H9	Shaft d9	Fit[†]	Hole H8	Shaft f7	Fit[†]	Hole H7	Shaft g6	Fit[†]	Hole H7	Shaft h6	Fit[†]
25	Max	25.130	24.890	0.370	25.052	24.935	0.169	25.033	24.980	0.074	25.021	24.993	0.041	25.021	25.000	0.034
	Min	25.000	24.760	0.110	25.000	24.883	0.065	25.000	24.959	0.010	25.000	24.980	0.007	25.000	24.987	0.000
30	Max	30.130	29.890	0.370	30.052	29.935	0.169	30.033	29.980	0.074	30.021	29.993	0.041	30.021	30.000	0.034
	Min	30.000	29.760	0.110	30.000	19.883	0.065	30.000	29.959	0.020	30.000	29.980	0.007	30.000	29.987	0.000
40	Max	40.160	39.880	0.440	40.062	39.920	0.204	40.039	39.975	0.089	40.025	39.991	0.050	40.025	40.000	0.041
	Min	40.000	39.720	0.120	40.000	39.858	0.080	40.000	39.950	0.025	40.000	39.975	0.009	40.000	39.984	0.000
50	Max	50.160	49.870	0.450	50.062	49.920	0.204	50.039	49.975	0.089	50.025	49.991	0.050	50.025	50.000	0.041
	Min	50.000	49.710	0.130	50.000	49.858	0.080	50.000	49.950	0.025	50.000	49.975	0.009	50.000	49.984	0.000
60	Max	60.190	59.860	0.520	60.074	59.900	0.248	60.046	59.970	0.106	60.030	59.990	0.059	60.030	60.000	0.049
	Min	60.000	59.670	0.140	60.000	59.826	0.100	60.000	59.940	0.030	60.000	59.971	0.010	60.000	59.981	0.000
80	Max	80.190	79.850	0.530	80.074	79.900	0.248	80.046	79.970	0.106	80.030	79.990	0.059	80.030	80.000	0.049
	Min	80.000	79.660	0.150	80.000	79.826	0.100	80.000	79.940	0.030	80.000	79.971	0.010	80.000	79.981	0.000
100	Max	100.220	99.830	0.610	100.087	99.880	0.294	100.054	99.964	0.125	100.035	99.988	0.069	100.035	100.000	0.057
	Min	100.000	99.610	0.170	100.000	99.793	0.120	100.000	99.929	0.036	100.000	99.966	0.012	100.000	99.978	0.000
120	Max	120.220	119.820	0.620	120.087	119.880	0.294	120.054	119.964	0.125	120.035	119.988	0.069	120.035	120.000	0.057
	Min	110.000	119.600	0.180	120.000	119.793	0.120	120.000	119.929	0.036	120.000	119.966	0.012	120.000	119.978	0.000
160	Max	160.250	159.790	0.710	160.100	159.855	0.345	160.063	159.957	0.146	160.040	159.986	0.079	160.040	160.000	0.065
	Min	160.000	159.540	0.210	160.000	159.755	0.145	160.000	159.917	0.043	160.000	159.961	0.014	160.000	159.975	0.000
200	Max	200.290	199.760	0.820	200.115	199.830	0.400	200.072	199.950	0.168	200.046	199.985	0.090	200.046	200.000	0.071
	Min	200.000	199.470	0.240	200.000	199.715	0.170	200.000	199.904	0.050	200.000	199.956	0.015	200.000	199.971	0.000
250	Max	250.290	249.720	0.860	250.115	249.830	0.400	250.072	249.950	0.168	250.046	249.985	0.090	250.046	250.000	0.075
	Min	250.000	249.430	0.230	250.000	249.115	0.170	250.000	249.904	0.050	250.000	249.956	0.015	250.000	249.971	0.000
300	Max	300.320	299.670	0.970	300.130	299.810	0.450	300.081	299.944	0.189	300.052	299.983	0.101	300.052	300.000	0.084
	Min	300.000	299.350	0.330	300.000	299.680	0.190	300.000	299.892	0.056	300.000	299.951	0.017	300.000	299.968	0.000
400	Max	400.360	399.600	1.120	400.140	399.790	0.490	400.089	399.938	0.208	400.057	399.982	0.111	400.057	400.000	0.093
	Min	400.000	399.240	0.400	400.000	399.650	0.210	400.000	399.881	0.063	400.000	399.946	0.018	400.000	399.964	0.000
500	Max	500.400	499.520	1.280	500.155	499.770	0.540	500.097	499.932	0.228	500.063	499.980	0.123	500.063	500.000	0.103
	Min	500.000	499.120	0.480	500.000	499.615	0.230	500.000	499.869	0.068	500.000	499.940	0.020	500.000	499.960	0.000

All dimensions are in millimeters.

Preferred fits for other sizes can be calculated from data given in ANSI B4.2–1978 (R1984).

[†]All fits shown in this table have clearance.

Source: Courtesy of The American Society of Mechanical Engineers.

American National Standard Preferred Hole Basis Metric Transition and Interference Fits (ANSI B4.2–1978, R1984)

Basic Size		Locational Transition			Locational Transition			Locational Interference			Medium Drive			Force		
		Hole H7	Shaft k6	Fit†	Hole H7	Shaft n6	Fit†	Hole H7	Shaft p6	Fit†	Hole H7	Shaft s6	Fit†	Hole H7	Shaft u6	Fit†
1	Max	1.010	1.006	+0.010	1.010	1.010	+0.006	1.010	1.012	+0.004	1.010	1.020	−0.004	1.010	1.024	−0.008
	Min	1.000	1.000	−0.006	1.000	1.004	−0.010	1.000	1.006	−0.012	1.000	1.014	−0.020	1.000	1.018	−0.024
1.2	Max	1.210	1.206	+0.010	1.210	1.210	+0.006	1.210	1.212	+0.004	1.210	1.220	−0.004	1.210	1.224	−0.008
	Min	1.200	1.200	−0.006	1.200	1.204	−0.010	1.200	1.206	−0.012	1.200	1.214	−0.020	1.200	1.218	−0.024
1.6	Max	1.610	1.606	+0.010	1.610	1.610	+0.006	1.610	1.612	+0.004	1.610	1.620	−0.004	1.610	1.624	−0.008
	Min	1.600	1.600	−0.006	1.600	1.604	−0.010	1.600	1.606	−0.012	1.600	1.614	−0.020	1.600	1.618	−0.024
2	Max	2.010	2.006	+0.010	2.010	2.010	+0.006	2.010	2.012	+0.004	2.010	2.020	−0.004	2.010	2.024	−0.008
	Min	2.000	2.000	−0.006	2.000	2.004	−0.010	2.000	2.006	−0.012	2.000	2.014	−0.020	2.000	2.018	−0.024
2.5	Max	2.510	2.506	+0.010	2.510	2.510	+0.006	2.510	2.512	+0.004	2.510	2.520	−0.004	2.510	2.524	−0.008
	Min	2.500	2.500	−0.006	2.500	2.504	−0.010	2.500	2.506	−0.012	2.500	2.514	−0.020	2.500	2.518	−0.024
3	Max	3.010	3.006	+0.010	3.010	3.010	+0.006	3.010	3.012	+0.004	3.010	3.020	−0.004	3.010	3.024	−0.008
	Min	3.000	3.000	−0.006	3.000	3.004	−0.010	3.000	3.006	−0.012	3.000	3.014	−0.020	3.000	3.018	−0.024
4	Max	4.012	4.009	+0.011	4.012	4.016	+0.004	4.012	4.020	0.000	4.012	4.027	−0.007	4.012	4.031	−0.011
	Min	4.000	4.001	−0.009	4.000	4.008	−0.016	4.000	4.012	−0.020	4.000	4.019	−0.027	4.000	4.023	−0.031
5	Max	5.012	5.009	+0.011	5.012	5.016	+0.004	5.012	5.020	0.000	5.012	5.027	−0.007	5.012	5.031	−0.011
	Min	5.000	5.001	−0.009	5.000	5.008	−0.016	5.000	5.012	−0.020	5.000	5.019	−0.027	5.000	5.023	−0.031
6	Max	6.012	6.009	+0.011	6.012	6.016	+0.004	6.012	6.020	0.000	6.012	6.027	−0.007	6.012	6.031	−0.011
	Min	6.000	6.001	−0.009	6.000	6.008	−0.016	6.000	6.012	−0.020	6.000	6.019	−0.027	6.000	6.023	−0.031
8	Max	8.015	8.010	+0.014	8.015	8.019	+0.005	8.015	8.024	0.000	8.015	8.032	−0.008	8.015	8.037	−0.013
	Min	8.000	8.001	−0.010	8.000	8.010	−0.019	8.000	8.015	−0.024	8.000	8.023	−0.032	8.000	8.028	−0.037
10	Max	10.015	10.010	+0.014	10.015	10.019	+0.005	10.015	10.024	0.000	10.015	10.032	−0.008	10.015	10.037	−0.013
	Min	10.000	10.001	−0.010	10.000	10.010	−0.019	10.000	10.015	−0.024	10.000	10.023	−0.032	10.000	10.028	−0.037
12	Max	12.018	12.012	+0.017	12.018	12.023	+0.006	12.018	12.029	0.000	12.018	12.039	−0.010	12.018	12.044	−0.015
	Min	12.000	12.001	−0.012	12.000	12.012	−0.023	12.000	12.018	−0.029	12.000	12.028	−0.039	12.000	12.033	−0.044
16	Max	16.018	16.012	+0.017	16.018	16.023	+0.006	16.018	16.029	0.000	16.018	16.039	−0.010	16.018	16.044	−0.015
	Min	16.000	16.001	−0.012	16.000	16.012	−0.023	16.000	16.018	−0.029	16.000	16.028	−0.039	16.000	16.033	−0.044
20	Max	20.021	20.015	+0.019	20.021	20.028	+0.005	20.021	20.035	−0.001	20.021	20.048	−0.014	20.021	20.054	−0.020
	Min	20.000	20.002	−0.015	20.000	20.015	−0.028	20.000	20.022	−0.035	20.000	20.035	−0.048	20.000	20.041	−0.054

Basic Size		Locational Transition			Locational Transition			Locational Interference			Medium Drive			Force		
		Hole H7	Shaft k6	Fit[†]	Hole H7	Shaft n6	Fit[†]	Hole H7	Shaft p6	Fit[†]	Hole H7	Shaft s6	Fit[†]	Hole H7	Shaft u6	Fit[†]
25	Max	25.021	25.015	+0.019	25.021	25.028	+0.006	25.021	25.035	-0.001	25.021	25.048	-0.014	25.021	25.061	-0.027
	Min	25.000	25.002	-0.015	25.000	25.015	-0.028	25.000	25.022	-0.035	25.000	25.035	-0.048	25.000	25.048	-0.061
30	Max	30.021	30.015	+0.019	30.021	30.028	+0.006	30.021	30.035	-0.001	30.021	30.048	-0.014	30.021	30.061	-0.027
	Min	30.000	30.002	-0.015	30.000	30.015	-0.028	30.000	30.022	-0.035	30.000	30.035	-0.048	30.000	30.048	-0.061
40	Max	40.025	40.018	+0.023	40.025	40.033	+0.008	40.025	40.042	-0.001	40.025	40.059	-0.018	40.025	40.076	-0.035
	Min	40.000	40.002	-0.018	40.000	40.017	-0.033	40.000	40.026	-0.042	40.000	40.043	-0.059	40.000	40.060	-0.076
50	Max	50.025	50.018	+0.023	50.025	50.033	+0.008	50.025	50.042	-0.002	50.025	50.059	-0.018	50.025	50.086	-0.045
	Min	50.000	50.002	-0.018	50.000	50.017	-0.033	50.000	50.026	-0.042	50.000	50.043	-0.059	50.000	50.070	-0.086
60	Max	60.030	60.021	+0.028	60.030	60.039	+0.010	60.030	60.051	-0.002	60.030	60.072	-0.023	60.030	60.106	-0.057
	Min	60.000	60.002	-0.021	60.000	60.020	-0.039	60.000	60.032	-0.052	60.000	60.053	-0.072	60.000	60.087	-0.106
80	Max	80.030	80.021	+0.028	80.030	80.039	+0.010	80.030	80.051	-0.002	80.030	80.078	-0.029	80.030	80.121	-0.072
	Min	80.000	80.002	-0.021	80.000	80.020	-0.039	80.000	80.032	-0.051	80.000	80.059	-0.078	80.000	80.102	-0.121
100	Max	100.035	100.025	+0.032	100.035	100.045	+0.012	100.035	100.059	-0.002	100.035	100.093	-0.036	100.035	100.146	-0.089
	Min	100.000	100.003	-0.025	100.000	100.023	-0.045	100.000	100.037	-0.059	100.000	100.071	-0.093	100.000	100.124	-0.146
120	Max	120.035	120.025	+0.032	120.035	120.045	+0.012	120.035	120.059	-0.002	120.035	120.101	-0.044	120.035	120.166	-0.109
	Min	120.000	120.003	-0.025	120.000	120.023	-0.045	120.000	120.037	-0.059	120.000	120.079	-0.101	120.000	120.144	-0.166
160	Max	160.040	160.028	+0.037	160.040	160.052	+0.013	160.040	160.068	-0.003	160.040	160.125	-0.060	160.040	160.215	-0.150
	Min	160.000	160.003	-0.028	160.000	160.027	-0.052	160.000	160.043	-0.068	160.000	160.100	-0.125	160.000	160.190	-0.215
200	Max	200.046	200.033	+0.042	200.046	200.060	+0.015	200.046	200.079	-0.004	200.046	200.151	-0.076	200.046	200.265	-0.190
	Min	200.000	200.004	-0.033	200.000	200.031	-0.060	200.000	200.050	-0.079	200.000	200.122	-0.151	200.000	200.236	-0.265
250	Max	250.046	250.033	+0.042	250.046	250.060	+0.015	250.046	250.079	-0.004	250.046	250.169	-0.094	250.046	250.313	-0.238
	Min	250.000	250.004	-0.033	250.000	250.031	-0.060	250.000	250.050	-0.079	250.000	250.140	-0.169	250.000	250.284	-0.313
300	Max	300.052	300.036	+0.048	300.052	300.066	+0.018	300.052	300.088	-0.004	300.052	300.202	-0.118	300.052	300.382	-0.298
	Min	300.000	300.004	-0.036	300.000	300.034	-0.066	300.000	300.056	-0.088	300.000	300.170	-0.102	300.000	300.350	-0.382
400	Max	400.057	400.040	0.053	400.057	400.073	+0.020	400.057	400.098	-0.005	400.057	400.244	-0.151	400.057	400.471	-0.378
	Min	400.000	400.004	-0.040	400.000	400.037	-0.073	400.000	400.062	-0.098	400.000	400.208	-0.244	400.000	400.435	-0.471
500	Max	500.063	500.045	+0.058	500.063	500.080	+0.023	500.063	500.108	-0.005	500.063	500.292	-0.189	500.063	500.580	-0.477
	Min	500.000	500.005	-0.045	500.000	500.040	-0.080	500.000	500.068	-0.108	500.000	500.252	-0.292	500.000	500.540	-0.580

All dimensions are in millimeters.

Preferred fits for other sizes can be calculated from data given in ANSI B4.2–1978 (R1984).

[†]A plus sign indicates clearance; a minus sign indicates interference.

Source: Courtesy of The American Society of Mechanical Engineers.

ANSI PREFERRED SHAFT BASIS METRIC CLEARANCE FITS–METRIC UNITS

American National Standard Preferred Shaft Basis Metric Clearance Fits (ANSI B4.2–1978, R1984)

Basic Size		Loose-Running Hole C11	Loose-Running Shaft h11	Fit†	Free-Running Hole D9	Free-Running Shaft h9	Fit†	Close-Running Hole F5	Close-Running Shaft h7	Fit†	Sliding Hole G7	Sliding Shaft h6	Fit†	Locational Clearance Hole H7	Locational Clearance Shaft h6	Fit†
1	Max	1.120	1.000	0.180	1.045	1.000	0.070	1.020	1.000	0.030	1.012	1.000	0.018	1.010	1.000	0.016
	Min	1.060	0.940	0.060	1.020	0.975	0.020	1.006	0.990	0.006	1.002	0.994	0.002	1.000	0.994	0.000
1.2	Max	1.320	1.200	0.180	1.245	1.200	0.070	1.220	1.200	0.030	1.212	1.200	0.018	1.210	1.200	0.016
	Min	1.260	1.140	0.060	1.220	1.175	0.020	1.206	1.190	0.006	1.202	1.194	0.002	1.200	1.194	0.000
1.6	Max	1.720	1.600	0.180	1.645	1.600	0.070	1.620	1.600	0.030	1.612	1.600	0.018	1.610	1.600	0.016
	Min	1.660	1.540	0.060	1.620	1.575	0.020	1.606	1.590	0.006	1.602	1.594	0.002	1.600	1.594	0.000
2	Max	2.120	2.000	0.180	2.045	2.000	0.070	2.020	2.000	0.030	2.012	2.000	0.018	2.010	2.000	0.016
	Min	2.060	1.940	0.060	2.020	1.975	0.020	2.006	1.990	0.006	2.007	1.994	0.002	2.000	1.994	0.000
2.5	Max	2.620	2.500	0.180	2.545	2.500	0.070	2.520	2.500	0.030	2.512	2.500	0.018	2.510	2.500	0.016
	Min	2.560	2.440	0.060	2.520	2.475	0.020	2.506	2.490	0.006	2.502	2.494	0.002	2.500	2.494	0.000
3	Max	3.120	3.000	0.180	3.045	3.000	0.070	3.020	3.000	0.030	3.012	3.000	0.018	3.010	3.000	0.016
	Min	3.060	2.940	0.060	3.020	2.975	0.020	3.006	2.990	0.006	3.002	2.994	0.002	3.000	2.994	0.000
4	Max	4.145	4.000	0.220	4.060	4.000	0.090	4.028	4.000	0.040	4.016	4.000	0.024	4.012	4.000	0.020
	Min	4.070	3.925	0.070	4.030	3.970	0.030	4.010	3.988	0.010	4.004	3.992	0.004	4.000	3.992	0.000
5	Max	5.145	5.000	0.220	5.060	5.000	0.090	5.028	5.000	0.040	5.016	5.000	0.024	5.012	5.000	0.020
	Min	5.070	4.925	0.070	5.030	4.970	0.030	5.010	4.988	0.010	5.004	4.992	0.004	5.000	4.992	0.000
6	Max	6.145	6.000	0.220	6.060	6.000	0.090	6.028	6.000	0.040	6.016	6.000	0.024	6.012	6.000	0.020
	Min	6.070	5.925	0.070	6.030	5.970	0.030	6.010	5.988	0.010	6.004	5.992	0.004	6.000	5.992	0.000
8	Max	8.170	8.000	0.260	8.076	8.000	0.112	8.035	8.000	0.050	8.020	8.000	0.029	8.015	8.000	0.024
	Min	8.080	7.910	0.080	8.040	7.964	0.040	8.013	7.985	0.013	8.005	7.991	0.005	8.000	7.991	0.000
10	Max	10.170	10.000	0.260	10.076	10.000	0.112	10.035	10.000	0.050	10.020	10.000	0.029	10.015	10.000	0.024
	Min	10.080	9.910	0.080	10.040	9.964	0.040	10.013	9.985	0.013	10.005	9.991	0.005	10.000	9.991	0.000
12	Max	12.205	12.000	0.315	12.093	12.000	0.136	12.043	12.000	0.061	12.024	12.000	0.035	12.018	12.000	0.029
	Min	12.095	11.890	0.095	12.050	11.957	0.050	12.016	11.982	0.026	12.006	11.989	0.006	12.000	11.989	0.000
16	Max	16.205	16.000	0.315	16.093	16.000	0.136	16.043	16.000	0.061	16.024	16.000	0.035	16.018	16.000	0.029
	Min	16.095	15.890	0.095	16.050	15.957	0.050	16.016	15.982	0.016	16.006	15.989	0.006	16.000	15.989	0.000
20	Max	20.240	20.000	0.370	20.117	20.000	0.169	20.053	20.000	0.074	20.028	20.000	0.041	20.021	20.000	0.034
	Min	20.110	19.870	0.110	20.065	19.948	0.065	20.020	19.979	0.020	20.007	19.987	0.007	20.000	19.987	0.000

Basic Size		Loose-Running			Free-Running			Close-Running			Sliding			Locational Clearance		
		Hole C11	Shaft h11	Fit[†]	Hole D9	Shaft h9	Fit[†]	Hole F5	Shaft h7	Fit[†]	Hole G7	Shaft h6	Fit[†]	Hole H7	Shaft h6	Fit[†]
25	Max	25.240	25.000	0.370	25.117	25.000	0.169	25.053	25.000	0.074	25.028	25.000	0.041	25.021	25.000	0.034
	Min	25.110	24.870	0.110	25.065	24.948	0.065	25.020	24.979	0.020	25.007	24.987	0.007	25.000	24.987	0.000
30	Max	30.240	30.000	0.370	30.117	30.000	0.169	30.053	30.000	0.074	30.028	30.000	0.041	30.021	30.000	0.034
	Min	30.110	29.870	0.110	30.065	29.948	0.065	30.020	29.979	0.020	30.007	29.987	0.007	30.000	29.987	0.000
40	Max	40.280	40.000	0.440	40.142	40.000	0.204	40.064	40.000	0.089	40.034	40.000	0.050	40.025	40.000	0.041
	Min	40.120	39.840	0.120	40.080	39.938	0.080	40.025	39.975	0.025	40.009	39.984	0.009	40.000	39.984	0.000
50	Max	50.290	50.000	0.450	50.142	50.000	0.204	50.064	50.000	0.089	50.034	50.000	0.050	50.025	50.000	0.041
	Min	50.130	49.840	0.130	50.080	49.938	0.080	50.025	49.975	0.025	50.009	49.984	0.009	50.000	49.984	0.000
60	Max	60.330	60.000	0.520	60.174	60.000	0.248	60.076	60.000	0.106	60.040	60.000	0.059	60.030	60.000	0.049
	Min	60.140	59.810	0.140	60.100	59.926	0.100	60.030	59.970	0.030	60.010	59.981	0.010	60.000	59.981	0.000
80	Max	80.340	80.000	0.530	80.174	80.000	0.248	80.076	80.000	0.106	80.040	80.000	0.059	80.030	80.000	0.049
	Min	80.150	79.810	0.150	80.100	79.926	0.100	80.030	79.970	0.030	80.010	79.981	0.010	80.000	79.981	0.000
100	Max	100.390	100.000	0.610	100.207	100.000	0.294	100.090	100.000	0.125	100.047	100.000	0.069	100.035	100.000	0.057
	Min	100.270	99.780	0.170	100.120	99.913	0.120	100.036	99.965	0.036	100.012	99.978	0.012	100.000	99.978	0.000
120	Max	120.400	120.000	0.620	120.207	120.000	0.294	120.090	120.000	0.125	120.047	120.000	0.069	120.035	120.000	0.057
	Min	120.180	119.780	0.180	120.120	119.913	0.120	120.036	119.965	0.036	120.012	119.978	0.012	120.000	119.978	0.000
160	Max	160.460	160.000	0.710	160.245	160.000	0.345	160.106	160.000	0.146	160.054	160.000	0.079	160.040	160.000	0.063
	Min	160.210	159.750	0.210	160.145	159.900	0.145	160.043	159.960	0.043	160.014	159.975	0.014	160.000	159.975	0.000
200	Max	200.530	200.000	0.820	200.285	200.000	0.400	200.122	200.000	0.168	200.061	200.000	0.090	200.046	200.000	0.075
	Min	200.240	199.710	0.240	200.170	199.885	0.170	200.050	199.954	0.050	200.015	199.971	0.015	200.000	199.971	0.000
250	Max	250.570	250.000	0.860	250.285	250.000	0.400	250.122	250.000	0.168	250.061	250.000	0.090	250.046	250.000	0.075
	Min	250.280	249.710	0.280	250.170	249.885	0.170	250.050	249.954	0.050	250.015	249.971	0.015	250.000	249.971	0.000
300	Max	300.650	300.000	0.970	300.320	300.000	0.450	300.137	300.000	0.189	300.069	300.000	0.101	300.052	300.000	0.084
	Min	300.330	299.680	0.330	300.190	299.870	0.190	300.056	299.948	0.056	300.017	299.968	0.017	300.000	299.968	0.000
400	Max	400.760	400.000	1.120	400.350	400.000	0.490	400.151	400.000	0.208	400.075	400.000	0.111	400.057	400.000	0.093
	Min	400.400	399.640	0.400	400.210	399.860	0.210	400.062	399.943	0.062	400.018	399.964	0.018	400.000	399.964	0.000
500	Max	500.880	500.000	1.280	500.385	500.000	0.540	500.165	500.000	0.228	500.083	500.000	0.123	500.063	500.000	0.103
	Min	500.480	499.600	0.480	500.230	499.845	0.230	500.068	499.937	0.068	500.020	499.960	0.020	500.000	499.960	0.000

All dimensions are in millimeters.

Preferred fits for other sizes can be calculated from data given in ANSI B4.2–1978 (R1984).

[†]All fits shown in this table have clearance.

Source: Courtesy of The American Society of Mechanical Engineers.

American National Standard Preferred Shaft Basis Metric Transition and Interference Fits (ANSI B4.2–1978, R1984)

Basic Size		Locational Transition			Locational Transition			Locational Interference			Medium Drive			Force		
		Hole K7	Shaft h6	Fit†	Hole N7	Shaft h6	Fit†	Hole P7	Shaft h6	Fit†	Hole S7	Shaft h6	Fit†	Hole U7	Shaft h6	Fit†
1	Max	1.000	1.000	+0.006	0.996	1.000	+0.002	0.994	1.000	0.000	0.986	1.000	-0.008	0.982	1.000	-0.012
	Min	0.990	0.994	-0.010	0.986	0.994	-0.014	0.984	0.994	-0.016	0.976	0.994	-0.024	0.972	0.994	-0.028
1.2	Max	1.200	1.200	+0.006	1.196	1.200	+0.002	1.194	1.200	0.000	1.186	1.200	-0.008	1.182	1.200	-0.012
	Min	1.190	1.194	-0.010	1.186	1.194	-0.014	1.184	1.194	-0.016	1.176	1.194	-0.024	1.172	1.194	-0.028
1.6	Max	1.600	1.600	+0.006	1.596	1.600	+0.002	1.594	1.600	0.000	1.586	1.600	-0.008	1.582	1.600	-0.012
	Min	1.590	1.594	-0.010	1.586	1.594	-0.014	1.584	1.594	-0.016	1.576	1.594	-0.024	1.572	1.594	-0.028
2	Max	2.000	2.000	+0.006	1.996	2.000	+0.002	1.994	2.000	0.000	1.986	2.000	-0.008	1.982	2.000	-0.012
	Min	1.990	1.994	-0.010	1.986	1.994	-0.014	1.984	1.994	-0.016	1.976	1.994	-0.024	1.972	1.994	-0.028
2.5	Max	2.500	2.500	+0.006	2.496	2.500	+0.002	2.494	2.500	0.000	2.486	2.500	-0.008	2.482	2.500	-0.012
	Min	2.490	2.494	-0.010	2.486	2.494	-0.014	2.484	2.494	-0.016	2.476	2.494	-0.024	2.472	2.494	-0.028
3	Max	3.000	3.000	+0.006	2.996	3.000	+0.002	2.994	3.000	0.000	2.986	3.000	-0.008	2.982	3.000	-0.012
	Min	2.990	2.994	-0.010	2.986	2.994	-0.014	2.984	2.994	-0.016	2.976	2.994	-0.024	2.972	2.994	-0.028
4	Max	4.003	4.000	+0.011	3.996	4.000	+0.004	3.992	4.000	0.000	3.985	4.000	-0.007	3.981	4.000	-0.011
	Min	3.991	3.992	-0.009	3.984	3.992	-0.016	3.980	3.992	-0.020	3.973	3.992	-0.027	3.969	3.992	-0.031
5	Max	5.003	5.000	+0.011	4.996	5.000	+0.004	4.992	5.000	0.000	4.985	5.000	-0.007	4.981	5.000	-0.011
	Min	4.991	4.992	-0.009	4.984	4.992	-0.016	4.980	4.992	-0.020	4.973	4.992	-0.027	4.969	4.992	-0.031
6	Max	6.003	6.000	+0.011	5.996	6.000	+0.004	5.992	6.000	0.000	5.985	6.000	-0.007	5.981	6.000	-0.011
	Min	5.991	5.992	-0.009	5.984	5.992	-0.016	5.980	5.992	-0.020	5.973	5.992	-0.027	5.969	5.992	-0.031
8	Max	8.005	8.000	+0.014	7.996	8.000	+0.005	7.991	8.000	0.000	7.983	8.000	-0.008	7.978	8.000	-0.013
	Min	7.990	7.991	-0.010	7.981	7.991	-0.019	7.976	7.991	-0.024	7.968	7.991	-0.032	7.963	7.991	-0.037
10	Max	10.005	10.000	+0.014	9.996	10.000	+0.005	9.991	10.000	0.000	9.983	10.000	-0.008	9.978	10.000	-0.013
	Min	9.990	9.991	-0.010	9.981	9.991	-0.019	9.976	9.991	-0.024	9.968	9.991	-0.032	9.963	9.991	-0.037
12	Max	12.006	12.000	+0.017	11.995	12.000	+0.006	11.989	12.000	0.000	11.979	12.000	-0.010	11.974	12.000	-0.015
	Min	11.988	11.989	-0.012	11.977	11.989	-0.023	11.971	11.989	-0.029	11.961	11.989	-0.039	11.956	11.989	-0.044
16	Max	16.006	16.000	+0.017	15.995	16.000	+0.006	15.989	16.000	0.000	15.979	16.000	-0.010	15.974	16.000	-0.015
	Min	15.988	15.989	-0.012	15.977	15.989	-0.023	15.971	15.989	-0.029	15.961	15.989	-0.039	15.956	15.989	-0.044
20	Max	20.006	20.000	+0.019	19.993	20.000	+0.006	19.986	20.000	0.001	19.973	20.000	-0.014	19.967	20.000	-0.020
	Min	19.985	19.987	-0.015	19.972	19.987	-0.028	19.965	19.987	-0.035	19.952	19.987	-0.045	19.946	19.987	-0.054

ANSI PREFERRED SHAFT BASIS METRIC TRANSITION AND INTERFERENCE FITS—METRIC UNITS

Basic Size		Locational Transition			Locational Transition			Locational Interference			Medium Drive			Force		
		Hole K7	Shaft h6	Fit†	Hole N7	Shaft h6	Fit†	Hole P7	Shaft h6	Fit†	Hole S7	Shaft h6	Fit†	Hole U7	Shaft h6	Fit†
25	Max	25.006	25.000	+0.019	24.993	25.000	+0.006	24.986	25.000	-0.001	24.973	25.000	-0.014	24.960	25.000	-0.027
	Min	24.985	24.987	-0.015	24.972	24.987	-0.028	24.965	24.987	-0.035	24.952	24.987	-0.048	24.939	24.987	-0.061
30	Max	30.006	30.000	+0.019	29.993	30.000	+0.006	29.986	30.000	-0.001	29.973	30.000	-0.014	29.960	30.000	-0.027
	Min	29.985	29.987	-0.015	29.972	29.987	-0.028	29.965	29.987	-0.035	29.952	29.987	-0.028	29.939	29.987	-0.061
40	Max	40.007	40.000	+0.023	39.992	40.000	+0.008	39.983	40.000	-0.001	39.966	40.000	-0.018	39.949	40.000	-0.035
	Min	39.982	39.984	-0.018	39.967	39.984	-0.033	39.958	39.984	-0.042	39.941	39.984	-0.059	39.914	39.984	-0.076
50	Max	50.007	50.000	+0.023	49.992	50.000	+0.008	49.983	30.000	-0.001	49.966	50.000	-0.018	49.939	50.000	-0.055
	Min	49.982	49.984	-0.018	49.967	49.984	-0.033	49.938	49.984	-0.042	49.941	49.984	-0.059	49.914	49.984	-0.086
60	Max	60.009	60.000	+0.028	59.991	60.000	+0.010	59.979	60.000	-0.002	59.958	60.000	-0.023	59.924	60.000	-0.087
	Min	59.979	59.981	-0.021	59.961	59.981	-0.039	59.949	59.981	-0.051	59.928	59.981	-0.072	59.894	59.981	-0.106
80	Max	80.009	80.000	+0.028	79.991	80.000	+0.010	79.979	80.000	-0.002	79.952	80.000	-0.029	79.909	80.000	-0.072
	Min	79.979	79.981	-0.021	79.961	79.981	-0.039	79.949	79.981	-0.051	79.922	79.981	-0.078	79.879	79.981	-0.121
100	Max	100.010	100.000	+0.032	99.990	100.000	+0.012	99.976	100.000	-0.002	99.942	100.000	-0.036	99.889	100.000	-0.089
	Min	99.975	99.978	-0.025	99.955	99.978	-0.045	99.941	99.978	-0.059	99.907	99.978	-0.093	99.854	99.978	-0.146
120	Max	120.010	120.000	+0.032	119.990	120.000	+0.012	119.976	120.000	-0.002	119.934	120.000	-0.044	119.869	120.000	-0.109
	Min	119.975	119.978	-0.025	119.955	119.978	-0.045	119.941	119.978	-0.059	119.899	119.978	-0.101	119.834	119.978	-0.166
160	Max	160.012	160.000	+0.037	159.988	160.000	+0.013	159.972	160.000	-0.003	159.915	160.000	-0.060	159.825	160.000	-0.150
	Min	159.972	159.975	-0.028	159.948	159.975	-0.053	159.932	159.975	-0.068	159.875	159.975	-0.125	159.785	159.975	-0.213
200	Max	200.013	200.000	+0.042	199.986	200.000	+0.015	199.967	200.000	-0.004	199.895	200.000	-0.076	199.781	200.000	-0.190
	Min	199.967	199.971	-0.033	199.940	199.971	-0.060	199.921	199.971	-0.079	199.849	199.971	-0.151	199.735	199.971	-0.265
250	Max	250.013	250.000	+0.042	249.986	250.000	+0.015	149.967	250.000	-0.004	249.877	250.000	-0.094	249.733	250.000	-0.238
	Min	249.967	249.971	-0.033	249.940	249.971	-0.060	249.921	249.971	-0.079	249.831	249.971	-0.169	249.687	249.971	-0.313
300	Max	300.016	300.000	+0.048	299.986	300.000	+0.018	299.964	300.000	-0.004	299.850	300.000	-0.118	299.670	300.000	-0.298
	Min	299.964	299.968	-0.036	299.934	299.968	-0.066	299.912	299.968	-0.088	299.798	299.968	-0.202	299.618	299.968	-0.382
400	Max	400.017	400.000	+0.053	399.984	400.000	+0.020	399.959	400.000	-0.005	399.813	400.000	-0.151	399.586	400.000	-0.378
	Min	399.960	399.964	-0.040	399.927	399.964	-0.073	399.902	399.964	-0.098	399.756	399.964	-0.244	399.529	399.954	-0.471
500	Max	500.018	500.000	+0.058	499.983	500.000	+0.023	499.955	500.000	-0.005	499.771	500.000	-0.189	499.483	500.000	-0.477
	Min	499.955	499.960	-0.045	499.920	499.960	-0.080	499.892	499.960	-0.108	499.708	499.960	-0.292	499.420	499.960	-0.580

All dimensions are in millimeters.

Preferred fits for other sizes can be calculated from data given in ANSI B4.2–1978 (R1984).

†A plus sign indicates clearance; a minus sign indicates interference.

Source: Courtesy of The American Society of Mechanical Engineers.

INDEX

Isometric grid paper

Rectangular grid paper

VP

C

D

A

B

One-point perspective

Two-point perspective